Y0-CAA-097

■ Encyclopedia of Human-Animal Relationships

■ Encyclopedia of Human-Animal Relationships

A Global Exploration of Our Connections with Animals

Volume 2: Con–Eth

Edited by
Marc Bekoff

GREENWOOD PRESS
Westport, Connecticut • London

Library of Congress Cataloging-in-Publication Data

Encyclopedia of human-animal relationships : a global exploration of our connections with animals / edited by
Marc Bekoff.
    p. cm.
  Includes bibliographical references and index.
  ISBN-13: 978-0-313-33487-0 (set : alk. paper)
  ISBN-13: 978-0-313-33488-7 (vol 1 : alk. paper)
  ISBN-13: 978-0-313-33489-4 (vol 2 : alk. paper)
  ISBN-13: 978-0-313-33490-0 (vol 3 : alk. paper)
  ISBN-13: 978-0-313-33491-7 (vol 4 : alk. paper)

  1. Human-animal relationships—Encyclopedias. I. Bekoff, Marc.
  QL85.E53 2007
  590—dc22   2007016552

British Library Cataloguing in Publication Data is available.

Library of Congress Catalog Card Number: 2007016552

ISBN–13: 978–0–313–33487–0 (Set)
ISBN–13: 978–0–313–33488–7 (vol. I)
ISBN–13: 978–0–313–33489–4 (vol. II)
ISBN–13: 978–0–313–33490–0 (vol. III)
ISBN–13: 978–0–313–33491–7 (vol. IV)

First published in 2007

Greenwood Press, 88 Post Road West, Westport, CT 06881
An imprint of Greenwood Publishing Group, Inc.
www.greenwood.com

Printed in the United States of America

The paper used in this book complies with the
Permanent Paper Standard issued by the National
Information Standards Organization (Z39.48-1984).

10 9 8 7 6 5 4 3 2 1

For my parents, who always encouraged the animal in me.

■ Contents

■ Alphabetical List of Entries

■ List of Entries that Feature Specific Animals

This list includes those entries that focus on particular animals. Other entries also include information about these animals but not to the extent that the following essays do. Please see the index for all mentions of the large number of different animals that are included in these volumes.

■ Conservation and Environment
From Anthropocentrism to Anthropoharmonism

Ecological concerns have begun to weave their way into our collective consciousness. Systematic animal abuse, unprecedented and rapid species extinction, poisoned air, polluted water systems, tropical forest burning, rampant militarism, ozone depletion, toxic waste sites, and global climate change all weigh heavily on our personal and professional lives.

Our current ecological moment is so grave and on such a profound scale that it compels us to rethink the basic question of what it means to be human. The ecological challenge to reexamine what it means to be human touches us at the core of our being, a being which is not only rational but psychological, emotional, phenomenological, and, indeed, spiritual.

As we slow-cook our climate with fossil-fuel emissions, delete thousands of species annually, cut down our remaining old growth forests, and watch as thousands of children, from Walkerton, Ontario, to Cairo, Egypt, die owing to contaminated water, we are confronted with sobering questions, questions of a deeply philosophical and spiritual vintage: What is our role here? What kind of world do we wish to leave future generations? What is the goal of "civilization," if its thrust forward leaves behind the world's ecosystems, a vast swath of the animal species, and a large percentage of our children in the wake of our destruction and consumption? What *on earth* are we doing?

All of these questions are critical in coming to grips with the enormity of the ecological challenge before us; they constitute a summons to fashion a new human-nonhuman relationship. This provides a context also for reassessing and critiquing human-centered, utilitarian approaches to animals, particularly in the West.

Biologist Marc Bekoff (2006) has explored the "intersubjectivity" of animals, or how they *experience* and relate to one another, noting that "many animals have subjective and intersubjective communal lives" as well as "a personal point of view on the world that they share with other individuals." Likewise, we also, in this moment of such profound ecological concern, are prompted to shift from an anthropocentric (human-centered) perspective to an "anthropoharmonic" perspective on human-nonhuman relationships. As advances in systems theory, quantum physics, and the life sciences are revealing, intersubjectivity is increasingly being discerned as a basic reality of all human and animal existence. This corresponds to the traditional knowledge of many of the worlds' indigenous communities as well.

From this perspective, harmonic means "of an integrated nature." Like an anthropocentric position, the anthropoharmonic perspective acknowledges the importance of the human and makes the human fundamental. Unlike anthropocentrism, however, it does not view the human as exclusively focal. Anthropoharmonism suggests that we are not rugged individual human monads but rather, as process theologian John B. Cobb (1972, 1994) has argued, "persons in community." Moreover, the community in which we share intersubjectivity is not just human, but nonhuman as well. Working towards an anthropoharmonic understanding of human-nonhuman relationships suggests that we as humans are in a type of "dialectical contingency" with both animals and the rest of the created world.

Our perceived role as master and lord over nature, which in some cases has ancient religious wellsprings but was given a hefty leg-up through modern industrialization, is now paired with an increasing sense of our interrelationship with and dependence upon the Earth's natural systems, and with the nonhuman animals which form an integral part of those systems. What is our proper place, then, between the rock of our technological prowess and the hard place of our biological vulnerability and profound intersubjectivity? Caught within this space, we are experiencing an ecological impetus to a new ontology, or nature of our being. This quest for a new ontology speaks to our worldviews and how these worldviews shape our interaction with culture, economics, politics, and the environment.

Many, such as cultural historian Lynn White Jr. (1967), have pointed to human anthropocentrism, aided by both modern technologies and religious and cultural heritages, as the provenance of our destructive environmental habits and our rapacious arrogance toward animals. In our hubris, we as a species have considered ourselves masters over rather than, as Aldo Leopold (1949) noted, "just plain citizens" of the life community on the planet. Such a stance has contributed to a most radical moment of destruction of the world's ecosystems, one that is, according to "geologian" Thomas Berry (2006), jeopardizing the last 65 million years of earth development, and through global climate change, altering the very foundations upon which life has evolved and flourished.

While the anthropocentric debate compels us to sift through and reinterpret human-centered texts and traditions, and "biocentric" debates, sparked by Aldo Leopold's land ethic and subsequent deep ecology perspectives, prompt an attempt to take nature and the created world seriously into ethical discourse, a "cosmocentric" conversation has emerged that posits a different ethical context. The framework adopted here is not human-to-human or human-to-nonhuman relations, but rather the human in relationship to the entire cosmos itself. Developed principally by Thomas Berry and mathematical cosmologist Brian Swimme (1992), the cosmocentric perspective suggests that unless we understand our role as humans within the greater unfolding of the universe itself, not only our imaginations but also our ethics will be truncated, and any ecological stratagems of healing earth-human, and human-animal relations will be incomplete.

For Berry and Swimme, the universe is primary; the human species is derivative. Emphasizing the awe, wonder, and celebrative, joyful beauty of the unfolding universe, they argue that the primary source of revelation lies not in scriptural texts but in the emergent cosmos. Consequently, the cosmocentric perspective embraces contemporary physics and astronomy to help ascertain the nature of the universe. While accenting the awe, mystery, and wonder of the cosmos, Berry and Swimme in a sense call for a new wisdom tradition, one that brings the world's religious traditions and contemporary scientific exploration together in a unified quest for meaning and ecological integrity—a healing of human-earth relations. For them, the universe story is the primary story that must be taught from kindergarten through doctoral programs and form the organizing narrative of our economic, political, cultural, religious, and ethical life. Significantly, both Berry and Swimme include a critique of consumer culture in their efforts to establish a cosmocentric ethics of the environment. They claim that a consumer cosmology has become the world's reigning worldview and that we are trained from an early age to view the universe not as a communion of subjects, but a collection of objects, to be bought, sold, used, and discarded. This is particularly evident in our widespread commodification of animals in factory farming, genetic modification for research and entertainment, and international trade.

In the cosmocentric perspective, humans, animals, and indeed all life and matter have a psychic-spiritual as well as a material dimension. Building on the work of Jesuit paleontologist Pierre Teilhard de Chardin (1881–1955), this perspective claims that the universe has consciousness, and thus all matter and creatures also have a psychic-spiritual dimension, that is, a consciousness. For Berry, because the universe is a singular reality, consciousness, from its origin, is a necessary dimension of reality, even a dimension of "the primordial atom that carries within itself the total destiny of the universe."

Recognition of intersubjectivity, intimacy, and communion between humans and animals and among all aspects of the created world, however, does not inevitably lead to justice. The shift from anthropocentrism to anthropoharmonism presents a challenge to move from a simple recognition of the radically intersubjective nature of reality to an ethically informed notion of justice, or, more colloquially, of a "right relationship" among humans, animals, and the rest of creation.

See also

Conservation and Environment—*Extinction of Animals*
Conservation and Environment—*Global Warming and Animals*
Ethics and Animal Protection—*Interest Conflicts between Animals and Humans*
Human Anthropogenic Effects on Animals

Further Resources

Bekoff, M. (2006). The public lives of animals: A troubled scientist, pissy baboons, angry elephants, and happy hounds. *Journal of Consciousness Studies, 13*(5), 115–31.

Berry, T. (2006). *Evening thoughts: Reflecting on Earth as sacred community*. M. E. Tucker (Ed.). San Francisco: Sierra Club Books.

Berry, T., & Swimme, B. (1992). *The universe story: from the primordial flaring forth to the Ecozoic Era, a celebration of the unfolding of the cosmos*. San Francisco: HarperCollins.

Cobb, J. B., Jr. (1972). *Is it too late? A theology of ecology*. Beverly Hills, CA: Bruce.

———. (1994). Ecology and process theology. In C. Merchant (Ed.), *Ecology: Key concepts in critical theory* (pp. 322–26). Atlantic Highlands, NJ: Humanities Press.

Gore, A. (2006). *An inconvenient truth: The planetary emergency of global warming and what we can do about it*. Emmaus, PA: Rodale.

Leopold, A. (1949). The land ethic. In *A Sand County Almanac* (pp. 237–63). New York: Oxford University Press.

Scharper, S. (1997). *Redeeming the time: A political theology of the environment*. New York: Continuum.

Scharper, S., & Cunningham, H. (2006, Winter). The genetic commons: Resisting the neoliberal enclosure of life. *Social Analysis: The International Journal of Cultural and Social Practice, 50*(3), 193–202.

White, L., Jr. (1967). The historical roots of our ecologic crisis. *Science, 155*, 1203–7.

Stephen Bede Scharper

■ Conservation and Environment
Global Warming and Animals

Over the last 100 years, the average global surface temperature has increased approximately 0.8°C. Even with this amount of warming, which seems small compared to what might occur in the relatively near future (2° to 6°C or more increase by 2100), wild animals are already exhibiting discernible changes. This is because all living things are affected by temperature in one way or another. Several types of changes have already been seen in the wild, including

1. Shifts in range boundaries (e.g., moving north in the Northern Hemisphere) and/or shifts in the density of individuals from one portion of their range to another (e.g., the center of the abundance pattern moving up in elevation)

2. Shifts in the timing (i.e., phenology) of various events primarily occurring in spring and/or autumn
3. Changes in the genetics, behavior, morphometrics (e.g., body size or egg size), or other biological parameter
4. Extirpation or extinction, the latter of which is the final irreversible change

Given what is known about physiological requirements of species, these changes are consistent with those expected with increasing ambient temperatures.

Changes in Ranges and Shifting Densities

The temperature occurring at any given location can greatly influence whether a particular animal is present or absent in the habitat around that location. As the globe warms we find that species are extending their ranges poleward and up in elevation. The reason for these shifts is that habitats previously avoided by various species have now warmed sufficiently to allow colonization. For example, thirty-five of fifty-six (65 percent) nonmigratory butterflies in Europe have been found to be expanding their ranges northward this century by 35 to 240 kilometers, while only 3 percent (two out of fifty-six) are shifting southward.

These types of changes in species ranges are of concern for at least three reasons. First, the synergistic effect between rapid global warming and habitat modification is probably the most important problem facing species today. This will become an even larger problem in the future because temperatures will continue to rise quickly and, as the human population grows, we will continue to modify more landscape. The movements (dispersal) of species forced by rising temperatures are frequently slowed and often blocked by human modifications of the landscape. Dispersing individuals must not only find suitable habitat through which to travel, but appropriate habitat in which to colonize. This is relatively easy for highly mobile species like butterflies, birds, and bats, but certainly scorpions, salamanders, shrews, and the like will have trouble navigating across highways and through farm fields or cities. When temperatures were increasing in prehistoric times, the rate that the temperature was rising was significantly lower and the habitat that species were moving through was not modified in the manner that it is today. Second, species near the poleward side of continents (e.g., South Africa fynbos) will have no habitats into which they can disperse. Additionally, the same will be true for species occurring in high mountain habitats. Once they reach the top of the mountain there is nowhere else for them to go—unless humans intervene and translocate them to some remaining suitable habitat. Third, in addition to not having habitat into which they can disperse, species living at the tops of mountains and the

Pikas are currently living in montane habitats in western North America where the ambient temperature is quite close to the maximum this small mammal can endure. Courtesy of Shutterstock.

poleward side of continents will have the added stress of species from lower latitudes and altitudes invading their habitats and thereby creating new competitive situations.

All three of these reasons can significantly increase the danger of extinction. For example, pikas are currently living in montane habitats in western North America, where the ambient temperature is quite close to the maximum this small mammal can endure. Moving up in elevation to cooler regions is not possible because the type of habitat needed by pikas is not available. Another example is a subspecies of a checkerspot butterfly in Baja California. It will probably go extinct in the near future because it too has a low tolerance to hot temperatures and cannot shift in to cooler regions because Tijuana and San Diego are blocking its way. Fourth, species movements will be independent of shifts of other species. The reason for this is that the physiological constraints of temperature, food availability, and many other factors are unique for each species. This independence will become more and more evident the higher the temperature goes. The independent movement will likely cause the disruption of species interactions. As an example, if, on the one hand, the range of a predator shifts and the range of its prey does not, then this could be a benefit if the prey is an endangered species. Without the predator, the abundance of the endangered prey could increase. If, on the other hand, the prey is a pest on one of our food crops, then the increase in its population due to the movement of its predator could certainly become detrimental.

Changes in Timing

The timing of many different biological events is strongly influenced by warming temperature. Such events for animals include the timing of hatching or birth, arrival after spring or fall migration, arousal from hibernation, and sexual receptivity. Using the timing of spring events as an example, Root and her colleagues (2003) found three different patterns. First, we found that species are acting progressively earlier. Over the last thirty years, around 700 species (plants and animals together) from locations around the globe were found to be changing the timing of a spring event by around five days earlier per decade. Second, species are acting later, but this is quite unusual, being that only six out of the same 700 species (less than 1 percent) showed this type of change. Third, species are exhibiting no change in the timing of their spring events. The number of species in this group is very difficult to determine. Scientist and journal editors rarely report on species not changing, unless this information is by chance reported along with information addressing change in another species. For migrating birds, unlike the birds showing change, the unchanging species are probably cueing on day length rather then climate as a migration trigger. More study is needed, however, on the topic of migration triggers.

Rapid phenological changes of species are of concern because for over tens of thousands of years or more, animals have been adjusting to the timing of those species around them. For example, as the globe warms farmers may have to change the timing of their planting and might even change the type of crop grown. Either of these changes could provide an insect with a food resource that was previously limited, thereby allowing the population size to grow. If the insect feeds on the nectar from the flowers of the crop, then the farmer could experience a benefit owing to the plants being pollinated. If, however, the insect feeds on the tissue of the crop plant, then the increasing size of the insect population could be seen as a detriment that must be countered in some manner (e.g., pesticide). In wild communities, changes in timing could mean that a food source of a species is not available at the time it is needed. This in turn could cause the species stress, either in time and energy looking for food or in competitive interactions over the little food available. Such stress may lead to lower fecundity rates, which, if not rectified could lead to extinction.

Changes in Genetics, Behavior, Morphometrics, or Other Parameters

Most studies concerning how rapid global warming has influenced animals focus on range and phenological shifts. The number of studies, however, are increasing that show other factors, such as genetics and behavior, are also changing along with rapidly warming temperatures. For example, Bradshaw, Fujiyama, and Holzapfel (2000) found that a day-length cue that initiates dormancy in a mosquito in eastern North America is genetically controlled. With global warming the habitats where this mosquito is found are staying warmer longer in the fall. Based on temperature, dormancy can occur later in the season, when the day length is shorter. As a result the genetic control of the day-length cue has changed to a shorter day length. Normally a very large number of generations is required to observe genetic changes. Consequently, enough time has not yet passed for us to observe genetic changes in most wild species. Indeed, the concern is that the genetic changes needed for the survival of a species may not happen quickly enough to "keep up" with the rapidity at which the planet is warming. Without a change in gene frequency and not being able to disperse into hospitable habitat, a species could be heading toward extinction.

An example of a behavioral change associated with global warming is the foraging habits of polar bears (*Ursus maritimus*). As the globe has warmed, these bears are increasingly foraging in garbage dumps rather than on seals. This change in behavior arose in response to the bears being less able to capture seal, their regular prey, because to hunt seals requires bears to be standing on sea ice. With global warming the ice is thinning and melting earlier in the spring, before the bears have put on enough depot fat to get them through the winter. By necessity bears now forage, unfortunately, much more at dumps. The type of food and the quantity are not sufficient to sustain the previous number of bears. Hence, numerous bears have been found dead or lacking the nutrients needed for reproduction. The population size of these bears has dropped. Additionally, other animals that depend on the polar bear as a keystone species (e.g., arctic fox [*Alopex lagopus*] and ivory gull [*Pagophila eburnean*]) may also be in significant trouble as the bears catch fewer and fewer seals, leaving fewer and fewer carcasses on the ice for these other scavengers.

Extirpation and Extinction

The size of a species' range, the density of individuals within their range, and how rare or common a species' preferred habitat is within its range are all important factors influencing a species risk of extinction. All of these factors can be directly affected by global warming. Species with small ranges, low densities, and few preferred habitats have a much higher probability of extinction. Extirpation (a population in a given location disappears) of populations on the hotter side of its ranges reduces the overall range size—a step in the direction of extinction. Additionally, extirpation decreases the density of individuals, which is again a step in the direction of extinction. For species able to extend their ranges—appropriate habitat is available and they are not at the poleward edge of continents or at the tops of mountains—rapid global warming will most likely only cause extirpation of populations in the warmest portions of their ranges. At the same time the populations dispersing and colonizing in high latitudes and altitudes could easily counter this decrease in population size.

Because temperatures are increasing rapidly around the globe, basically all of the species in the world are going to be affected. With "only" about 0.8°C increase in the average global temperature thus far, numerous species are already speeding toward extinction. These "facultatively extinct" species are those that cannot move to a different location as the temperature increases. This includes species occurring at the poleward edge of continents, on tops of mountains, or on islands, whether they are surrounded by water or inhospitable land (e.g., desert or city). For example, the Mallee emu-wren (*Stipiturus malle*) is quite sedentary (rarely moving farther than 5 or 6 kilometers) and has a small fragmented range that is threatened by fires. It cannot move until its habitat moves, which will likely be much slower than the speed at which the emu-wren will need, given the rate of temperature increase. Unless humans intervene and translocate individuals to a suitable habitat farther south, this bird will most likely go extinct within the next twenty-five to fifty years. Unfortunately only about 2,000 square kilometers of suitable native habitat is available today, and creating suitable habitat farther south will require two- to three-year old spinifex grass (*Triodia irritans*). After the birds are moved to a new habitat, then we would need to ensure that both the habitat and emu-wren survive, which will necessitate prohibiting a fire cycle that is less than ten to fifteen years. Such interventions by humans would be difficult and likely not practical for this species or for other species in the same circumstances. We do not appear to have the money, land, personnel, or political will with long-term commitment to accomplish such an enormous task. This is why many scientists are predicting that we are standing at the brink of a mass extinction that will be caused by one very careless species. Projections are that between 15 and 40 percent of all species alive today will become extinct because of human enhancement of atmospheric greenhouse gases. Given that there are somewhere between 5 million and 50 million species (only around 1.75 million are described), that means somewhere between 750,000 and 2 million or 7.5 million to 20 million species could go extinct primarily because of human use of fossil fuels and the dumping of their combustion products into the atmosphere.

Further Resources

Bradshaw, W. E., Fujiyama, S., & Holzapfel, C. M. (2000). Adaptation to the thermal climate of North America by the pitcher-plant mosquito, *Wyeomyia smithii*. *Ecology, 81,* 1262–72.

Derocher, A. E., Lunn, N. J., & Stirling, I. (2004). Polar bears in a warming climate. *Integrative and Comparative Biology, 44,* 163–76.

Grayson, D. K. (2000). Mammalian responses to Middle Holocene climatic change in the Great Basin of the western United States. *Journal of Biogeography, 27,* 181–92.

Parmesan C., & Yohe, G. (2003). A globally coherent fingerprint of climate change impacts across natural systems. *Nature, 421,* 37–42.

Reynolds, A. (2002). *Warnings from the bush*. Climate Action Network Australia. Retrieved February 27, 2007, from http://www.cana.net.au/bush

Root, T. L., Price, J. T., Hall, K. R., Schneider, S. H., Rosenzweig, C., & Pounds, J. A. (2003). Fingerprints of global warming on wild animals and plants. *Nature, 421,* 57–60.

Terry L. Root

Global Warming and Oceans

Terry L. Root

Global Warming

The Earth's climate is different from what it was only 20,000 years ago, when ice sheets covered much of the Northern Hemisphere. Since the industrial revolution humans have been dumping exhaust from burning fossil fuels into the atmosphere, thereby significantly increasing the amount of carbon dioxide (CO_2) in the atmosphere. This "extra" CO_2 increases the natural capacity of the atmosphere to trap radiant heat near the Earth's surface. This trapping is called the greenhouse effect, and it works in the following manner. Solar energy reaching the Earth's surface warms the lower atmosphere. Gases such as water vapor and CO_2 trap a large fraction of this heat near the Earth's surface. The natural greenhouse effect, not aided by human emissions, is responsible for keeping our planet at a livable temperature—around 33°C on average at the surface. The concentration of greenhouse gases in the atmosphere determines how much heat is trapped. Even seemingly small human-induced additions of greenhouse gases have already resulted in the average global temperature increasing around 0.8°C in the last 100 years. Without technology that allows us to switch off our use of fossil fuels and still maintain or improve our typical lifestyle, within the next century the global average temperature could continue to warm by 1°C if we are quite lucky, 6°C if we are unlucky, or 12°C or more if we are very unlucky. This could result in ecologically significant changes, which are why climatic considerations are fundamental in the discussion of possible ecological consequences of wildlife.

Acidification of Oceans

Besides increasing the average global temperature, the added CO_2 in the atmosphere is discernibly changing the pH balance of the oceans. The CO_2 concentration is greater in the atmosphere than in the oceans, meaning that at the air-water boundary CO_2 is diffusing into the water. The carbonic acid that forms in the water has already discernibly changed the oceans pH from 8.0–8.3 (before the industrial revolution) to 7.9–8.2. If we do not change the way that we currently use fossil fuels, The Royal Society of the United Kingdom has estimated by 2100 the pH range could be 7.5–7.8. Coral reefs and other structure built by animals using calcium carbonate (e.g., shells) will not only be difficult to build, but those that are built will dissolve. Obviously this will be detrimental to marine organisms, but we will also be affected because ocean fisheries will most likely be negatively affected and reefs that protect the shores during large storms could be so weakened or eroded to be of little use. Scientists in The Royal Society warn that even if we curtailed our CO_2 emissions in 2010, the pH change already evident in the oceans will take tens of thousands of years to return to preindustrial levels. What we are doing today will truly impact the lives of people over 500 generations from now.

Further Resources

Intergovernmental Panel on Climate Change (IPCC). (2001). *Climate change 2001: Impacts, adaptations, and vulnerability.* New York: Cambridge University Press.

The Royal Society. (2005). *Ocean acidification due to increased atmospheric carbon dioxide.* Policy document, December 2005.

■ Conservation and Environment
Global Warming and the Spread of Disease

Heat stress is probably the most obvious thing people think of when the idea of global warming comes up. A heat wave in Europe during the summer of 2003 killed more than 10,000 people in France alone. Many of the dead were elderly—the group most likely to live alone and most susceptible to heat-related health problems.

Other climatic effects are more subtle, but no less deadly. Higher rates of ground-level ozone are a major respiratory irritant, and vector-borne diseases thrive in warmer temperatures. And that is the problem that is keeping New York City's public health officials up nights.

In the summer of 1999—the hottest and driest in a century—sixty-two cases of West Nile encephalitis were reported in New York, and seven people died (see "Beyond the Bite," *Your Health,* November/December 2003). A general health warning was issued, and city residents began to get used to helicopters overhead spraying clouds of malathion and pyrethriod pesticides.

According to Dr. Dickson Despommier, a professor of public health at New York's Columbia University, the disease is spread by *Culex pipens* mosquitoes, whose favorite prey is birds. But periods of high heat and drought send such common urban-dwelling species as crows, blue jays, and robins out of the city in search of fresh water. City bird populations are further reduced as unlucky individuals are bitten and killed by West Nile infection.

"By reproductive imperative the mosquitoes are forced to feed on humans, and that's what triggered the 1999 epidemic," Despommier says. "Higher temperatures also trigger increased mosquito biting frequency. The first big rains after the drought created new breeding sites."

Despommier says that this same pattern is also discernible in recent West Nile outbreaks in Israel, South Africa, and Romania. In Bucharest, his investigation turned up abandoned buildings whose basements were full of water, a perfect Culex breeding ground.

Another prominent proponent of the West Nile–global warming connection is Dr. Paul Epstein of Harvard University. "Droughts are more common and prolonged as the planet warms," he says. "Warm winters intensify drought because there's a reduced spring runoff. The cycle seems to rev up in the spring, as catch basin water dries up and what is left becomes organically rich and a perfect mosquito-breeding place. The drought also reduces populations of mosquito predators."

In 2002, West Nile spread across the country, appearing in forty-four states and the District of Columbia. Five provinces of Canada were also affected. It spread to forty-six states in 2003, when there were 9,862 cases in the United States. As the *New York Times* reported, the mosquitoes that carry the disease can breed effectively even in dry cities like Phoenix, where abandoned swimming pools serve as effective incubators. West Nile has killed more than 500 Americans. "Because West Nile is apparently carried by migrating birds, wildlife managers in North America have few options but to watch and wait for its arrival, and hope that stricken bird populations can recover in time," says Jeff Burgett of the U.S. Fish and Wildlife Service.

Adapted from "Too Darn Hot: Global Warming Accelerates the Spread of Disease," by Jim Motavalli with research assistance by Aaron Midler, in *E/The Environmental Magazine*; November/December 2004, Vol. XV, no. 6. Norwalk, CT 06851. www.emagazine.com. Used by permission.

In a growing scientific consensus, public health officials believe the next drought will give this serious virus even a wider reach. Spraying certainly has not stopped these infectious bugs. Researchers at France's University of Montpellier said in mid-2003 that a mutation in the West Nile mosquitoes' genetic code resulted in their singular resistance to pesticides.

A Gathering Force

It is plain that global warming will have a major impact on human health and that new threats will emerge. "With vector-borne diseases, it's pretty well established that increased temperatures cause higher and higher loads of viruses," says Peter Daszak, executive director of the Consortium for Conservation Medicine. "You also get an increase in biting rates of carriers like mosquitoes."

We have so far identified only 1 percent of the bacteria and 4 percent of the viruses on the planet, according to *Harper's*. Some twenty-nine previously unknown pathogens emerged between 1977 and 1994. In 2001, the American Public Health Association cited such probable effects as increased heat-related mortality, more rapid spread of vector-borne disease, greater incidence of water-borne diseases because of more intense precipitation, and threatened agriculture to recommend that more attention be paid to climate change.

The Congressionally mandated "National Assessment of Climate Change and Health," issued in 2001, foresaw greater incidence of heat stroke, malaria, yellow fever, and respiratory disease as a result of global warming. As countries like the United States become more "tropical," it said, insect- and rodent-borne diseases may be seen more often in the First World. Increased flooding will also breed more of these carriers. A 2004 Pentagon report on the possibility of abrupt climate change envisioned grave threats to national security arising from loss of agricultural productivity and widespread famine.

Some of the most enlightening studies arise from new technology. Rita Colwell, a professor at the University of Maryland-College Park, has studied public health in Bangladesh for twenty-five years, and recent work with satellite imaging and computer modeling has helped her make some new correlations between cholera outbreaks, water-borne pathogens, and climate that would not have been possible without such modern tools. "It gives us the ability to actually predict the number of cases based on the histor- ical record and climate variables," she says, adding that she is "hesitant to say flat out" that global warming is a major player. But she certainly sees it as a strong possibility.

Research published in the *American Journal of Public Health* in 2001 by Dr. Jonathan Patz and others reported that 68 percent of waterborne disease outbreaks studied came after major precipitation events (which are predicted to occur more frequently because of global warming). A growing pile of studies like these are making it clear that climate change will present a major challenge to our already overburdened global public health system.

Jim Motavalli, with research assistance by Aaron Midler

■ Conservation and Environment
The Great Apes of Asian Forests, "The People of the Forest"

The uninterrupted human activities in forest habitats threaten the lives of our closest rel- atives and "people of the forest," who are the great apes such as the gorilla, chimpanzee, and orangutan, and we have to make wise actions to protect them from extinction.

Nearly two-thirds of the world's original forests have been destroyed, and if the current rate of deforestation continues, we will have virtually no natural forest left in a centaury. Tropical forests are being destroyed at a rate of 17 million hectares per year, and the Asian timber reserve may last for only forty years.

The great apes comprise the families *Pongidae* (orangutan) and *Hominidae* (gorilla, chimpanzee, and human). Taxonomists in the past placed all great apes in a single family *Pongidae* and humans in a separate family *Hominidae*. However, recent molecular genetics data and new fossil evidence led experts to conclude that the African apes (gorillas and chimpanzees) diverged from each other some 5 million years ago and are more similar to each other than to orangutans. We, the humans, or *Homo sapiens*, share 98.4 percent of our DNA with chimpanzees. Although we do not look like chimpanzees or gorillas, we are similar enough in many ways to belong to the same family. The Great Ape Project, an ape-rights group, calls for certain civil rights to be granted to all great apes, including the rights to life, liberty, and freedom from human torture.

The great apes have larger bodies and bigger brains than other primates and are sexually dimorphic— the males are larger than the females. Generally, the great apes are less arboreal and more terrestrial than the lesser apes (gibbons). The social structure of each genus of great apes is different. Orangutans lead a solitary life, while gorillas live in one-male, multi-female groups. Chimpanzees live in large multi-male

and female groups often spilt into subgroups, which periodically rejoin. Humans often live in monogamous pairs, which is similar to the gibbons and one male, multi-female family groups, such as gorillas. The great apes are in many ways emotionally, physiologically, anatomically, and behaviorally close to us. They can recognize themselves when they look into a mirror. They have the ability to learn and use sign language designed by anthropologists in laboratory conditions. They use tools and they also can break and throw large sticks toward intruding rivals or humans to scare them away. All the great apes construct their own nests for sleeping; they have a lifespan between fifty and sixty years.

Local people in Africa and Asia consider the great apes people of the forest, and Asia's only red ape, "Orang Utan," literally means "the people of the forest" in the native Malay language. Wild populations of orangutans are now restricted to two separate enclaves, the islands of Borneo and

Although protected by international law, large numbers of smuggled orangutans can be seen in public and private zoos, in recreational parks, and as house pets. Courtesy of G. Agoramoorthy.

Sumatra, and exist in two forms, the Bornean *Pongo pygmaeus pygmaeus* and the Sumatran *Pongo pygmaeus abelii*. They have lost 80 percent of their rainforest habitat in the last few decades. The Sumatran orangutan has been listed as one of the top twenty-five highly endangered species of primates in the world—which means that they have been picked for extinction! The forest fires deliberately caused by humans in Borneo and Sumatra not only destroy large areas of rainforest each year, but also kill unknown numbers of orangutans. The rehabilitation centers in Borneo and Sumatra continue to receive large numbers of orphaned baby orangutans as a result of logging, forest fires, and pet trade. Poachers usually shoot the mothers to retrieve young babies to be sold in local market. These rare apes are still being smuggled out of Borneo and Sumatra to be sold in pet and zoo markets around the world. In Taiwan, for example, hundreds of orangutans were smuggled for the pet trade not long ago, and large numbers of orangutans can still be seen in public and private zoos, recreational parks, and as house pets. Orangutans are protected by law in habitat countries such as Indonesia and Malaysia and also by international law, including the Convention of International Treaty on Endangered Species, which prohibits trade of live specimens or body parts.

In several Southeast Asian countries, such as Taiwan, exotic pets are viewed as status symbols among the rich upper class, even though it is against the law to keep them. Orangutans suffer from this chronic trade because of weak law enforcement in the region. There is no doubt that wildlife is under threat in Southeast Asia because of the intensity of illegal trade. Snakes, monkeys, apes, turtles, birds, and all sorts of endangered animals are being taken out of their habitat to meet the growing demand in wildlife trade. Furthermore, rainforest is being lost to logging with poorly planned development, and porous borders make smuggling a lot easier. And while supply is dropping, demand remains strong, pushed by poverty and hunger. Asia and the Pacific have 23 percent of the world's land area, but 58 percent of its people, and millions live below the poverty line. It appears that countries such as China, Vietnam, Indonesia, Cambodia, and Laos are the vacuum cleaners for various species of wildlife including orangutans, while Hong Kong, Taiwan, Singapore, Thailand, and Malaysia are major transit points or destinations for the illicit trade. It is therefore essential to strengthen law enforcement, promote public education and eventually eradicate poverty in developing countries to put an end to this horrific trade before more species are added to the extinction list.

The fate of rainforests in Southeast Asia will largely determine the fate of endangered species such as the orangutans. We already know that several species of amphibians, reptiles, birds, and mammals are declining in numbers around the world, and the situation is chronic in Southeast Asia. We must not forget that the world's biological diversity not only supports our economic wealth but also our own survival on this delicate planet. We must realize that reckless human activities are endangering the survival of our closest relative, the people of the forest, and it is time to take some meaningful action to save them from extinction.

Further Resources

Agoramoorthy, G. (1998). Status of orangutans in Taiwan. *International Zoological Yearbook, 36,* 118–22.

Galdikas, M. F. (1995). *Reflections of Eden: My years with the orangutans of Borneo.* Boston: Little, Brown and Company.

Gisela, K., & Rogers, L. (1994). *Orang-utans in Borneo.* Armidale: The University of New England Press.

Govindasamy Agoramoorthy

■ Conservation and Environment
Humane Sustainable Development

Over the past twenty years there have been intense national and international debates over the limits of the dominant economic model and the meaning of a possibly more effective alternative called "sustainable" development.

The currently dominant model of development was consolidated immediately following World War II, largely to aid in the reconstruction of a devastated Europe and to ensure that such evils as fascism, communism, destructive nationalism, and religious fanaticism would be replaced by a common pursuit of economic prosperity. This approach to development has in some ways been remarkably successful (as in the case of Europe). But by discounting the future, "externalizing" social and ecological costs, and failing to recognize goals for development other than short-term economic gain, this single-minded pursuit of economic growth is destroying the social and ecological fabric necessary for the well-being of humans and other living beings both now and especially in the future.

A major problem in the current approach to fostering global economic development is the assumption that nonhuman animals and nature are "objects" that have no intrinsic worth or moral claim on us and that human societies can exploit for even trivial human ends. Our economic, legal, and political systems embody this anthropocentric and mechanistic premise in their principles and practices.

The worldviews of both modern science and European Christianity that shaped our globalizing economies have encouraged human beings to exploit animals with little regard for their suffering. Francis Bacon and René Descartes, founders of modern science, believed that only humans have souls and that nature is just a great machine. Descartes regarded the screams of animals being vivisected as no more than the noise a machine makes as it breaks down, a mere grinding of gears. So too has the Christian church emphasized human dominion over the Earth and all its creatures. Too often dominion has been interpreted to mean domination—implying that the only value of the rest of creation is its utility to humans. Hence, our economic task is only to exploit these natural resources efficiently.

Such worldviews must change if we are to create a humane and sustainable path of development. Scientists, religious leaders, educators, and politicians must begin to recognize that the community of life on Earth, as Thomas Berry (2000) says, is a "communion of subjects, not a collection of objects." With such a shift in worldviews, our economics, science, and education would cultivate compassion for all sentient beings and contribute to lifestyles and business practices that are ecologically sound, socially just, and humane, as well as economically beneficial.

The following sections describe two major shifts we must make to achieve humane sustainable development.

Recognize That Existence Is a Living and Interconnected Community of Diverse Subjects Who Deserve Our Respect and Care

Animals, plants, rivers, and stars—all beings and natural systems—are enlivened by a mysterious presence. All partake of this life and feeling. Evolution and development are processes that must be understood in their psychic and spiritual as well as material/physical dimensions.

Diversity is a central value. Protecting the diversity of life forms, cultures, and languages, as well as rights and opportunities for each individual, is fundamental. When we lose these different expressions of life, we lose sources of essential knowledge, wisdom, and technologies. We also lose the richness of our souls, for at some deep level our ecological and ethical selves require the enhancement of diversity against both the violence of exploitation and the efficiency of monoculture.

We must also accept a world of material limits. A correlate of extending rights to others is recognizing the limits of human technological and intellectual capacities. This new worldview demands that we humans accept a set of restraints on our exploitation of the natural world, based on respect for the integrity of the life community, as well as healthy skepticism about human ability to control and manage extraordinarily complex natural systems benignly. Human humility is essential.

Create a New "Bottom Line" for Economics and Development That Differentiates between Constructive and Destructive Economic Activities

The globalized market focuses on increasing financial transactions through fostering material production and consumption. All religions and philosophies recognize qualitative ends beyond enhancing quantitative, material well-being. So should our key measures of progress and development. The nonmonetized human exchanges in the social economy, the contributions of the natural economy, and the fundamental spiritual ends of human life all need to be factored into our calculations of the costs and benefits of economic development.

The only viable strategy for promoting an economy that supports humane sustainable development is to differentiate between "good" and "bad" growth. The difference between the two is the desirability of what is growing and its effects on the natural and social systems that support the monetized economy. If growth enhances natural and social capital as well as long-term (and spiritual) consequences, then it is good. If it liquidates and undermines these, then it is bad. Just as cancer is "bad" growth in the body, so too are cancerous economic activities. The fundamental task is to refocus the economy on "good" growth and to reward those who produce it. We must quit subsidizing destructive economic activities.

What is needed is a shift in our model of the good life. Rather than fostering mobility and greed, this new approach encourages rootedness and caring. Policies to enhance healthy social and natural economies require a commitment to low input, sustainable local enterprises that strengthen the bonds of community, and responsiveness to ecosystematic constraints. Of course, such local initiatives are only possible within a strong international framework of "good globalization," in which nations demand and enforce ecologically sound, socially just, spiritually awakening, and humane standards as well as "free" trade agreements. Examples of local enterprises and national policies that support this shift are the "best practices" for humane sustainable development.

Examples of Humane Sustainable Development

How are communities, governments, and individuals implementing humane sustainable development at the local, regional, and national levels? The following three cases highlight promising initiatives in regions of the world where development pressures are high, biological and cultural diversity is being destroyed, and people and animals are suffering. These will show how financial and social benefits can be achieved by communities that create sustainable livelihoods, protect animals, and practice humane sustainable agriculture.

The Global Alliance for Humane and Sustainable Development (GAHSD)

The Global Alliance for Humane and Sustainable Development is a nongovernmental organization (NGO) based in Costa Rica that works in partnership with international institutions, donor organizations, academia, the public sector, private companies, and associations to support local economic development initiatives and to establish trade and environmental programs that serve as community development tools. The GAHSD's first major initiative is the Central American Free Trade Agreement (CAFTA) Alliance. It is also exploring opportunities to address environmental, humane, and habitat issues in the context of the World Trade Organization negotiations, the Free Trade of the Americas negotiations, and the U.S.–Southern African Customs Union Free Trade Agreement negotiations (Colombia, Ecuador, and Peru). Partners include a variety of governmental, industrial, and nongovernmental groups, including wildlife protection and rehabilitation groups, organic agriculture and other industry associations, a natural cosmetics company, and cooperative producer programs. The CAFTA Alliance effort to "link aid with trade" works to provide better opportunities for third world countries participating in trade. Activities within these program areas have included beef and pork sector competitiveness and animal welfare, organic cacao production and commercialization, promotion of organic agriculture and sustainable ecotourism, providing organic products and economic opportunity for threatened communities, research and education for humane agricultural systems, rural agricultural tourism, and curriculum reform for agriculture and livestock training. RAVS has held workshops in most of the countries to help reduce the number of stray dogs and improve the health of all companion animals. The Alliance has also worked with the government on international and domestic legislative issues including animal welfare regulations, outreach for the CAFTA, CITES implementation support, and rescue-center improvement.

The Jane Goodall Institute's Lake Tanganyika Catchment Reforestation and Education (TACARE) Project

The TACARE Program seeks to restore and conserve African forest habitats for chimpanzees, promote conservation, and address the education, economic, and health-care needs of the local people. The primary target area of TACARE is the remaining indigenous forests of the Kigoma region in western Tanzania, and it is currently coordinating with over thirty villages on or near the shore of Lake Tanganyika. The program was designed as a pilot project to address poverty and support sustainable livelihoods in villages around Lake Tanganyika while halting the rapid degradation of natural resources, especially in the remaining indigenous forest. The program focuses on community socioeconomic development and offers training and education in sustainable natural resource management. TACARE offers an innovative model of a community-centered conservation approach, which effectively addresses human needs while promoting conservation values. TACARE's activities are divided into five primary project areas: community development, forestry, agriculture, health, and Roots & Shoots, an education program.

Integrated Human and Conservation Development at Tanjung Puting National Park, Kalimantan

This program of the Friends of the National Parks Foundation (FNPF) is aimed at supporting local communities in developing comprehensive approaches to living humanely and sustainably to the benefit of the local people, their agriculture, draft and companion animals, and the forest/habitats and wildlife alongside which they live. The FNPF works with local and international partners, officials, and stakeholders to identify and attempt to address the needs and issues around Tanjung Puting National Park. This project aims to foster greater self-sufficiency in local communities via participatory processes. It also promotes community care and management of natural and farmed resources using humane animal husbandry practices and sustainable agriculture and land management policies. The Agriculture and Livestock Management program will design pilot projects and support local research drawing from indigenous processes as well as modern science. The FNPF works with local and international partners, officials, and stakeholders to identify and attempt to address the needs and issues around Tanjung Puting National Park. This project aims to foster greater self-sufficiency in local communities via participatory processes. It also promotes community care and management of natural and farmed resources using humane animal husbandry practices and sustainable agriculture and land management policies. The Agriculture and Livestock Management program will design pilot projects and support local research drawing from indigenous processes as well as modern science.

FNPF assists with the provision of health and education services to ensure a sound base for local capacity building, and to foster love of nature via conservation and humane education. This includes improving access to education for elementary school children in villages around the park and contributing to addressing the health concerns of villagers, especially women and children.

These three cases illustrate ways in which different stakeholders are working together in order to help local, regional, and national communities develop both humanely and sustainably. A development scheme that fails to recognize the interconnectedness of all species will ultimately fall short of its goals. Peace, security, and economic viability in human society depend on a healthy ecosystem, in which all forms of life are protected and respected. Humane sustainable development ensures that all generations of life have a viable future.

Further Resources

Berry, T. (2000). *The great work: Our way into the future*. New York: Harmony/Bell Tower.

TACARE. http://www.janegoodall.org/africa-programs/programs/tacare.asp (accessed February 27, 2007).

Waldau, P., and Patton, K. (Eds.) (2006). *A communion of subjects: Animals in religion, science, and ethics*. New York: Columbia University Press.

Richard M. Clugston

■ Conservation and Environment
India's Bats and Human Attitudes

Bats are the only mammals capable of true flight. These fascinating creatures of the night belong to the order *Chiroptera*, meaning "hand-wing." The order is further divided into two suborders: *Megachiroptera*, which comprises fruit eaters—these large bats have large, keen eyes and an acute sense of smell to locate fruits—and *Microchiroptera,* which comprises insect eaters—these small bats have small eyes and large ears, and they produce ultrasound over 15 kHz, which human ears cannot hear. *Microchiroptera* rely on echolocation to navigate and to catch insects.

Although people squirm at the very mention of the word "bat," bats are rather clean animals that groom frequently. The myth that all bats carry the rabies virus persists. However, statistics say that only 0.5 percent of bats contract rabies. And bats, almost as a rule, only bite in self-defense. They pose no threat to people. Worryingly, being one of the slowest reproducing mammals of their size—bats produce one young a year—bats are extremely vulnerable to extinction. That these gentle, beneficial creatures have been widely misunderstood and neglected further adds to the danger.

India is home to about 100 species of bats, including twelve species of fruit bats. Data on the conservation status, population density, and ecology of many bat species are limited, however, because of a lack of field studies. Recently, I conducted a rapid survey in India and focused my attention on two species in particular: the common Indian flying fox, *Pteropus giganteus,* and the much lesser-known Salim Ali's fruit bat, *Latidens salimalii.* Covering 720 square kilometers in the south Indian state of Tamil Nadu, I came across only nine colonies of the Indian flying fox. Each colony's population varied greatly: some had about 100 bats, whereas others had over 1,200 bats. I monitored two colonies near the town of Sirkali for a year. Because of intensive hunting, the population in these colonies dropped from 1,000 to 700 in one and from 650 to 450 in the other.

With no protected natural forests suitable to sustain them, a large number of the flying fox colonies are located in unprotected agricultural land around villages. The flying fox prefer to roost in large fig trees (*Ficus benghalensis*) that are located in sacred groves and protected by local villagers. However, the very same bats become fair game once they leave their roost to forage for food away from the sacred groves. Within a mere two months, I saw hunters selling forty-six bats in the local markets. These giant bats weigh about a kilogram, and their meat is considered a delicacy. There is also a belief among some in India that bat meat is capable of curing asthma. The population decline in the two colonies I observed is an indication that hunting is beginning to take its toll on one of the largest bats in India. If hunting continues, these intriguing bats will become locally extinct in many villages.

The Salim Ali's fruit bats are far more elusive than the flying fox, which I spent about a year with. This bat entered the Guinness Book of World Records in 1993 as one of the three rarest bats in the world. The other two are the small-toothed fruit bat from Sulawesi and a hipposiderid bat from Vietnam. I discovered for the first time the roosting site of these elusive bats in a cave in the rainforest near a coffee and cardamom estate adjacent to the Megamalai forest reserve, at an altitude of about 1,000–1,175 meters. I counted about 250 Salim Ali fruit bats when they emerged from their cave at dusk. I saw some of them plucking forest fruits and carrying them back into the caves. In the past, estate workers killed these bats along with the endangered Nilgiri langurs and lion-tailed macaques for food. There is a popular local belief that these rainforest mammals have medicinal value and cure diseases, including respiratory illnesses. Recently, the plantation workers have realized the importance of these rare species and vowed not to disturb them. This is especially significant in view of the fact that these bats live on a private coffee and cardamom estate. They are, therefore, not covered by any legal protection. The future, indeed the very survival, of these bats depends on the estate workers' goodwill.

Part of the antipathy toward bats is owed to the fact that fruit bats eat fruits and thus harm the output of the fruit industry. Although there are no authenticated scientific reports available on the extent of damage caused by fruit bats to cash crops in India, it appears that such damage is minimal. A few simple steps could change this trend. First, people can be educated about bats, including their role in the ecosystem, thus dispelling myths about them and fear of them. Bats are also important for biodiversity. Second, bat conservation projects can be planned, along with bat roosting sites. There is so little known about the population status of so many species of bats that one could start a local "bat-watching" society. Third, local legislators and parliamentarians could be asked to provide legal protection for bats.

Further Resources

Agoramoorthy, G., & Hsu, M. J. (2002). Biodiversity surveys are crucial for India. *Current Science, 82,* 244–45.
———. (2005). Population size, feeding, forearm length and body weight of a less known Indian fruit bat, *Latidens salimalii. Current Science, 88,* 354–56.
Bates, P. J. J., & Harrison, D. L. (1997). *Bats of the Indian Subcontinent.* Kent: Harrison Zoological Museum.

Govindasamy Agoramoorthy

■ Conservation and Environment
Mollusk Extinction

Human-caused animal extinctions may be the ultimate of all human-animal interactions. Extinction is forever, eliminating any other associations with humans. Since 1500, 643 animal species have been declared extinct, almost all from human causes, including hunting and harvesting, pollution, and habitat degradation.

A woman snorkeling in a pearl farm off the coast of Japan. Mussels have been harvested for seeds in pearl cultivation. ©Westend 61/Alamy.

When we think of extinct animals, we think of charismatic megafauna such as passenger pigeons, dodoes, Tasmanian tigers, or Steller's sea cows. When we think of threatened or endangered species, we think of chimpanzees, gorillas, Siberian tigers, rhinoceroses, pandas, whooping cranes, or California condors, each of which has campaigns and dedicated advocates devoted to saving them from that ultimate death. Yet 87 percent of all animal species are invertebrates. They are not cute and fuzzy, and they have suffered the most sweeping extinctions.

Going, Going, Gone!

Mollusks—notably freshwater snails and mussels—comprise 42 percent of the extinct animals today, more than all the extinct mammals and birds combined. In addition, 1,930 nonmarine mollusks are listed as threatened (only forty-one marine species are so listed).

Freshwater mussels are particularly threatened. The southeastern United States is the center of diversity of the group, with over 300 species described from that area. Freshwater mussels were living in the streams there when the Appalachian Mountains were being formed and are there now, even as the mountains have eroded to nubs of their former towering glory. Endangered mussels with colorful names like purple wartyback, fluted kidneyshell, elktoe, orange nacre mucket, and pink heelsplitter grow in their own particular stretches of the Tennessee and Mississippi River systems and, because of human activities, they are losing the battle of survival.

Human Causes

First, Native Americans ate these mussels, which are well-represented in ancient middens, the trash piles of the tribes. Second, European settlers discovered that these mussels made pearls, and countless millions were harvested solely for the pearl trade. Some mussel pearls were so highly valued that one was sold in 1902 for $65,000. Third, mussel shells were used to make mother-of-pearl buttons in the early 1900s, and millions more were harvested for this industry. At the peak of production, over 9,700 harvesters collected mussels for the button industry. Fourth, 125 dams have been built since the 1920s on the waterways of the American Southeast, isolating stocks of mussels, degrading their habitats with silt, and slowing down river flows. Some

European zebra mussels covering a stick. These tenacious mussels siphon food from native bivalve species and smother them to death. Courtesy of Shutterstock.

100-year-old mussels are isolated between dams with no way to reproduce because the fish they need to disperse their young no longer live there.

Mussels were again harvested beginning in the 1970s for seeds in the Japanese pearl oyster trade. A tiny piece of a freshwater mussel shell is placed inside the mantle of a pearl oyster, which the oyster feels is an irritant. It lays down layers of nacre around the irritant to make it smoother, and the result is a cultivated pearl. Ironically, Japanese pearl oysters recently suffered a 60 percent death rate due to a virus, dramatically cutting down the demand for these mussel shells.

In addition to those threats to their existence—and the increasing pollution from humans' cohabitation—freshwater mussels are now facing yet another threat from two of their own relatives. The European zebra mussel was first found in the Great Lakes in 1985, probably first arriving there in ships' ballast water, and it has since spread dramatically into the Tennessee River. It is so prolific that it thickly coats all substrate layers with its tiny shells, siphoning food from native bivalve species and smothering them to death. Another invader, introduced from China—the tiny Corbicula clam—also competes for food in lakes and streams.

A few people are working to save freshwater mussels by transplanting them to clean waters, raising them in labs, and educating farmers about fertilizing fields close to rivers or grazing cows near streams where they could crush mussel shells with their hooves. Some efforts are being made to legislate protection of the threatened species, much like those used to save the Banff Springs snail, which is only found in five hot springs near Calgary in Canada. (Three of the five are now protected, and snails have been successfully transplanted into hot springs where they had formerly been eradicated by human bathing activity.)

But the future of many of the still-living freshwater mussels looks bleak. Unlike the millions of dollars donated for preserving condors, whooping cranes, or pandas, little money is generated for saving freshwater mussels or other threatened mollusks. When they go extinct, who will cry for the lost pink heelsplitters? Who will mourn what Lydeard and his colleagues called the "silence of the clams"?

See also

Conservation and Environment—*Extinction of Animals*
Ecotourism

Further Resources

Lydeard, C., et al. (2004). The global decline of nonmarine mollusks. *Bioscience, 54,* 321–30.

Roland C. Anderson

■ Conservation and Environment
The Passenger Pigeon

The passenger pigeon (*Ectopistes migratorius*) is legendary as a symbol of unbelievable abundance and of human-caused extinction. These birds, slightly larger than contemporary mourning doves were an important food source for Native Americans as well as for both rural and urban residents in America before the twentieth century. The species occurred only in North America, primarily east of the Rocky Mountains. They nested almost exclusively in the deciduous forest that stretched from the Mississippi River to the Atlantic Ocean and from southern Canada to the mid-Atlantic and Midwest states. There were some 3 to 5 billion passenger pigeons in America prior to the arrival of the Europeans; they comprised perhaps a quarter of the continent's bird life. Yet, entirely due to human activities, greed, and ignorance, the passenger pigeon was extinct in the wild by the end of the nineteenth century, and its last representative died in the Cincinnati Zoo in 1914.

The Passenger Pigeon in Abundance

Reports of huge numbers of passenger pigeons, in passing flocks so large that they obscured the sun, might appear to be beyond belief were they not so consistent among independent observers for three centuries. Flocks were described as "having neither beginning nor end" and the birds themselves as being "in such prodigious numbers, as almost to surpass belief" and similar superlatives. These flocks had major impacts on the landscape, moving great quantities of nutrients, breaking tree limbs with the combined weight of roosting birds, consuming the food supply of some birds, mammals, and human settlers, and becoming the food supply of people and other predators.

Fossil records extend back to 100,000 years before the present and include western states not part of the colonial-era range. There are many records from Native American sites during the Woodland period (early Holocene, c. 500 BCE–1200 CE), particularly in what are now the Great Lake states. Pigeon bones have been found at settlement sites dating to about 1300 CE, but never in great quantities. Anthropologist Stephen Williams speculated that pigeon populations rose beginning around 1450–1500 CE, coinciding with a major decline in human populations, from about 800 CE, in the agricultural Mississippian culture that had occupied much of what is now the eastern United States. The human decline occurred particularly in the lower Mississippi valley and partway up the Ohio River. The decline may have been caused by a cooling climate during the Little Ice Age (c. 1400–c. 1850). Climate change and abandoned agricultural land may have created a forested landscape suitable for a rising population of passenger pigeons.

The sightings of French explorer Jacques Cartier are the first surviving record of passenger pigeon numbers; he recorded seeing "an infinite number" on July 1, 1534, at Prince Edward Island. Although Spanish conquistadors explored southern North America from 1540 to 1580, their writings contain no mention of skies filled by flocks of passenger pigeons. The explorer Samuel de Champlain recorded "countless numbers" on July 12, 1605, along the coast of southern Maine.

The enormous nomadic flocks roamed the continent, traveling hundreds of kilometers daily in search of food. In most seasons they fed primarily on acorns and beechnuts. In certain years, oaks and beech trees produce superabundant crops of nuts, known as mast. The locations of the mast crops varied from year to year. When the pigeons located a rich source of mast in the spring, they established a huge nesting colony (at least hundreds of thousands of pairs), known as a "city." Often there would be dozens of nests on a single tree. Each nest contained a single egg. Colonies ranged from 50 hectares to thousands of hectares. The largest nesting described covered much of the southern two-thirds of Wisconsin in 1871. The colonies had time only for a single nesting before the food supply was exhausted and the birds moved on. It is unlikely that the food was sufficient to allow for more than a single nesting in a year. During summer, the birds fed on abundant berries in the Great Lakes states and provinces.

Passenger Pigeons and Humans

Passenger pigeons were a source of fresh or preserved meat, feathers, and sport for native people and later for European settlers and their descendants. The hunting practices of tribes varied in the degree to which they disturbed nesting colonies or waited until the nestlings had reached sufficient size before killing them for food.

After Europeans colonized North America, essentially every flock that was encountered by humans at any time of year was subject to shooting, netting, and other means of killing. Many of the earliest writings of European colonists mention hunting of Passenger Pigeons either in transit or at nesting colonies or night roosts. "When these roosts are first discovered, the inhabitants, from considerable distances, visit them in the night with guns, clubs, long poles, pots of sulfur, and various other engines of destruction. In a few hours, they fill many sacks and load their horses with them" (Wilson, 1812). Fresh pigeon meat was a welcome food source on the frontier, particularly when the spring flocks arrived after the pioneers had spent the winter eating dried provisions.

The flocks of pigeons were agricultural pests. They ate all types of cultivated grains, particularly at planting time, when the seed was sown on the top of the field prior to harrowing it into the ground. Buckwheat was preferred, followed by wheat, corn, and barley. Rye and oats were eaten less commonly. Cultivated peas were eaten eagerly. There is conflicting information as to whether or not pigeons ate sprouting grains, but they were reported to have eaten tender shoots of varying kinds. After a field had been visited by pigeons, it frequently was ruined for the year. However, despite their abundance, pigeons apparently had a relatively small impact on agriculture, except very locally.

Commercial trade in pigeon meat, fat, and feathers began in the eighteenth century. Professionals used massive nets (2 m wide by 6–10 m long) to capture live pigeons for sale to be used in shooting matches and to capture birds to be killed for market. A large area known as "the bed" was cleared and prepared either by creating mud mixed with saltpeter and anise or by baiting a dried area with grain or mast. Natural salt licks were also used because pigeons were very fond of salt. The net was laid alongside the bed and set by an adjustment of ropes and a powerful spring pole. The operator sat in a blind made of boughs. A captured "stool pigeon" was used to help attract passing flocks. It was attached to a perch or box and manipulated to flap its wings or was tossed from its tethered post into the air when a flock passed by. When the birds came to the bait, the net was sprung. As many as 10 "strikes" of the net were made at a single bed on one day, with "forty or fifty dozen" birds being a good haul for one strike, although twice as many could be captured at once.

Often tens of thousands of adults and young were killed at a single colony. Yet this extensive killing probably had little, if any, impact on pigeon populations until the infrastructure (especially the railroad) was available to transport pigeons and carcasses from nesting colonies to markets in urban areas, beginning around 1840. In 1851 nearly 2 million pigeons (dead and alive) were shipped from a single nesting. In the late nineteenth century, from 600 to 1,200 men worked as professional pigeon trappers, using massive nets to capture live pigeons for shooting matches and using guns, poles, fire, and other means to kill birds for market. Wherever the pigeons nested or roosted (nonbreeding aggregations), hundreds of local people joined the professionals in wreaking havoc on the pigeons. "As settlement advanced, as railroads were built, spanning the continent, as telegraph lines followed them, as markets developed for the birds, an army of people, hunters, settlers, netters and Indians, found in the pigeons a considerable part of their means of subsistence, and the birds were constantly pursued and killed whenever they appeared, *at all seasons of the year*" (Forbush, 1927). In 1842, 3,000 live pigeons were transported by rail from Michigan to Boston. In 1851, an estimated 1.8 million pigeons were sent to New York City from a nesting in northern New York (Schorger, 1955, p. 145). By the time the Civil War ended, most of the United States east of the Mississippi was connected by railroad. Only a handful of nesting colonies were too far from rail or ship for market exploitation. Even a nesting in 1881 in Oklahoma, 176 km from the railroad, was pillaged by commercial trappers.

Pigeon populations, although reduced, were still enormous in the early 1870s. The nesting in Wisconsin in 1871 apparently included nearly the entire population of the species, estimated to be more than 135 million adults (less than 10 percent of what had been estimated only decades earlier). The population declined drastically during the 1870s. The last huge nesting, "something like 100,000 acres," according to ornithologist Edward Forbush, took place in 1878 in Michigan, near Petosky. There was a tremendous slaughter of perhaps 10 million birds, which were shipped out by the barrel. No enormous mass nestings were reported subsequently.

However, wherever the pigeons tried to nest or roost they were persecuted. In spring 1883, all of the young were reportedly taken. One man was said to have 60,000 and several others 10,000 young each. Over 5,000 birds were reportedly killed at a roost in Missouri the following winter. Over 1,000 carcasses were shipped to Boston in 1891. Market hunting continued until at least 1893, and shooting was reported to the end. Although the population had probably declined beyond point of recovery by the late 1880s, as many as 2,000 pigeons were reportedly taken for market in the early 1890s. By the mid-1890s, flocks of hundreds were noteworthy. By 1895, a flock of just ten pigeons drew attention.

Without the mass nestings, which had provided protection from predators, nestings of small groups and isolated pairs apparently produced insufficient numbers of offspring to maintain the species. The once dominant species trickled away to extinction. The end of the century brought the end of the passenger pigeon. The last reliable specimen was taken in 1900 in Ohio.

A few aviculturalists had taken wild pigeons into captivity for their private collections. Studies of these captive birds by early ethologists, such as Charles O. Whitman at the University of Chicago, provided much information on the habits of passenger pigeons. There were few recorded attempts to breed them in captivity, but there were some successes. A flock kept by David Whittaker of Milwaukee in the late 1880s apparently produced some young. These birds were reportedly the source of Professor Whitman's flock and perhaps the source of a flock of about two dozen pigeons at the Cincinnati zoo around 1875. The last of the descendents of this flock, the fabled Martha, died on September 1, 1914. With Martha's passing, this once noble species was no more.

The passenger pigeon had been successful due to its nomadic behavior and its colonial habits, which allowed it to exploit superabundant food sources that were unpredictable in space and time. Additionally the numbers were so enormous that nonhuman predators could not reduce the population. Ironically, these same factors attracted the human predators from whose relentless persecution the passenger pigeon was unable to recover.

Major Factors in the Passenger Pigeon's Decline

Two factors predominated in causing the decline: habitat destruction and direct exploitation by humans for food. Other explanations have been proposed, including climate, disease, and weather-related catastrophe, but without evidence.

Every passenger pigeon colony that was accessible to humans was exploited. I, along with fellow ornithologist Harrison B. Tordoff, have contended that the development of the transcontinental railroad and the telegraph in the nineteenth century were key factors leading to extinction. Railroads allowed access to nesting colonies, and the telegraph provided a way for scouts who located colonies to inform the professional pigeon trappers.

In addition to those killed for commerce, large numbers of birds were destroyed by locals or otherwise killed but not transported. A million birds could be lost at a single nesting. Yet, even the enormous numbers of birds killed were probably not sufficient to cause the precipitous decline in the population. Overhunting did not exterminate the passenger pigeon, as is commonly believed. Rather, the disturbance of the nesting colonies over a period of almost thirty years, well over twice the lifetime of the average bird, led the birds to abandon the colonies before they had raised young. This, coupled with slaughter of the squabs (nestlings)—more than 200 grams, full of fat, and apparently very tasty—and adults largely eliminated replacement of the population.

Another nineteenth-century technology, the portable saw mill (introduced in the 1870s), sped the destruction of what had once been a completely forested landscape, with disastrous effects on the pigeons. By 1880, about 80 percent of the original forest of New England had been cleared. Deforestation in the major nesting area of north central Pennsylvania began in 1872 but did not reach full speed until 1892. Michigan and Wisconsin were still well wooded in 1883, and although they were being logged rapidly, the trees being removed were largely pine, whose loss had less impact on the passenger pigeon than the loss of deciduous trees.

Deforestation was a major factor in the decline of the species because it reduced the opportunities for nesting and roosting colonies. The nomadic passenger pigeons roamed over enormous areas to find some conditions suitable for nesting colonies. Even when eastern North America was largely forested, there were relatively few areas with sufficient mast for a nesting colony. After the forest cover was decimated, there were many fewer suitable areas even for a reduced population of pigeons. In some years in the last decades of the nineteenth century there was no nesting at all.

The extinction of the passenger pigeon predated any conservation movement in America. The extinction of the once "limitless" flocks of pigeons along with the near extermination of the American bison (*Bison bison*) introduced Americans to the concept of human-induced extinction. Previously, the idea that a species, so abundant, could disappear was apparently inconceivable to the human mind. Some states belatedly approved laws prohibiting disturbance of the nesting colonies, but these laws were passed only after the birds were well on their way to extinction, and the laws were rarely enforced. The abundance of the pigeons was such that few people recognized that there were any risks to the species. Arguments that there was no need for protection generally doomed any proposed legal protection. In the words of ecologist and ethicist Aldo Leopold,

> There will always be pigeons in books in museums, but these are effigies and images, dead to all hardships and to all delights. Book-pigeons can not dive out of a cloud to make the deer run for cover, or clap their wings in thunderous applause of mast-laden woods. Book-pigeons cannot breakfast on new-mown wheat in Minnesota, and dine on blueberries in Canada. They know no urge of seasons, no lash of wind and weather. They live forever by not living at all. (Leopold, 1947, p. 3)

Further Resources

Bendire, C. (1892). Passenger pigeon. In C. Bendire (Ed.), *Life histories of North American birds: With special reference to their breeding habits and eggs* (U.S. National Museum Special Bulletin No. 1). Washington, DC: National Museum of Natural History.

Blockstein, D. E. (2002). Passenger Pigeon (*Ectopistes migratorius*). In A. Poole & F. Gill (Eds.), *The birds of North America* (No. 611). Philadelphia: The Birds of North America, Inc.

Blockstein, D. E., & Tordoff, H. B. (1985). A contemporary look at the extinction of the passenger pigeon. *American Birds, 39,* 845–52.

Forbush, E. H. (1913). The last passenger pigeon. *Bird Lore, 15,* 99–103.

——— (Ed.) (1927). Passenger pigeon. In *Birds of Massachusetts and other New England states* (Vol. 2, pp. 54–82). Boston: Massachusetts Department of Agriculture.

Goodwin, D. (1983). *Pigeons and doves of the world* (3rd ed.). London: British Museum of Natural History.

Leopold, A. (1947). On a monument to the pigeon. In W. E. Scott (Ed.), *Silent wings: A memorial to the passenger pigeon* (pp. 3–5). Madison, WI: Wisconsin Society for Ornithology.

Mershon, W. B. (1907). *The passenger pigeon.* New York: Outing Publishing Co.

Mitchell, M. H. (1935). *The passenger pigeon in Ontario* (Contribution No. 7 of the Royal Museum of Zoology). Toronto: University of Toronto Press.

Schorger. A. W. (1955). *The passenger pigeon: Its natural history and extinction.* Madison: University of Wisconsin Press.

Wilson A. (1812). *American ornithology* (Volume 5, pp. 102–12). Philadelphia: Bradford and Inskeep.

David E. Blockstein

■ Conservation and Environment
Pesticides' Effects on Honeybees

Honeybees contribute substantially to the pollination of various wild plants and food crops. The annual value of agricultural crops benefiting from honeybee pollination is estimated at as much as $20 billion/year in the United States alone. Studying the influence of agrochemicals on honeybee behavior is important for the survival of honeybees, public policy issues, honeybee population regulation, environmental degradation, and the use of biological controls.

This entry briefly surveys the literature on what is known about the effect of agrochemicals on honeybee behavior, including learning, foraging, and physiology. We will focus on the effects of toxic chemicals, sublethal effects, and chemicals considered harmless to honeybees.

Toxic Chemicals

The use of toxic chemicals to control insect pests has a long history. Chemicals such as DDT, sevin, rotenon, diazionon, methoxychlor, and imidacloprid have been used to control such pests as the Colorado potato beetle, cabbageworm, and gypsy moth. What has not always been known is how these chemicals affect honeybee behavior. Data generated over the past 50 years have shown that pesticides disrupt the functioning of the central nervous system, metabolic processes, and some physiological processes, such as molting and reproduction. Pesticides that are specially formulated to kill target insects usually do so by influencing receptor molecules in central nervous system, mechanical, photo, and/or chemical receptors. Pesticides have also been developed that are synthetic analogs of enzyme substrates that interfere with metabolic pathways.

As a case study, consider the Africanized honeybee in Brazil. The Africanized bee is important to the economy of Brazil in two main ways. Aside from the production of honey as a major agricultural product, bees serve as pollinators of the cotton crop as well as many others crops in the Brazilian economy.

Cotton is an important crop for the agrarian sector and development of the textile industry in Brazil. Cotton production in Brazil was adversely affected soon after the appearance of the cotton boll weevil in 1983 and has led, for example, to unemployment, depreciated land value, and the closing of cotton gins and oil mills. The major strategy to combat the boll weevil is the use of pesticides. The use of pesticides such as endosulfan, decis, baytroid, and sevin to control the boll weevil has had adverse effects on the honeybee population. When bees were exposed to baytroid and sevin, death quickly resulted. Interestingly, bees exposed to endosulfan could acquire a learned response, but over the course of training, the learning became unstable and soon disappeared. Those exposed to decis showed a pattern of learning indistinguishable from untreated controls.

Sublethal Effects of Agrochemicals

The study of toxic chemicals on honeybee behavior has extended to the area of sublethal effects. When a toxic chemical is released into the environment it can be degraded, for example, by rain or ultraviolet rays from the sun. The result is that honeybees can be exposed to sublethal levels of agrochemicals that normally would be lethal. Evidence exists that sublethal doses of pesticides may be decreasing the number of honeybee colonies available for pollination and reducing the effectiveness of honeybees as pollinators. Sublethal doses of deltamethrin, for example, disrupt the homing flight of honeybees, whereas parathion disrupts the communication dance of foragers. In addition to the disruption of natural behavior, it is known that sublethal exposure to permethrin, coumaphos, and diazinon retards learning.

Effects of Agrochemicals Considered Harmless

Recently, a new line of investigation has begun on agrochemicals considered harmless to honeybees. These compounds may include some of the new generation pyrethroids, insect growth regulators, and metabolic by-products, all of which are currently used in formulation of new products. Many of these new products are considered by the Environmental Protection Agency, and other regulatory bodies, to be user-friendly, target-specific, and environmentally safe. However, little is known about their effects, if any, on honeybee behavior. In order to use these chemicals effectively and without injuring these important pollinators, it is necessary to know what effects these agrochemicals have on honeybee behavior.

The first experiment in the study of chemicals considered "not harmful" to honeybees was an investigation of dicofol. Dicofol is a chlorinated hydrocarbon pesticide. It is considered nontoxic to most insects and is used primarily to control mites. However, honeybees pretreated with dicofol exhibited significantly lower levels of learning than honeybees not pretreated.

Recently, other experiments have been conducted using insect growth regulators tebufenozide and diflubenzuron. The results of these experiments were similar to those with dicofol and equally unexpected. The learning ability of honeybees was again disrupted by agrochemicals that were once thought to be harmless.

Another example is *imidacloprid*. Imidacloprid is a novel insecticide that mimics nicotine. It is applied to the seeds of crops and, as the plant develops, is transported to the stem and leafs of the plant. Aphids and other pests, such as the Colorado potato beetle, will die if they ingest imidacloprid. Imidacloprid is also used on sunflower seeds. Sunflowers are an excellent source of nectar for honeybees, and sunflowers depend on bees for pollination. Although it is toxic to honeybees, honeybees are not in direct contact with imidacloprid. It is known from the plant data that the average values of imidacloprid contained in the pollen of sunflowers and of corn was found to be around 3 parts per billion, which is one-fifth of the dose known to cause changes in waggle dance communication in honeybees. The French government decided to prohibit use of imidaclopride on sunflower seeds because of its effect on honeybees.

Pesticides are the primary weapon against insect pests. Unless carefully monitored, the use of agrochemicals can be ecologically unsound, leading to problems such as insect pest resistance, outbreaks of secondary pests, adverse effects on nontarget organisms, pesticides residues, and direct hazards to those individuals applying the chemicals.

Further Resources

Abramson, C. I., Aquino, I. S., Ramalho, F. S., & Price, J. M. (1999). Effect of insecticides on learning in the Africanized honeybee (*Apis mellifera* L.). *Archives of Environmental Contamination and Toxicology, 37,* 529–35.

Abramson, C. I., Squire, J., Sheridan, A., & Mulder, Jr., P. G. (2004). The effect of insecticides considered harmless to honeybees (*Apis mellifera* L.): Proboscis conditioning studies using the insect growth regulators Confirm F (Tebufenozide) and Dimilin 2L (Diflubenzuron). *Environmental Entomology, 33,* 378–88.

Devillers, J., & Pham-Delègue, M. H. (Eds.). (2002). *Honeybees: Estimating the environmental impact of chemicals.* London: Taylor & Francis.

Winston, M. L. (1997). *Nature wars: People vs. pests.* Cambridge: Harvard University Press.

Charles I. Abramson and Janko Bozic

■ Conservation and Environment
Rewilding

Rewilding is a science-informed conservation strategy pioneered by conservation biologist Michael Soulé and the Wildlands Project in the 1990s. It is based on the important role of large carnivores for maintaining and restoring health and integrity in ecological communities as shown by field research around the world. In turn, large carnivores need large roadless "core" habitats where they are relatively safe from human killing and habitat destruction, along with safe movement linkages between core habitats. In other words, rewilding protects and restores large carnivores, big areas of roadless habitat, and wildlife movement linkages between roadless areas. Far from a radical approach, rewilding is a logical evolution of conservation thought and practice.

At the end of World War I, two new conservation approaches appeared in the United States. Forest rangers, led by Aldo Leopold, were increasingly worried about the inroads "motor-cars" were making into the backcountry of national forests. They called for "wilderness areas" to prevent motor roads and motor-cars in such wild areas. The Ecological Society of America established a Committee on Natural Areas, led by Victor

Shelford, to identify representative samples of all of America's plant communities to be protected in their natural state. In the 1920s, these were seen as separate efforts, with wilderness areas being set aside to protect "primitive conditions of travel" (horsepacking and canoeing) and natural areas being set aside to preserve "primitive conditions of growth" (in the words of the day). During the 1930s, however, Aldo Leopold and the other founders of The Wilderness Society began to see the ecological and recreational values of natural areas and wilderness areas, respectively, as compatible values to protect in wilderness areas. So, for many decades, both conservationists and scientists have seen protected areas (national parks, wildlife refuges, wilderness areas, and so on) as the most important tool for protecting wild nature.

In the 1960s, Robert MacArthur and E. O. Wilson developed the theory of island biogeography, which led other biologists to recognize the "area-species relationship"— all else being equal, a larger area will support a larger number of species than a smaller area. Concurrently, field biologists and other researchers began to publish claims that the Earth was in a mass extinction event, greater than any since the demise of the dinosaurs 65 million years ago, and that it was caused by human beings. Based on this research, Michael Soulé organized a conference in San Diego to create a new discipline— conservation biology. The task of conservation biology was to halt the human-caused mass extinction, and it recognized that protected areas were the best tool to do so. Conservation biologists asked how protected areas could be better selected, designed, and managed to protect all species. By 1985, another conference had been held, a Society for Conservation Biology created, and the *Conservation Biology* journal planned.

In 1986, Reed Noss, at the University of Florida, proposed a new model for protected areas in Florida. Instead of the old approach of isolated, islandlike reserves in a sea of development, Noss called for a network of wildlife movement corridors linking protected areas, which would then be surrounded by multiple-use buffers. Noss's proposal was revolutionary, in that his model of a network linking protected areas has now replaced the previous model of isolated protected areas alone. At least it has become the new model in how scientists, biologists, and land managers think, if not always in how conservation is practiced.

Also in the late-1980s, James Estes, working with sea otters in kelp forests of the North Pacific Ocean, and Michael Soulé, working with coyotes and native birds in the San Diego suburbs, published research showing the importance of large carnivores to the health and integrity of ecosystems. In the 1990s, additional research, including that on wolves in Yellowstone National Park and jaguars in Venezuela, backed this theory up. Wolves, big cats, and other large carnivores exercise "top-down regulation" of ecosystems in five ways: control of numbers of their prey, behavior of their prey, numbers of smaller predators, behavior of smaller predators (mesocarnivores), and through competitive exclusion, whereby in a guild of burrowing rodents, for example, the elimination of their predator may allow one species to outcompete and replace the others. Research in Yellowstone National Park, in particular, also showed that when large carnivores were repatriated to an ecosystem, where they had been previously eliminated, the ecosystem began to heal itself.

In the 1980s and 1990s, many conservationists and biologists recognized that few ecosystem types still had substantial undisturbed areas. They realized it was not enough to protect what was left; ecosystems needed to be restored. In most cases, such ecological restoration was on the small scale—local wetlands or streams, for example. Other conservationists, however, including the original Earth First!, called for the restoration of large areas of wilderness by closing roads and removing dams.

In 1992, a small group of conservation biologists and wilderness conservationists formed The Wildlands Project to begin ecological restoration on a large scale. This was called "rewilding." Michael Soulé soon developed a specific scientific approach of rewilding based on the necessary role of large native carnivores. In 1997, he and John Terborgh convened a workshop of thirty leading conservation biologists to consider protection and restoration of wild Nature on the continental scale. In 1998, Soulé and Noss wrote that rewilding was based on the "Three Cs": large carnivores, core protected areas (roadless), and connectivity between the cores. In 2000, the Wildlands Project and cooperating groups published an ambitious rewilding vision for the Sky Islands region of southeastern Arizona and southwestern New Mexico, with further connections to the Sierra Madre of Mexico and to the Southern Rocky Mountains. This Sky Islands Wildlands Network called for the recovery of the Mexican wolf, jaguar, and other species in a protected network of federal, state, county, and private (with cooperating owners) lands. It was followed by rewilding visions for the New Mexico Highlands and Southern Rockies, stretching from the Mexican border to southern Wyoming. Similar visions are in the works for the so-called Spine of the Continent MegaLinkage from Alaska's Brooks Range to Mexico's Sierra Madre.

Dave Foreman called for a rewilding-based North American Wildlands Network using Four Continental MegaLinkages—Artic/Boreal, Pacific, Spine of the Continent, and Appalachian—in his 2004 book, *Rewilding North America.* At the October 2005 8th World Wilderness Congress in Anchorage, Alaska, conservationists from Latin America, Europe, Africa, Asia, and Australia discussed rewilding projects in their regions—many crossing international boundaries.

Rewilding, then, is large-scale ecological restoration that stresses the vital role of large carnivores in maintaining healthy ecosystems and the need large carnivores and other species have for large roadless core wildernesses tied together across continents by wildlife-movement linkages.

Further Resources

Foreman, D. (2004). *Rewilding North America: A vision for conservation in the 21st century.* Washington, DC: Island Press.

Dave Foreman

■ Conservation and Environment
Whale and Dolphin Culture and Human Interactions

Culture is seen by many as a uniquely human attribute. But if we define culture in a way so that it includes the generally accepted forms of human culture, such as religion, language, art, technology, symbolism, social conventions, political structures, and "pop" culture, then non-humans have culture too. The key to culture is learning behavior from others—social learning. Once behavior is being imitated, emulated, taught, or transferred between individuals by any other form of social learning, culture can happen. With culture, the processes of genetically driven evolution become changed. Behavior can sweep through a population, or be entrained in it by "cultural conservatism." Group-specific badges, such as ethno-linguistic markers, can evolve and drive cooperation within, and competition between, culturally marked groups. These processes have dominated the recent history of humans, but they occur in other species, including oceanic species, and they can affect how the species interact with humans.

In the centuries since humans became marine mammals, interactions between humans and whales have mostly involved humans intentionally killing whales. The scale of the slaughter was extraordinary—most large whales that lived in the twentieth century were killed by whalers. But as whaling ran its course in the 1970s, the human-caused deaths did not cease. Whales are killed, often slowly and painfully, by entanglement in fishing gear, by ship strikes, and as we have recently discovered, by noise.

Humans can affect whales in ways other than through a fast or slow death. We can injure them, disturb them, and affect their behavior. Our profound alterations of the marine habitat have closed some niches and opened others. In the North Pacific, gray whale calves seem to be an important food for some killer whales. In the North Atlantic there have been no gray whales since their extirpation several hundred years ago. During the course of whaling, killer whales in all oceans scavenged the carcasses of other species of whale killed by whalers. But when whaling virtually stopped in the 1970s, the killer whales moved on. In many parts of the world they have started removing fish from longlines, to the consternation of fishermen. The destruction of the sea otter populations along the Alaskan Aleutian archipelago in the 1980s, and consequent restructuring of almost the entire nearshore ecosystem, seems to have been the result of a prey shift by just a few killer whales, perhaps some of those who had subsisted largely on whale carcasses in the heyday of whaling.

That diet shifts by just one non-human predator should have such significant conservation and management consequences is partially a tribute to the killer whale's power, size, and intelligence. But, as with another voracious predator, the human, there is another important factor: culture.

Culture is defined in many ways but the essence is that individuals learn their behavior from each other in such a way that groups of individuals acquire distinctive behavior. When behavior becomes determined by cultural, rather than by genes or individual learning, then it can take some unusual forms and have immense consequences. Humans are the prime example. Human culture includes some wonderfully useful features that enrich our lives. These include language, technology, art, and music. But some forms of culture, such as Kamikaze cults, guns, and fast-food restaurants, are harmful to individual humans, and others, such as nuclear weapons, fundamentalist religious beliefs, and fossil-fuel burning, threaten us as a species and in many cases other species, too.

Because of the abilities of our brains and the opposability of our thumbs, human culture has reached extraordinary heights and depths, literally and figuratively. But other animals have culture. It has been found in fish, rats, and many other species, but it is best known in songbirds, primates, and cetaceans (whales and dolphins). The cultures of the different species vary characteristically. For instance, song-birds seem to be cultural primarily in their songs, whereas culture has a particular role in the foraging and social behavior of chimpanzees. In one important respect, whale and dolphin culture seems closest to humans. In several species of whale and dolphin, social groups that use the same habitat behave differently, in an analogous fashion to multicultural human societies.

How Dolphins and Whales Are Affected by Humans

Just as human culture profoundly affects our interactions with others species, whale and dolphin cultures may also influence the interspecific relationships. Here are some examples that have arisen over recent years during our dealings with whales and dolphins, starting with dolphins.

The bottlenose dolphin ("Flipper") is the most-studied cetacean. It is found in many parts of the world and has been studied in some of them. The site of one of the longest and most detailed studies is Shark Bay, Western Australia. The dolphins in Shark Bay have a wide diversity of feeding strategies, including using sponges as tools to probe beneath the surface, stranding intentionally on beaches, and attacking very large fish. It seems as though these strategies are largely passed on through social learning, perhaps principally from mother to offspring, and so are a form of culture. One of the strategies, begging for fish from beach goers, has important negative consequences: the calves of the dolphins who do this have higher mortality. This only involves a few animals, but on the other side of Australia, in Moreton Bay, there are two communities of bottlenose dolphins. They use the same waters, but one regularly feeds on discards from prawn-trawlers, probably a cultural behavior, and the other does not. The communities rarely interact. They will be differentially affected by human activities, such as changes in trawling activity due to overexploitation of the prawns.

On a more positive note, 25–30 bottlenose dolphins in Laguna, Brazil, run a fishing cooperative with local human fishers, in which the dolphins and fishers follow a strict protocol with the dolphins herding the fish into the nets and feeding on the entrapped fish, to the benefit of both. This has been going on for generations, the cooperative fishing culture apparently passed from mother to daughter in the dolphins, and father to son in the humans. There are other dolphins in the Laguna area who do not participate in the cooperative fishing and sometimes try to disrupt it. There are reports of similar human-dolphin fishing cooperatives in other places, including one involving a different species, the Irrawaddy dolphins in Burma.

In a similar vein, there was for many decades a whaling cooperative in Twofold Bay, Australia. Generations of killer whales would herd baleen whales into the hunting areas of shore-based whalers and then scavenge the dead animals once the whalers had done their work.

From a cultural perspective, the most interesting whales and dolphins may be those which form permanent matrilineal groups. In such species, most female whales swim in the same social unit as their mothers while both are alive. In killer whales and pilot whales this often extends to the males, so that there is no dispersal from the natal social unit by either sex. In such cases the female units can develop distinctive cultures. The most easily studied parts of these cultures are vocal repertoires. Pods of killer whales have distinctive dialects and are grouped into clans that are recognized by vocal similarity but seem to be based on common ancestry. Sperm whale social units associate preferentially with other units from their own clan, even though units from two or more clans may share particular waters. In humans, dialects are markers for rich cultural differences between ethno-linguistic groups, and so it seems to be in the whales. The nonvocal cultural differences are those that are most likely to interact with anthropogenic effects on the ocean habitat.

The two principal sperm whale clans off the Galapagos Islands can be distinguished by their codas—Morse code–like patterns of clicks. But they use the waters differently. Groups of the "Regular clan" ("click-click-click-click") primarily use the waters close to the islands and have convoluted paths as they search these waters for deep water squid. In contrast, the groups of the "Plus-one" clan ("click-click-click-click-pause-click") are generally further from shore and move in straight lines. In most conditions the groups of the "Regular clan" appear to have greater feeding success, but in the years when El Niño strikes and the waters warm, losing much of their productivity, all whales do worse, but the "Plus-one" groups are less affected and do relatively better. Global warming seems likely to increase the frequency and strength of El Niño's as well as the prevalence of El-Niño–like conditions. Preserving the cultural inheritance of the "Plus-one" clan may be crucial to the survival of the sperm whales in the waters.

More is known about killer whale cultures. Cultural differences have been recognized across several tiers of social structure—matrilineal units, pods, clans, communities, and type—and span a wide variety of behavior. Apart from the vocal dialects, evident within all tiers, there are differences in foraging behavior, social behavior, and play behavior. The "southern resident community" has a ritualized greeting ceremony when pods meet; is known for breaching, leaping from the water; and for a short while its members had a strange distinctive fad: pushing dead salmon around. In contrast the "northern residents" do not show the greeting ceremony and rarely breach but have a "rubbing beach" that they use regularly.

Some of these differences interact with human behavior. Most dramatically, when killer whales were captured for the display industry, they were fed fish. This was fine for the "residents" that eat fish. But a "transient" killer whale who was also caught died from starvation rather than eat fish. The transients primarily eat mammals.

This is an example of cultural conservatism taken to the extreme. But culture can play it either way. Sometimes it promotes conservative behavior, preventing adaptive responses to changed circumstances, but in other situations, culture can allow a species to quickly adapt to new environments as animals learn new ways of life from one another. The spread of scavenging from whalers and feeding from long lines by killer whales mentioned at the start of this article are two examples in which social learning likely helped spread an activity that some humans found extremely annoying.

When culture becomes a major determinant of behavior, as it appears to have with killer whales and sperm whales, it can take dramatic forms, as a look at human behavior so clearly shows. Cultural conservatism and cultural opportunism are joined by group-specific cultural "badges" and maladaptive adaptive. We do not know why groups of apparently healthy whales and dolphins mass strand on beaches, but it seems likely that a usually sensible cultural imperative, such as "stay with the group whatever happens," has a part. Thus we need to view the behavior of cultural animals with a different perspective, and this carries through to their conservation biology.

This came to a head with the trans-border "southern resident" killer whales, whose small population was declining. They differ from the "northern residents," a healthier population, by only one known base-pair in the genetic code, but, as mentioned previously, a host of cultural traits. Should the "southern residents" be specifically protected under endangered species legislation? The Canadian listing committee (COSEWIC) thought so and listed them as "endangered." The United States equivalent, the National Marine Fisheries Service (NMFS), thought not and listed all killer whales in the area as "depleted." However, after protests and legal challenges, NMFS changed its perspective and upgraded the "southern residents" to "endangered."

Cultural species may, through a new and rapidly spreading form of behavior, quickly become embroiled in a conflict with humans, or through their cultural conservatism they may not react appropriately when we change their environment. But cultures, like genes, have evolved through natural selection, and they mostly have an important role in allowing the animals to live their lives. Just as we seek to preserve genetic biodiversity, we must preserve the cultural diversity of such species, so that the cultured species of the ocean, such as sperm and killer whales, have the knowledge to survive when we change their habitat.

See also

Ecotourism—*The Whale and the Cherry Blossom Festival*
Ecotourism—*Whales, Dolphins, and Ecotourism*
Geography—*Animal Geographies*
Living with Animals—*Dolphins and Humans: From the Sea to the Swimming Pool*

Further Resources

Chilvers, B. L., & Corkeron, P. J. (2001). Trawling and bottlenose dolphins' social structure. *Proceedings of the Royal Society of London B, 268*, 1901–05.

Rendell, L., & Whitehead, H. (2001). Culture in whales and dolphins. *Behavioral and Brain Sciences, 24*, 309–24.

Richerson, P. J., & Boyd, R. (2004). *Not by genes alone: How culture transformed human evolution.* Chicago: University of Chicago Press.

Whitehead, H., Rendell, L. Osborne, R. W., & Würsig, B. (2004). Culture and conservation of non-humans with reference to whales and dolphins: Review and new directions. *Biological Conservation, 120*, 431–41.

Hal Whitehead

■ Conservation and Environment
Whale Watching: A Sustainable Use of Cetaceans?

Watching whales and dolphins is fascinating more and more people around the world. Whale watching—the observation of cetaceans (the biological order of whales and dolphins) in their natural environment—today is promoted as an important means of increasing environmental awareness. It is also regarded as a sustainable use of cetaceans—as long as it is conducted in an ecologically sustainable manner.

This branch of the ecotourism sector is growing more rapidly than any today. Already we can hear voices calling out for stronger regulation or even the ceasing of whale-watching activities as they have become a negative influence on the cetaceans' lives in some places.

The following is an overview of the different types of whale watching as well as its (positive and negative) effects, both on humans and the cetaceans. Also, I will outline how whale watching can be conducted so as to make it a truly sustainable use of cetaceans.

Types of Whale Watching

Let us define whale watching as the observation of cetaceans in the wild. This includes all cetacean species, from the small Hector's dolphins (*Cephalorhynchus hectori*), to medium sized pilot whales (*Globicephala macrorhynchus*), to orcas (*Orcinus orca*), to beaked whales (*Ziphius* sp.), and up to the blue whale (*Balaenoptera musculus*), the largest animal that ever lived on earth. In general, whale watching has a strong commercial component or is even a purely commercial venture. However, the definition given here comprises commercial and noncommercial activities in all types of cetacean habitats (coastal and offshore waters, estuaries, and rivers) but excludes whales and dolphins in captive (dolphinaria) or "semi-captive" settings (enclosed bays). Furthermore, we must not forget the numerous private boaters as well as observers from land who actively or passively come close to whales and dolphins.

When thinking about whale watching, we intuitively imagine people on a boat. In fact, the great majority of whale watchers worldwide use vessels to approach cetaceans. Vessel types vary widely in terms of size and passenger number. In some places kayaks are used to come close to marine mammals; elsewhere rubber boats without boarders are in use. Larger motor vessels, yachts, and catamarans recommend themselves in areas where sea state and swell may be changing quickly. Small former fishing boats (even whalers) are used, or specially designed whale-watching boats. Even ferries may be employed to watch cetaceans. As an example, ferry lines connecting the United Kingdom and Spain are advertising the chance to see a variety of cetaceans on their passage through the Bay of Biscay. Finally, whales and dolphins may be observed from cruise ships and/or their helicopters.

Besides seagoing trips, there are uncountable sites worldwide where you can watch cetaceans from elevated coastlines, from watch towers, and even from beaches. From a cliff twenty or thirty meters high you can see as far as several miles offshore. As an example, along the Pacific coast of California are many sites with special watch-towers. In fact, whale watching started as a land-based activity with the establishment of the world's first grey whale watch post close to San Diego (which still exists).

Yet another way of whale watching is swimming with cetaceans. This activity has grown in significance and is witnessing a fast expansion all over the globe. More and more operators are including swims in their trips or are specializing in this kind of nature tourism. Scientists are increasingly worried about swim-with-cetacean programs because they hold a considerable potential to disturb the animals. As a consequence, some nations have now prohibited such activities or put strong regulations on them.

Finally, there are so-called sociable dolphins: single animals turning up on coasts worldwide performing astonishingly human-directed behavior on a regular basis. The phenomenon is not well understood, although there have been reports of such incidents for many centuries. Sociable dolphins are mostly bottlenose dolphins (*Tursiops truncatus*). However, there are well-documented cases involving belugas, orcas, and other species. In some areas, these dolphins become local attractions, sometimes transforming those locations into tourist destinations. Famous examples are Fungie on the Dingle Peninsula in southwestern Ireland and Jojo in the Bahamas.

Development of the Industry

There are remainders of cave paintings showing whales in Norway dating back thousands of years. Many myths and legends deal with a special relationship between humans and cetaceans. Ancient Greek and Roman stories report of friendships between boys and dolphins; the god Apollo was said to transform into a dolphin regularly. Contrastingly, despite a long history of whaling, whale watching is a recent phenomenon. Organized trips were first established in the 1950s in California after it was recognized that the (newly protected) grey whales were returning to their nursing and calving ground, thereby using migration routes close to the coast.

Erich Hoyt, a renowned expert in the field of whale watching, has thoroughly documented the development of the whale-watching industry, and most of the numbers in this paragraph are taken from his comprehensive "Whale Watching 2001" report. During the 1970s, with the increasing environmental awareness, whaling became an issue to the broad public. Subsequently, whales turned into global symbols for the environmental movement, and whale watching spread over the United States. One may speculate that the newly developing image of whales and with this the rising public pressure led the International Whaling Commission (IWC) to implement the ban on commercial whaling in 1986, which is still valid today. After a moderate growth until the late 1980s, in 1995 there were 45 nations and overseas territories offering whale watching, with more than 5 million whale watchers.

In 2001 this number had risen steeply to 87 countries and overseas territories with more than 500 whale watching locations and an estimated total of no less than 9 million whale watchers per year. The (exponential) growth in recent years has slowed down only a little, if at all. The industry now is worth several billion US$ and has witnessed yearly growth rates of up to 100 percent or more in some countries. Thus, whale watching has become one of the fastest growing branches of the tourism industry. Today, the number of whale watchers worldwide is estimated at 10–12 million per annum. In Canada, whale watching rivals the current income of major fisheries, and Iceland's income through whale watching recently was higher than the earnings of the nation's whaling industry at its best times.

Erich Hoyt found that since the 1990s whale watching has experienced worldwide growth rates of more than 10 percent per year. To date, fascinated humans can go whale watching on every continent of the globe. Hence, the rapidly expanding business of watching whales and dolphins has made the question of the protection of the animals a crucial issue.

Bow-riding dolphins: do they watch us, too?
©Fabian Ritter/MEER e.V.

In contrast, whale-watching regulations have been established in only a few countries and territories. Where rules or codes of conduct exist, they often lack scientific background. Studies dealing with growing human-cetacean interactions are still relatively rare, and the majority of operators are still purely commercial.

How Does Whale Watching Affect Humans? The Benefits

The most obvious value of whale-watching tourism is its economic power. Whale watching has become an important regional source of income, at times being one of the primary earners. In several locations around the world, whale watching has virtually transformed the economic structure. In many areas the income generated from whale watching by now has surpassed any other source of income, often resulting in higher profits than the previous whaling industry. Whale watching thus has potential to become the driver for regional, possibly even national, economic development. Examples are Iceland, where whale-watching tourism is spreading rapidly nationwide, and Kaikoura, a small village on the east coast of New Zealand and now the "capital of whale watching."

One eminent aspect of whale watching is the intense and direct effect on our own nature. Seeing a large whale surfacing close to the boat, listening to a dolphin's whistles while it is bow-riding, or watching mothers with calves from shore affects us in a way that sometimes cannot even be put into words. This can lead us to think differently about our relationship to these animals (or to nature in general). Thus, cetaceans can encourage us to support them.

Moreover, whale watching can significantly contribute to knowledge about cetaceans. With a little training, paper and pencil, and camera at hand, any operator can collect basic sighting data in a simple, cheap, and effective way and thereby contribute to the understanding of abundance, distribution, habitat use, or behavior. As a result of improved knowledge through whale watching it is possible to accomplish scientifically based public education. This will be highlighted further later. For governments, whale watching has become a learning field for management in the field of tourism. In this relation, the protection of natural resources through wise management can be a contribution to the reputation of nations, regions, and communities.

Finally, whale watching can be seen as an alternative to keeping cetaceans in captivity. Natural behaviors can be observed, not trained animals. A lot can be learned about these fascinating animals, first of all that wild animals cannot be separated from their environment without losing some of the central aspects of their lives. Almost automatically, whale watchers will be sensitized also for the threats that marine mammals have to face today. However, this requires some form of educational program.

How Does Whale Watching Affect Cetaceans? The Risks

In the light of the well-known detrimental effects of anthropogenic activities already putting cetaceans under pressure on a global scale (habitat destruction, pollution, overfishing, noise, bycatch, and direct hunt), it is essential that whale watching does not become another curse to the animals.

Although the least disturbing type of whale watching is the observation from land, the majority of whale watchers use boats to approach cetaceans. Today we know that vessels can influence cetacean behavior in various ways. When assessing the impact of whale watching (and other) boats on cetaceans, two broad categories can be identified: short-term effects and long-term effects.

Short-term effects are usually relatively easy to observe, as they show up immediately when cetaceans and vessels interact: for example, instant changes of behavior, swimming direction and/or speed, prolonged dive times, or the separation of groups (especially mothers and calves) due to insensitive conduct. The underwater noise of vessel engines and propellers may also interfere with the communication skills of dolphins and whales. The worst scenario is a collision or propeller strike. And as Paul Forestell, a researcher studying the humpback whales off Hawaii, has said, another often underestimated effect is the creation of a misleading image of the animals due to marketing based on competitive claims, leading to false expectations on the side of the tourists (Forestell & Kaufman, 1995).

If disturbances become frequent, repetitive, and persistent, long-term effects such as changes of behavioral budget, or stress due to permanent (physical and noise) disturbance may be witnessed. This may result in elevated susceptibility to diseases, lower reproduction rate and/or changes in distribution, and even the disappearance of animals due to migration to other, undisturbed areas. In a study conducted by M.C. Allen in Florida in 2000, it was found that bottlenose dolphins showed a decrease in use of their foraging sites during periods of high boat density. In that sense, cetaceans with a high degree of site fidelity are much more vulnerable to whale watching than those species that do not constantly live in the same area. Habituation and sensitization of individual animals, groups, or populations may also be observed. Habituation may, for example, lead to risky behaviors such as close approaches to propellers, which I have often

Atlantic spotted dolphins riding the bow of a small whale watching boat. ©Fabian Ritter/MEER e.V.

observed in Atlantic spotted dolphins (*Stenella frontalis*). Thus, even apparently positive interactions could have negative effects on populations.

However, whale watching up to now could not be made responsible for population declines or major changes in habitat use. In most places where whale watching is taking place, the animals appear to be able to adapt to increasing boat traffic. Whale watching tourism alone would not be much of a problem in many locales if there weren't already other problems in place—for example, pollution and overfishing. Rather, we have to think about synergetic effects, in other words, the accumulation of human-caused interferences.

Long-term effects are much more difficult to detect because they require a constant monitoring of populations—which is realized only rarely. If performed in an unsustainable manner over longer periods, whale watching can pose a threat to cetacean populations. That is why some people ask if the use of whales for tourist purposes is just one more form of harmful exploitation of cetaceans. Hence, it is important that a precautionary approach is applied.

Preconditions for the Sustainability of Whale Watching

It is necessary to regulate the intensity of the use of cetaceans. A possible solution has been termed "ecological" or "sustainable" whale watching. With the incorporation of the Principles of Sustainability, it is possible to develop the industry in an ecologically orientated manner.

Code of Conduct

Conscientious whale watchers are not pushing anything. They are looking out patiently for cetaceans without racing around at high speed. If they meet cetaceans, they let the animals decide to come closer or not. Their way of approach is cautious and passive. They do not try to attract dolphins or instigate special behaviors by speeding up, making sounds, or throwing fish or other items overboard. Thus, ecological whale watching in the first place means avoiding effecting natural behaviors and acknowledging different types of responsiveness. Cetaceans sometimes do interact willingly; sometimes they don't. "We are guests at sea—and we should behave so!" is the whale watcher's golden rule.

Numerous guidelines, legal regulations, and non-binding codes of conduct for whale watching have been established worldwide. The number of nations developing their own sets of rules is increasing slowly but steadily. Carole Carlson (1996) in her global overview of whale watching regulations described that according to region and species, they cover different aspects. However, there are a number of general features, for example, reducing vessel speed, setting a maximum number of boats, maintaining minimum distances, regulating the way of approach (direction, speed), and so on. A list of typical aspects established in whale watching regulations worldwide is presented in Table 1.

Table 1. Aspects Usually Comprising Whale-Watching Guidelines/Regulations

1. Maximum number of boats within, e.g., 100/300/500 meters (usually 3 boats).

2. Minimum distance: do not approach closer than 50/100 meters (usually 100 m).

3. Maximum duration of encounters, e.g., 15 or 30 minutes.

4. Do not disturb the natural behavior of the cetaceans.

5. Approach slowly on an angle from laterally behind the group or animal—no "head-on" approach.

6. Reduce speed to "no wake" speed when close to animals.

7. Avoid sudden changes in speed and direction.

8. If animals approach is close, put in neutral gear and wait.

9. Do not separate mothers from their offspring.

10. Avoid loud sounds of any kind.

11. Do not throw anything overboard.

12. When more than one boat is present, coordinate activities via radio.

13. No swimming, snorkeling, or diving.

14. Maintain low speed when leaving the animals.

It is important to know about the characteristics of the species, as a number of studies have shown a high degree of inter- and intra-specific variation in responsiveness. The duration of the encounter, similar to the conduct, principally has to be put in relation to a species' group size, group formation, and behavior. Attention also should always be directed to the percentage of calves and juveniles present in a group because mothers with calves or newborns should not be approached at all.

Onboard research (here: acoustics) is a feature typical for operations combining tourism with scientific studies. ©Fabian Ritter/MEER e.V.

Research

The cooperation between whale-watching operators and researchers cannot be overemphasized. With the enormous success of this tourism branch also arises the possibility to use nonscientific platforms (whale-watching boats) for the collection of biological data. Whale-watching activities constitute an enormous potential to explore cetacean populations on a regular basis. As an example, the data assembled through our research project in the Canary Islands has been collected exclusively from whale-watching boats. Years ago, a simple sighting scheme was implemented and the operator willingly reserved a place on board to conduct behavioral research. Today, La Gomera's waters are exceptionally well known in relation to the status of cetacean populations. The tour operator in return can promote his engagement in research and the presence of a researcher on board, which also offers a good way to provide exposure of tourists to scientific settings.

Erich Hoyt found that research integrated into whale-watching activities also makes good financial sense because it increases the quality of the experience for tourists. What is more, a study conducted by Mark Orams in Australia found that tourists may prefer going on boats of operators who support research and conservation efforts. Nonetheless, the incorporation of scientific research into whale-watching operations still lags far behind the overall development.

Public Education

Whale watching can have substantial effects on cetaceans. Hence, stakeholders need to be educated about their potential negative influence, as well as how to adapt their behavior so as to avoid such impacts. Moreover, educational programs stimulate support and awareness for marine wildlife and conservation efforts.

There are several opportunities to inform trip participants about cetaceans: before, during, and after the trip. My own experience tells me that people become highly interested in receiving information just after a sighting. However, showing a film in advance of or after the excursion contributes greatly to sensitize tourists. Under no circumstances should a guarantee on a sighting be given!

It is recommended to introduce the presence of instructed personnel on all whale-watching boats. Having naturalists on board strongly contributes to removing wrong expectations on the side of the participants. Providing the local public with information about "their" whales and dolphins is crucial. A positive effect on the local economy and the environment mainly depends on how the public comprehends the need and shape of conservation measures. The meaning of whales and dolphins for the ecosystem and their possible economic importance to the region should be explained, for example, via free excursions for local and pupils. Materials such as brochures, films, lectures, exhibitions, Internet, and multimedia presentations can add to face-to-face distribution of educative messages. Ideally, some form of interpretation center should be established. Signs presenting cetaceans species encountered off the coast could be placed at highly visible locations, especially to promote land-based whale watching where appropriate.

The study by Mark Orams also found that close proximity to the animals may not be decisive for a successful whale watching trip, and a high degree of satisfaction for tourists may be achieved even in the absence of whales. This is confirmed by our own experiences. People returning from a trip without seeing cetaceans are rarely greatly disappointed because they were told that this could happen.

Management

In a 2001 "Good Practice Guide," written by B. Garrod, Wilson, and Bruce, it is said that "whale watching directly depends on a high quality marine environment in which

Numerous whale-watching boats and scheduled ferry traffic make the cetaceans' life hard off Tenerife (Canary Islands). ©Fabian Ritter/MEER e.V.

to operate. Poorly planned and managed ecotourism can contribute to its own demise since ecotourism requires the provision of an ongoing high quality resource for its successful operation." While managing the human behavior around cetaceans (it is not the whales that can be managed!), it is most important to regulate the intensity of the use of cetaceans, in other words, the period of time that the animals are subjected to boat presence. This can be achieved by controlling the number of boats, sighting duration, and temporal or spatial closures of biologically important areas.

Other fundamental aspects are the regulation of user groups, such as whale-watching operators, local fisheries, and general vessel traffic. All users of cetaceans should be managed so as to avoid strong competition between operators and to develop a sense of community within them. Another feasible measure is the dispersion of boats by time intervals (time of day, days per week, weeks per month, season, etc.).

A necessary feature of effectively regulated whale-watching activities is a licensing system. It is strongly recommended to connect the issuing of licenses and permits to qualitative features of the whale-watching trips, such as the obligation to present profound information and to conduct or facilitate the collection of scientific data.

The ultimate success of any management will depend less on the quantity of the regulations imposed than on willingness of each person—entrepreneur, politician, skipper, tourist, scientist, and resident—to recognize his or her own responsibility and act accordingly. Luckily, the cetaceans themselves fascinate people in a way that can make whale watching a deeply emotional, sometimes spiritual, experience—leading people to think off the track, becoming inspired. This effect on the human psyche, which I have experienced over and over again, is in itself a fascinating field well worth studying systematically.

Many questions arise for the future. What kind of effects of whale watching will be observable in the long term, and how will they look? Are habituation or sensitization

This rough-toothed dolphin, apparently not bothered, performs high breaches close to a whale watching boat. ©Fabian Ritter/MEER e.V.

processes involved? Cetaceans have shown remarkable regional and geographic adaptability to different habitats as well as to anthropogenic activities. What is more, there are substantial differences within species. Each species has its own characteristic way to deal with whale watchers. Comparing different study sites will show regional differences in the responsiveness of species. This dynamic context is difficult to explore and even more difficult to manage. However, we have to keep in mind that cetaceans may be able to adapt to our own activities in ways we still cannot even imagine. This is an enormous learning field for us as an intelligent mammalian species facing another form of life that has been termed "the crown of creation in the sea."

Our own curiosity to explore cetaceans, their habitats, and their interactions with us should be led by one premise: that we humans learn how to change our behavior so that cetaceans do not have to change theirs.

Further Resources

Allen, M. C. (2000). Habitat selection of foraging bottlenose dolphins in relation to boat density near Clearwater, Florida. *Marine Mammal Science 16* (4), 815–24.

Carlson, C. (1996). A review of whale watching guidelines and regulations around the world. *International Whaling Commission Scientific Committee, SC/48/025.*

Forestell, P., & Kaufman, G. D. (1995). Doing well by doing good: Need eco-tourism be an oxymoron? Proc. 6th Pacific Islands Seminar, Tokai University at Honolulu, Nov. 14–17, pp. 63–67.

Garrod, B., Wilson, J. C., & Bruce, D. M. (2001). *Planning for marine ecotourism in the EU Atlantic area: Good practice guidance.* Bristol: University of the West of England.

Hoyt, E. (2001). *Whale watching 2001: Worldwide tourism numbers, expenditures, and expanding socioeconomic benefits.* Yarmouth Port, MA: International Fund for Animal Welfare.

International Fund for Animal Welfare (IFAW). (1997). *The educational values of whale watching.* Provincetown, MA.

IFAW, Tethys Research Institute, & Europe Conservation. (1995). *Report of the workshop on the scientific aspects of whale watching.* Montecastello di Vibio, Italy. March 30–April 4, 1995.

Orams, M. (1998). *Marine tourism: Development, impacts and management.* New York: Routledge.

Ritter, F. (2003). *Interactions of cetaceans with whale watching boats—Implications for the management of whale watching tourism.* Berlin: M.E.E.R. e.V. Available at www.m-e-e-r.org/

Fabian Ritter

■ Conservation and Environment
Wolf and Human Conflicts: A Long, Bad History

Wolves and humans have a long history. Coexistence goes back tens of thousands of years but began in earnest with the agricultural revolution approximately 10,000 years ago. Around this time we domesticated some wild hoofed animals, called ungulates (initially these were sheep and goats in the Middle East), and our relationship with wolves changed forever. Before ungulate domestication, we admired wolves as a hunter, for we were hunters too, but once we husbanded wild animals, domestication bred out their natural defenses against wolf attack and they became easy prey to wolves. Domestic livestock vulnerability continues to this day. This has resulted in much conflict between humans and wolves, who when they discover it, find cows, sheep, and the like easy prey.

One reason for such a long history is that wolf and human distribution has overlapped over a wide geographic area. Wolves are Holarctic in distribution, meaning they live atop the globe, essentially everywhere north of the equator overlapping with areas humans have occupied as well. Some have pointed out how similar we are—usually monogamous, live in families, very social, communicate vocally, good predators—and this has placed wolves in our culture and minds. Yet as often happens, similarity does not befit coexistence and instead we compete, vie, for the position of top predator. U.S. president Teddy Roosevelt, for example, an avid hunter, called wolves "the beast of waste and desolation."

In short, our shared history is not a good one. We have been killing wolves since we domesticated ungulates. But it is more than that, we embellished our relationship with wolves, made them into something purposely evil because they got in the way of our farming and ranching. We considered them contrary to human settlement; references to this are found in the Christian bible. So we started killing them—first in Europe and then in North America. The first bounty on wolves was offered by the Massachusetts Bay Colony in 1630. Thousands were killed in the American West. It has only been within the last 30 or so years that anyone has even questioned this bloody relationship with wolves (there are some exceptions to this, one being a biologist from the 1930s and 1940s named Adolph Murie, who worked in Denali National Park, Alaska, and Aldo Leopold, who wrote an essay *Thinking Like a Mountain,* where he watched the "green fire" go out of a female wolf's eyes that he had shot, which caused him to question what he had done). So why is there a problem? Why is this animal so difficult for us to coexist with?

Three Major Areas of Wolf-Human Conflict

There are three major areas of wolf-human conflict: human safety, livestock (cows, sheep, etc.) killing, and competition for ungulates (deer, elk, moose, etc.). Whether based on fact or not, how we feel about wolves, and ultimately what we think should be done with them, revolves around one of these three issues.

Human Safety

For centuries people have perceived wolves as being dangerous. Their long white fangs, howling, nighttime activity, and ecological role as a large predator likely spawned this view of a horrific beast. If you live in wolf country, you

Myths and stories about wolves have contributed to a negative view of wolves by the public. This Arthur Rackham illustration from Little Red Riding Hood is a perfect example. ©Art Resource, NY.

will hear such statements as the following: watch out for your kids being snatched from the bus stop; don't go out at night or alone; walking the woods will never be safe again. In April 2006 the governor of Wyoming was quoted in the newspaper as saying that the number of wolves in Wyoming was a human safety threat. Where do these feelings come from? "Little Red Riding Hood"? "The Three Little Pigs"? "Peter and the Wolf"? "A wolf in sheep's clothing"? Few other animals have such a place in fable and human imagination as do wolves, and this affects our view of them from a young age.

During medieval times in Europe, plagues (often called "the black death") swept through, killing people in great numbers. Wolves were seen standing on the piles of the dead, and some they exhumed from their graves. This horrifying scene, combined with a skulky manner, did nothing to endear wolves to people. These images shaped our view of wolves and became embedded in human culture. But was this fear of wolves really warranted? And how often do wolves really kill people?

Evidence shows that wolves rarely kill people, and through the millennia there are very few verified human fatalities from wolves. In all of North America during the twentieth century only sixteen people have been attacked, and none were killed. Virtually all of these wolves had been accustomed (habituated) to people, usually first by being fed human food. Wolves are naturally shy of people and almost always need a period of time to acquaint themselves with humans to lose their natural fear and shyness, assuming they are not immediately killed first. This then can lead to attack, but usually it does not. This is unlike cougars and bears, which might attack the first time they encounter a human, not needing human exposure to overcome their natural fear. European history and Asian history have verified cases of wolves killing humans; most of these are from rabid wolves, but some were healthy wolves. Nonetheless, it is still very rare, given how long wolves have been exposed to people throughout history.

In the past few years a number of children have been killed by wolves in India, but upon closer inspection it seems the circumstances were unusual. For example, some children may have been left unattended while their parents were guarding the family livestock. Because livestock is closely watched, and the wolf's natural prey is much depleted in the area, hungry wolves have turned to the vulnerable and unattended children. This does not mean these children-killing wolves should not be removed; it is just that the root of the problem cannot be simply stated as a wolf problem.

In short, wolves do attack people and sometimes kill them, but it is blown vastly out of proportion to how often and probable an event it really is. Compared with other carnivores, wolves are the safest to live around and the least likely to attack and kill people, yet they are considered the most dangerous by the general public.

Livestock Depredation

If wolves rarely kill people, what about domestic livestock—animals such as cows, sheep, horses, pigs, goats, and dogs? Killing livestock is called depredation, and wolves may kill livestock. After all, livestock is essentially wolf prey without its natural defenses. Livestock is derived from wild ungulates but modified through captive breeding to better suit human husbandry and production. The end result of this is that they are largely unable to defend themselves against predators—essentially their defenses have been bred out of them to make them easier to live with. Living with us makes it hard to live side-by-side with a wild animal. This makes livestock especially vulnerable to wolves and other predators.

Another factor that contributes to livestock depredation is when humans move into a natural area formerly untouched by humans, and wild ungulates decline because of excessive hunting. This occurs because people view wild prey as food, perceive the resource as limitless (e.g., plains bison in the 1800s), or want to remove the wild prey to reduce competition with incoming livestock (e.g., save the grass for the cows). In the American West during settlement this is exactly what happened; wildlife was killed in excess, sent to markets for food (market hunting), and wildlife in general was viewed cheaply because there was so much of it. Numbers of wild ungulates, the wolf's natural food, fell dramatically. This left wolves with reduced opportunity to feed on wild prey, so to survive they turned to livestock. Conflicts resulted, and wolves in the west were killed in great numbers. From 1915 to 1942, 24,132 wolves were killed in Colorado, Wyoming, Montana, and the western Dakotas. Government trappers, called "wolfers," were hired specifically for killing wolves, and veterinarians introduced diseases to kill wolves (which is how the disease mange was introduced to North America).

When wolves were reintroduced to Yellowstone National Park in 1995 and 1996, data from livestock killing during the western settlement period in the 1800s fueled opposition to the reintroduction. A court case against restoring wolves to the park used statistics from this time period stating that wolf killing of livestock would be common. However, recent experience from other areas where wolves and livestock commingle with restored wildlife populations show that livestock killing by wolves is rare. Modern day wildlife management has restored natural populations of wild prey such as elk and deer, the wolf's natural prey; therefore, the need to prey on livestock is less. Countering these arguments, lawyers defending wolf reintroduction stated that 100-year-old data were inappropriate and predicted livestock killing would be uncommon—a prediction that has been borne out. In those days what was most available for a wolf to eat was the rancher's cow; that is no longer the case.

This pioneer era of wolf conflict added to the wolf's negative image, and a few clever wolves became adept at avoiding attempts to kill them. Settlers named them, and stories of these crafty wolves led to fables and legendary status for some individual wolves. Names such as "Old Three Toes" (probably because she was caught in a trap and had her toes pulled off escaping from the trap, hence the track only had three toes in it), "Old Lefty," and the "Custer Wolf" (the insinuation being it was the last of its kind) developed into local and then widespread mythology, which only enhanced the idea of cunning evil wolves that needed to be exterminated. This contributed to the zeal to kill them, and the people who did became heroes.

Today livestock depredation by wolves is rare, but it does happen and therefore is an important problem to be dealt with. Even though it is neither common nor widespread, when it is your cow, sheep, or family dog, it's a very big deal. Preventing attacks is the most effective strategy, but these methods are not always effective and usually cost money and time to implement.

Competition for Wild Prey

The last area of wolf and human conflict is the most legitimately debated of the three major issues: competition between wolves and humans over wild prey. Outside of live-stock country—in the continental United States and in southern Canada where the prairie province grasslands have been converted to agriculture—the big issue is wolves killing wild ungulates that people want to hunt. In some cases this hunting provides food for people living off the land in remote regions (subsistence hunting), but in most cases the hunting is recreational. The question of concern is this: Do wolves take food and/or hunting opportunities away from people? In ecological terms, we seek to know if wolves depress or cause declines in ungulate populations.

Many people have presumed wolves negatively affect prey numbers. "How could this not be?" the logic goes. "The wolf killed something; is it then not there for human use?" This conclusion has resulted in widespread wolf killing to reduce competition with human hunting. Called "wolf control," one such program was under way during the winter of 2005–06 in parts of Alaska. Typically, control is done through aerial gunning, either from helicopters or airplanes, but historically other means have been used, the most common being poison baits. This has been particularly controversial because poison is an indiscriminate killer; anything that eats the bait, and many things do, dies. The nonselective killing caused by poison is not favored by many people, but for others poison is preferred because of its low cost and efficiency (i.e., no expensive helicopters are needed).

Another reason wolf control is controversial is that for it to be effective, in other words for the moose or caribou population to increase, most of the wolves need to be removed. Typically this means 80–90 percent of the wolves need to be killed over several years, or it is not enough of a reduction in predation pressure on prey for them to recover from low population levels. To many this is too disruptive to the natural system, and for some it is terribly unethical to kill so many animals in a nonsporting (from aircraft) fashion. Proponents admit it is not a form of hunting, rather it is a means to an end, and thus fair chase considerations, which are normally adhered to for recreational hunting, are unnecessary. Viewpoints of both groups are strongly held; therefore, wolf-control debates have become highly polarized. Political solutions are often sought but tend to be only short term, in favor of one or the other constituency without addressing the root issue of proper wildlife management.

Whether or not wolves cause prey population declines is one of the most hotly debated topics in all of wildlife management, and a survey of studies will reveal multiple answers to the same question. In short, sometimes wolves do depress wild prey, and sometimes they don't; it depends on other factors, and this list of factors can sometimes be long and not correctly measured by scientists because we don't know what they are or they are too inconceivable to think of. As Frank Egler says, "ecosystems are not only more complex than we think, but more complex than we can think."

If there is any generalization to be made from this, it is that the difference between studies indicates that no wolf-prey system is the same; each has its own unique set of circumstances that make generalization difficult. But it is human nature to draw

Wolves are still met with hostility and fear, in spite of the fact they've been proven to be incredibly social and intelligent animals. Courtesy of Shutterstock.

conclusions from knowledge or experience of one situation—and it is easy to find a case supporting your viewpoint—and it becomes the solution for all. These solutions become fixed in people's minds as an opinion and are then held as a value and therefore are not subject to logical discourse. This further entrenches the viewpoint, which leads to greater polarization. Social science research in this area has shown that at this point more factual information, regardless of what side it supports, actually serves to harden or make one's viewpoint more extreme.

Solutions: Can We Live with Wolves?

So how do we live with wolves, or as some others have framed the question, "Can we live with wolves?" For most of human civilization the answer to this question has been, "No we cannot." It has only been in the last thirty or so years that we have decided to try. To some this is a ridiculous proposition—that we should in essence reject the wisdom of our ancestors. But now there are other viewpoints. Some feel all life, whether human or not, has intrinsic value even if it costs us economically. Research has found that the way one feels about wolves has a lot to do with your world view. For example, do you have a utilitarian concept of the world, it is here for us to use, or do you believe we should live in harmony with it and limit our use of resources? If your philosophy is the former then you will probably be in favor of controlling or eliminating wolves in areas of conflict. Your philosophy and values are not easily debated. This issue can be likened to the current debate on the legality/morality of human abortion.

Despite this, we, as managers of the natural world, have to decide (some would debate this statement but it is safe to say that there are no longer any environments on earth unaffected by humankind). We have learned in the last few decades some key information about wolves that can help us live with them. Myths and stories no longer have to be our sources of information.

Human Safety

We know that wolves are not a natural threat to people, but that under some circumstances they will attack and even kill people. We know too that most of the cases where this happens involve wolves that have lost their natural fear of people. Typically wolves lose their fear of people by being fed human food. This diminishes wolves' natural fear of humans and often leads to attacks. Keeping wolves from gaining exposure and familiarization with people and not feeding them can solve this problem. Strictly enforcing zero tolerance for wolves near people is a must. To accomplish this, people will have to watch their garbage in wolf country; they cannot feed wolves; and they must scare away any wolves that choose to be near people.

Dogs are another attractant for wolves because dogs are very similar to wolves (all dogs are derived from wolves), so wolves can be competitive with dogs. Therefore, we need to be knowledgeable in wolf country about the possibility of an attack on a dog. Tending dogs is as important as tending human food and behavior.

In essence we need to keep the space between wolves and humans, much as we should for any wild animal species. If this fails, we can negatively condition or teach the wolf that comes too close to humans to stay away. If that fails, removing the offending wolf (e.g., relocating it or killing it) may be the last solution, because if we do not, the animal will hurt somebody, the behavior will spread to other animals through observational learning, and the wolf population will suffer because of widespread publicity about wolf attacks on humans. Wolf pets and wolf/dog hybrids should also be discouraged because they are not good pets and only contribute to negative opinions. If you want a wolf, get a dog; they are very similar, except that dogs have had the wildness bred out of them, like other domestic livestock.

Livestock Depredation

To reduce conflicts between livestock and wolves, proper land management (keeping wolves out of heavily settled areas) and better animal husbandry are ways to reduce wolf-livestock conflict. There will be many areas of dispute as to whether or not wolves belong there, but at least this is a start. This means that further human developments into relatively pristine wildlife habitat may need to be limited or prohibited. In areas of legitimate livestock-wolf overlap, preventive measures can be employed. None of these are perfect, but by being proactive and accepting that better animal husbandry in areas with wolves and other native carnivores (bears, cougars, coyotes) is better than not adapting; otherwise, livestock will be lost and wolves will continue to be killed. More frequent human tending (e.g., shepherding, cowboying), livestock guarding dogs, adaptive fencing and corrals, scare devices, and rounding up at nighttime all offer ways to reduce, or if used in combination, eliminate, wolf attacks and kills. All of these techniques make life more difficult, but they offer ways for coexistence.

Europe poses an example to many ranching operations in North America. There most wolves live in more settled landscapes; as in the early American West after market hunting, the availability of livestock is greater than wild prey (in Italy and Spain for example, a large part of the wolf diet is obtained from garbage dumps), so livestock needs to be protected. With a combination of guard dogs and human shepherding, loss of livestock has been virtually eliminated in some areas. In North America such practices are rarely adopted, possibly because the initial solution to livestock loss was wolf killing, a practice that has become culturally engrained. Recently, predator control on public land (land owned by all Americans), where much grazing occurs, has become controversial because grazing is subsidized by charging lower-than-market prices because it is expected that there will be some losses to predators. Families using these grazing allotments for several generations are not happy with the increasing societal tolerance of predatory animals. Because they pioneered and settled the land with predator elimination, living with wolves, bears, cougars, and coyotes is not agreeable to most.

This acrimonious debate seems intractable, but by emphasizing the proper land use, adopting measures to prevent wolf attacks on livestock, paying the ranchers for losses, and when all else fails, killing the offending wolf or wolves, coexistence seems possible. Many individuals are willing to contribute financially to a fund that pays ranchers for confirmed losses to wolves; this lifts the burden of having wolves from individuals to society. Ranchers complain about the uncompensated losses of missing livestock and the extra time and trouble it takes to prevent a wolf attack. Having wolves on your ranch leads to worry and sleepless nights that are not compensated for, so it is impossible to price all costs, but nonetheless, some compensation is a start. In Sweden, sheep producers are paid a flat fee that is prorated, calculated by the size of the local predator populations. This is a form of a societal payment to coexist with large carnivores. This system eliminates the oftentimes contentious debates over how many livestock predators have killed. If you suffer no losses you keep all the money, whereas if you suffer more losses, you don't get more. This provides an economic incentive to prevent losses.

Competition for Wild Prey

Understanding the full range of factors that control animal populations and prioritizing land use offer possible solutions to reducing competition between wolves and humans. Prey populations are strongly affected by the number of predators; if wolves and humans are the only predators, then impacts of prey will not be as strong, whereas if bears and cougars are also present, stronger management intervention will be necessary because depressing prey populations becomes more likely. Limiting human hunting may also be necessary. In addition to how many predators there are, attributes of the prey's productivity play into decision making as well. Some prey reproduce at a higher rate than others; hunting can be more liberal for these productive species (such as white-tailed deer).

What about the habitat? Intact landscapes will offer more for all wildlife and therefore reduce the need for heavy-handed human management. Adequate food, water, and shelter will go a long way toward preserving all wildlife and provide for some human use. Also, having a variety of land use objectives will help. Some areas should be managed with minimum human intervention, whereas others should be managed to maximize human use (e.g., wildlife harvest). As long as a variety of areas are represented, and one use does not outweigh others, an assortment of land uses (conservation versus preservation) offers great hope.

Conclusion

No one claims wolves are easy to live with; virtually everyone acknowledges that life is more difficult with them. There are ways, however, to live with them and to allow them to live undisturbed in other places. Recognizing and preserving undisturbed areas is key for many species of wildlife—not just wolves. Knowing where wolves don't belong is just as important. Key areas will be at the human-wildland interface, where a variety of techniques to minimize wolf damage to the human economy will be necessary with an occasional wolf removed to prevent spread of the damage. Some areas should have no wolves. With some tolerance, work, and thoughtful planning, the answer to the question "Can wolves and humans live together?" will be "Yes."

See also

Conservation and Environment—*Coyotes, Humans, and Coexistence*
Ethics and Animal Protection—*Wolf Recovery*
Living with Animals—*Wolf Emotions Observed*

Further Resources

Egler, F. http://www.caves.org/section/ccms/wild01/lera_wild01.htm (accessed March 2, 2007).

Leopold, A. (1966). *A sand county almanac.* New York: Ballantine.

Lopez, B. (1978). *Of wolves and men.* New York: Charles Scribner's and Sons.

McIntyre, R. (1995). *War against the wolf.* Osceola, WI: Voyageur Press.

McNay, M. (2002). Wolf-human interactions in Alaska and Canada: A review of the case history. *Wildlife Society Bulletin, 30,* 831–43.

Mech, L. D., & Boitani, L. (2003). *Wolves: Behavior, ecology, and conservation.* Chicago: University of Chicago Press.

Robinson, M. (2005). *Predatory bureaucracy.* Boulder: University Press of Colorado.

Smith, D. W., and Ferguson, G. (2006). *Decade of the wolf: Returning the wild to Yellowstone.* Guilford, CT: The Lyons Press.

Douglas W. Smith

■ Cruelty to Animals
The Bear Bile Industry in China: A Personal Essay

The practice of farming endangered Asiatic Black Bears (also known as Moon Bears after the golden crescent of fur across their chests) has a relatively short history: introduced in Korea in the early 1980s, the procedure of caging bears and surgically implanting metal catheters into their gall bladders to "milk" them daily for their bile was soon adopted in China. It was hoped that bear farming would provide an easy solution to satisfy the local demand for bile, while reducing the number of bears taken from the wild. However, wild bears are still poached today for their whole gall bladders or as an illegal source of new stock for the farms. Other bears are captive-bred on the farms and taken from their mothers at three months of age. Confined in tiny wire cages, desolate cubs are seen suckling upon their own limbs, or even on the limbs or faces of their siblings in adjacent cages, while making "comfort" vocalizations reminiscent of cubs in the wild feeding on their mother's breast.

Disregarded on the farms as the sentient, intelligent creatures they are, both adults and cubs are at the mercy of people who see them not as vibrant beings deserving of life and freedom in their own right, but merely as tools of a trade—to be used and exploited until they have nothing else to give. I struggle to think of any other species with the ability to endure the physical and mental torture that farmed bears suffer for as long as twenty miserable years or more. Investigating undercover while working as a Consultant for the International Fund for Animals Welfare (IFAW), my first visit to a bear farm in 1993 left me in despair. Since then I have worked not only to rescue bears but also to end the practice of bear farming in Asia. In 1998 I founded the Animals Asia Foundation together with a team of highly dedicated and professional people, and my life has not been the same since.

Literally at death's door, the bears arrive at our Bear Hospital stacked in rusting metal cages with appalling physical and mental traumas as a result of their years on the farms. Bears such as Andrew, Freedom, Belton, and Frodo with severed limbs as a result of being trapped in the wild or Crystal and Gail with their canine teeth cut back to gum level, exposing pulp and nerves, and their paw-tips brutally sliced off to permanently declaw them—making them less dangerous to "milk" their bile.

The bears are gently unloaded and moved into quiet and cool recovery rooms, while our veterinary team assesses the condition of each one before prioritizing them for emergency health checks or life-saving surgery.

Traditional Chinese Medicine

According to the last official Chinese government figures, 7,002 bears are factory farmed and "milked" each day for their bile, which officials maintain is an essential component of Traditional Chinese Medicine (TCM).

Bear bile has been used in TCM for over 3,000 years, but today, doctors agree that it can be easily replaced with numerous herbal and synthetic alternatives, which are both cheap to obtain and effective to use. The challenge now is to escalate the voice and support of these doctors and publicize their concerns without denigrating the principles and effectiveness of TCM itself.

Eminent professors such as Professor Zhu Zheng Lin are dedicated both to the culture and usage of TCM as well as to the end of bear farming. Professor Zhu has written, "I believe that it is the time now and it is the responsibility of our new generation of

Traditional Chinese Medicine practitioners that we further develop the TCM theories left by our ancestors, and not rely on the old beliefs of bear bile or tiger bone" (cited in Zhiyong and Yanling, 1999).

Similarly, Professor Liu Zheng Cai's theory postulates the following: "If people don't use bear bile, the industry will have no reason to exist. I always tell my patients and students that bear bile is not necessary, and is replaceable. By not using bear bile, it complies with the TCM theory of 'harmony with nature'" (cited in Zhiyong and Yanling, 1999).

Ms. Zhou Jian Hua, a doctor who recently visited the Animals Asia Sanctuary, uses even stronger language: "I have being selling TCM tonic food for 13 years, including bear bile and bear bile wine, which have brought me much benefit. But, after visiting your Sanctuary, I am shocked and feel shamed of having hurt these animals. I used to buy bear bile, even bears, freely. But starting from today, I will never sell bear products again" (cited in Zhiyong and Yanling, 1999).

With supportive statements from doctors such as these, there is hope that, like the government in Vietnam, the Chinese government will seriously evaluate the true status of the industry and acknowledge that both local and international condemnation of this cruel trade further supports its closure forthwith.

For the bears suffering and dying on the farms right now, freedom cannot come soon enough. Day after day, their bile is drained through crude metal catheters implanted deep into their gall bladders or via permanent, open, and infected holes in their abdomens—known as the "free-dripping" technique, a technique that the Beijing department of the Convention on International Trade in Endangered Species (CITES) states is "hygienic" and "humane."

However, veterinary professionals view this technique somewhat differently. Their observations prove that the gall bladder fistulation surgeries and bile extraction performed on the bear farms significantly increases the risk of disease in the bears and must, by the nature of the wound, cause pain. Even a basic medical understanding recognizes that a permanent hole in the abdomen is a perfect vector for bacterial infection and can never be considered hygienic or humane, as the bear farmers maintain.

Additionally, the hole naturally tries to heal over; to prevent this, the farmers have devised a "fake free drip" technique whereby a Perspex catheter is discreetly embedded within the hole to prevent it from closing up. Knowing that the catheter is undetectable once the fur has grown over the wound, the farmers are able to circumvent the regulations and deceive government inspectors by maintaining that the hole is "naturally and permanently open," allowing bile to freely drip out as the regulations dictate.

The Five Freedoms of Animal Welfare

For a species that has won the hearts of children around the world who have grown up with their beloved teddy bear and then developed empathy for its wild counterpart, bear farming is a betrayal of a special relationship we share. It also clearly violates all five of the Five Freedoms of Animal Welfare upon which developed nations base their treatment of animals.

Freedom from Hunger and Thirst

The large majority of bear farms do not allow bears free access to drinking water, and those that do restrict access to certain times of the day or week. The rescued bears are usually underweight or even emaciated, eat ravenously, are frantic for water, and have abnormal hair coats. In animals that hibernate and are therefore uniquely adapted to withstand privation, this indicates chronic malnutrition and a severe restriction of drinking water.

Freedom from Discomfort

Chinese regulations state that bears are to be given space to move about at all times, except for the brief period each day when they are caged for bile extraction. The reality is that most bears spend their entire lives in the 'extraction cages' and are never let out. Some farms have a small, bare, concrete enclosure in which some of the bears may move about, however, there is no effort to provide environmental or psychological enrichment, to provide comfortable places for the bears to rest, or to allow the bears to engage in anything resembling normal behavior.

Freedom to Express Normal Behavior

Normal bear behavior does not occur on bear bile farms. Cubs are weaned unnaturally early for the species (three months or less, instead of 18 to 24 months as in nature), which deprives the mother of normal maternal behavior and the cub of normal development. Weaned cubs are often raised in isolation because of the inability on the part of the farm staff to manage infectious disease. Living space for the bears is insufficient, and socialization is nonexistent or poor. The space, environment, and stimulus necessary for bears to express normal behavior are therefore not found, nor will they be found, on bear farms.

Freedom from Fear and Distress

Fear of people is evident in the majority of the bears that Animals Asia has visited on bear farms and that arrive at our Bear Rescue Centre. Head wounds indicate the mental trauma they have endured over the years as they repeatedly bang their heads against the bars in a frantic attempt to stimulate their intelligent minds. Hunger, thirst, pain, physical abuse, deprived environments, and chronic illness constitute distress.

Freedom from Pain, Injury, or Disease

Of the bears received at our rescue center, 60 percent have missing limbs, mutilated digits from efforts at declawing, significant areas of scarring on the body, or other signs of major injury. Another 25 percent have severe dental problems associated with the practice of cutting canine teeth (to prevent injury to the farmer), malnutrition, and chewing on metal cage bars out of psychological distress. Farmed bears also appear to have a relatively high frequency of vision loss due to a variety of causes. Infections in bears received at our Hospital are resistant to nearly every antibiotic available in China, which indicates that the bears suffer chronic infections and that the use of antibiotics on bear farms is ubiquitous and inappropriate. Clearly, the bears on bear farms are not without pain, injury, and disease.

What We Have Learned from the Bears

But out of the dark comes light. With 205 bears rescued so far we continue moving slowly but surely toward our end goal. Our rescued bears are spokespersons for hope.

The bears' capacity to forgive is humbling. Initially displaying frightening aggression when they arrive at our Sanctuary from the farms, they lash out at the people who are showing them respect and love for the first time in their lives because they cannot immediately differentiate between those of us who care and their previous tormentors. However, the gradual transition in personality from an animal so violent and fearful of the human species to one that is finally trusting, inquisitive, and completely at ease with people can only be described as remarkable.

Our experienced team has the expertise and passion to care professionally and lovingly for these bears and to build world-class sanctuaries of hope, which give them something they have never had before—choices. The bears can choose to play and integrate with new friends or to lead a solitary lifestyle as many would choose in the wild.

In February 2006 we said goodbye to Andrew, a bear we rescued in October 2000, a bear whose majesty and kindness has left a lasting legacy in everyone's hearts. Our gentle, forgiving Andrew was euthanized after we discovered that a massive seven-kilogram liver tumor had ravaged his body and was bringing his life to an end. Everybody—Chinese and Western staff alike—was inconsolable.

Of over 1,000 tributes we received from around the world, someone wrote, "You are not weaker without Andrew, but stronger because of him." And we are. Bears like Andrew and those before him who have taught us so much about the individual, and who have guided our knowledge of the species as a whole, will never die in vain. The bears' story—and the terrible truth behind bear farming in China—will continue to be told, underlining the betrayal of a relationship between ourselves and animals with whom we share this earth.

What they teach us today is astonishing. Animals that have probably never before built nests are doing so today, as evidenced by skillfully created bamboo beds deep within their natural forest enclosure. Similarly, animals that have never before climbed trees are shinning up into branches several meters high with ease.

They remember and form close friendships with specific members of the group—often sleeping two or even three together in their hanging basket beds. They will often gang up together against another bear. In fact, we have one female group of bears we affectionately call the "knitting circle" because they remind us so much of elderly ladies who seemingly have nothing better to do but gossip the day away and who will join together to warn away another bear who comes too near to their exclusive circle. Such friendships for a species recognized to be solitary in the wild are extraordinary.

Our bears remember the source and the cause of pain. It is no secret that our veterinary surgeons, the staff who work to care for these bears day and night, are also loathed by many of the bears. They recognize individual people and will often huff in caution if a veterinarian just steps into the room. They then follow through with warning growls or explosions of rage if an anesthetic jab stick is seen.

They are closely connected with comforts and habits that we require and perform in our own everyday lives. Straw piles are carefully collected and transported from one end of the grassy enclosure, up the stairs, and into their basket beds inside the den. Hence it appears that they are preempting the pleasure they know that this mattress of straw will bring once it lines their bed.

They can be preemptive in other ways. For example, one bear will often wait until another is distracted and then steal toys or food away, in every sense like a greedy sibling, biding their time until the opportunity arises to steal from a too-trusting brother or sister.

Their inherent sense of fun and mischief finally overcomes all but the most severe legacy of pain and stress. Skittish and happy, many of the bears adore their new opportunity to play, either with the myriad of toys and enrichment enhancers specifically created within their enclosure, or within the natural surrounds of their enclosed bamboo forest zones, or simply, and perhaps most appreciatively, with each other. In the absence of all three, the bears will often throw themselves onto the ground in a deliberate, giddy somersault, simply because they can.

Our Sanctuary benefits more than bears. Farmers are compensated with funds to start a new livelihood if they close their premises immediately. The project sees significant local employment, the utilization of local food and construction materials, and the development of a facility that retains the integrity and culture of the surrounding area—promoting a community-based project for bears and people alike. As a result, the groundswell of encouragement from the general public within China continues to escalate and shows how supportive people are toward ending a shameful and unnecessary practice in the country of their birth.

Animals Asia's educational initiatives are central to the China Bear Rescue and successful Open Days see the Moon Bear Rescue Centre opening its doors to welcome hundreds of visitors each month, including busloads of school children who descend upon the center in a whirl of excitement, walking around wide-eyed with wonder as they watch bears such as Jasper and Aussie tumbling together in the forest or spy Sausage balancing high in the tree tops! Every visitor to the Centre leaves with a sense of being part of something enormous, a new beginning for animals in China, and promising to help in any way that they can.

"Friends of Animals Asia" Support Groups that are springing up in Universities as educational outreach programs build long-term cooperation between Animals Asia and the students who represent a compassionate new future for China. Together with their support, Animals Asia is increasing the impact of their official slogan "Rescue Black Bears, Give up Using Bear Bile" and educating a wider section of the general public about bear farming and the concept of animal welfare in China.

In Vietnam, where the industry is now illegal, we are preparing to welcome the first of 50 bears into a new Sanctuary in Hanoi before the end of the year.

Open wounds like the ones seen here on Jasper's stomach are just some of many horrifying wounds often found on bears rescued from "bear farms" in China. Courtesy of Annie Mather/Animals Asia.

Formerly a resident of a "bear farm" in China, Jasper is now happy and healthy residing at the Moon Bear Rescue Centre. Courtesy of Annie Mather/Animals Asia.

Legislation to Help the Bears

In July 2000, a landmark agreement between Animals Asia, the China Wildlife Conservation Association (CWCA), and the Sichuan Forestry Department pledged to rescue 500 bears and work toward ending the practice of bear farming. Since October 2000, more than forty bear farms have been closed and over 200 individuals have been released into the care of the Animals Asia team.

Adding a powerful voice from the international political arena are members of the European Parliament, who in December 2005 passed only the seventh Declaration in the history of the European Union, calling on China to end bear farming before the Olympic Games in Beijing in August 2008.

Political support also grows from within. Recent visits by several members of the National People's Congress saw their promise of offering "serious help to end bear farming by the 2008 Olympic Games in Beijing." Acknowledging the government's commitment to a "Green Olympics—embracing harmony and unity between mankind and nature" this is our perfect opportunity to position a statement of goodwill and hope for all animals, but especially a statement against the plight of 7,000 caged and suffering bears who remain on the farms.

Meanwhile, our own promise to the bears is that we will care for them for the rest of their thirty-year lifespan. They deserve nothing less. Our sanctuary and others following it will be a place of peace where majestic bears like Andrew can be remembered with love and respect for the first time in their lives; where hundreds of bears see release from a lifetime of torture, enjoy tender loving care, take their first steps into the forest and play with their friends; and where all bears wake up every single day with the sun on their backs and without fear in their hearts.

Further Resources

Animals Asia Foundation. www.animalsasia.org/

Loeffler, K., Robinson, J., & Cochrane, G. (2006). *Compromised health and welfare of bears subjected to bile extraction in China's bear bile farming industry with special reference to the free-dripping technique.* Hong Kong: Animals Asia Foundation.

Zhiyong, F., & Yanling, S. (1999). 3rd International Symposium on the Trade in Bear Parts National Institute of Environmental Research. (October 26–28). Seoul: TRAFFIC East Asia. http://www.traffic.org/

Jill Robinson

■ Cruelty to Animals
Harming Animals and Its Impact on People

It is generally assumed that humans will be adversely affected by harming animals. One argument, known as the "graduation" or "escalation" model, claims that those who deliberately mistreat animals, in culturally unsanctioned ways, will move on to subsequent aggression and violence toward people. A second argument, the "coarsening" model, claims that those who harm animals in institutional settings, where such acts are socially sanctioned, undergo some degree of desensitization as their moral and emotional sensitivity to suffering becomes blunted.

That harming animals affects people—but only in negative ways—is an old idea. As early as the seventeenth century, the philosopher John Locke suggested that harming animals had a destructive impact on those who inflict the harm. In later centuries, the psychologist Anna Freud and the anthropologist Margaret Mead followed suit, arguing that cruelty could be a symptom of character disorder. Children or adolescents who harmed animals were thought to be on a pathway to future violence because these acts desensitized them or tripped an underlying predisposition to aggression. By releasing destructive impulses and turning them on animals, the floodgates restricting violence opened and future targets were likely to be human, or so it was argued.

When studies tried to verify what is now known as the "link," results proved to be mixed and sometimes misinterpreted to support this idea. Starting with Macdonald's "triad" (1961), researchers had a hard time proving that animal abuse, in combination with firesetting and bedwetting, leads to subsequent violence. Macdonald himself failed to establish that violent psychiatric patients were significantly more likely than nonviolent psychiatric patients to abuse animals. In subsequent research, the evidence has been less than compelling, raising doubts about the validity of the link. For every study that purports to find a significant association between cruelty to animals and the impulse to violence, there is another that does not. And in studies reporting significant findings, there are methodological problems that cast doubt on their results because they rely on self-reports of people who, from the study's outset, were seriously troubled or disturbed, and they treat violence as the sole outcome, even though other kinds of problems might be subsequently linked to prior abuse.

Indeed, if the link were valid, then the reverse should be too, namely, that kindness toward animals should predict compassion toward people. However, there are examples of people who are kind to animals but cruel to fellow humans. For instance, some murderers show compassion toward animals. This was the case with Robert Stroud, the Birdman of Alcatraz, who shot a bartender, stabbed an inmate, and assaulted a prison guard while caring for the health of hundreds of canaries. And several members of the Nazi general staff, including Hitler, demonstrated extreme concern for animals in their personal lives as well as through the enactment of animal protection legislation.

Nevertheless, many people continue to believe the link exists, in part because the idea has strong commonsense appeal and resonates with cultural stereotypes and myths about the origins of violent behavior. For instance, in fiction writing one of the most effective ways to create a mean, unlikable character is to have the person ruthlessly brutalize an animal, because doing so must be a sign that humans are next in line to be harmed. Steven King confesses that he used this imagery to manufacture just this sort of person for his book, *The Dead Zone*. Speaking about his main character, Greg Stillson, King writes, "I wanted to nail his dangerous, divided character in the first scene of the book. . . . When he stops at one farm, he is menaced by a snarling dog. Stillson remains friendly and smiling. . . . Then he sprays tear gas into the dog's eyes and kicks it to death" (King, 2000, p. 193). Riding this commonsense appeal and cultural resonance, activists have argued that cruelty should be prevented because it is a nodal event leading to further violence. By the end of the twentieth century, the link became the dominant focus of organizational campaigns against cruelty, such as the First Strike program of the Humane Society of the United States (HSUS). Even those who do not care about animal welfare might now be concerned about preventing cruelty, given the urgency felt by many to identify adolescent "red flags" that signal a future violent adult.

Others argue that harming animals can negatively affect people who work in organizations where such acts are socially sanctioned. Unlike the link, here harming animals is not thought to lead to violence but to a general coarsening of one's personality. For instance, those who experiment on animals are thought to endure moral or emotional damage, even though their actions are institutionally approved. Presumed deleterious affects on humans formed the basis of anti-vivisection campaigns as early as the nineteenth century, when calls to prevent experimentation stressed injustice to animals as well as harm to scientists. They believed that using animals in painful experiments destroyed human sensitivities by forcing people to distance or coarsen themselves from the assumed suffering of lab animals. On the basis of this belief, there were calls to end experimentation because of its detrimental impact on human character.

Although most contemporary debate focuses on the moral basis for using or not using animals in experiments, some still claim that it negatively affects the identities of scientists and technicians. The latter suffer what is assumed to be a lasting and important moral damage by becoming insensible to the pathos of the lab animal's situation. Yet even those who make this assumption acknowledge that if there is a patent lowering of moral sensitivity, compared to our ordinary attitudes about how animals should be treated, it only occurs in the laboratory context. The damage, then, is at worst temporary and situational.

However, only a handful of studies have examined the impact of animal experiments on those conducting them, and irreparable moral or emotional harm seems unlikely. Indeed, even across-the-board situational coarsening is debatable. On the contrary, although such work can be stressful at times to those who have direct and sustained contact with certain kinds of lab animals, many escape or transcend these negative effects by relying on institutional coping techniques that shield their identities from lasting harm. Despite such findings, the belief still lingers and informs many pleas to end biomedical research.

Despite equivocal research support for the escalation and coarsening models, harming animals may still have a substantial impact on humans. Both models may be correct, at least in certain situations, for some people. Future research will hopefully refine these models and provide a more precise understanding of how and when harming animals influences human character and behavior.

Indeed, harming animals can be seen as having an enormous impact on people if we do not think about this relationship in traditional ways. For one, we need to consider groups of people who do not themselves harm animals but who are affected by such treatment of animals. In addition to those who abuse or neglect animals, there are many other groups of people who contend or deal with animal harm, whether to prevent cruelty, punish abusers, educate the public, or mourn the victims. For example, the owners of animals that are abused by others become secondary victims and are clearly affected by this harm. They sometimes go through a series of highly emotional stages of adjustment to this trauma, including feelings of rage and hopelessness.

We also need to consider the possibility that other effects, sometimes quite positive from the person's viewpoint, are possible in addition to or instead of more negative ones predicted by the escalating or coarsening models. The effects of such harm are not so simple or only negative because it is a transformative experience for animals and people. Being harmed can change the meaning of animals to victim, sacrificial object, target, plaything, or martyr, whereas encountering harm can shape the identities of humans in a positive or socially approved way. For example, college students who recall their "youthful indiscretions" with animals often view this harm as helping them attain an adultlike social status at a time when society prevented them from doing so because of their age. Another example can be seen in humane societies, when "big" cases of animal abuse are brought to their attention. Despite, or perhaps because, these cases are so disturbing, they often end up having positive affects for those who work in these organizations.

Critics will think it unsavory to propose that harming animals can have beneficial effects. Some may be troubled because this focuses on the human side of harm rather than on the animal's experience. Although it is understandable and proper to focus attention on animals, because they suffer and die, harm is also experienced by people—many of whom are not themselves causing harm. Taking the spotlight off the animal victim means that researchers and policy analysts can explore this topic without an ideological agenda by giving a voice to those who come face to face with the harm of animals and are forced to deal with it—asking themselves if what they see constitutes acceptable or unacceptable harm, if they or others are wrong to commit these acts, and if they can approach or use such harm in ways that make them feel better about themselves.

Others might be troubled because the expanded approach to understanding and studying the harm of animals suggests, at a social psychological level, that it can have a positive impact. This idea will be considered heretical if misconstrued, even implicitly, to mean that harm should be encouraged or at least tolerated. However, by asking how people interpret and use the harm of animals in beneficial ways, the goal is not to condone it, just as analysts seeking to understand "evil" are not forgiving it.

Further Resources

Arluke, A. (2000). Secondary victimization in companion animal abuse. In A. Podberscek, E. Paul, & J. Serpell (Eds.), *Companion Animals and Us* (pp. 275–91). New York: Cambridge University Press.

———. (2006). *Just a dog: Understanding animal cruelty and ourselves*. Philadelphia: Temple University Press.

Arluke, A., Levin, J., Luke, C., & Ascione, F. (1999). The relationship of animal abuse to violence and other forms of violence. *Journal of Interpersonal Violence, 14*, 963–75.

Birke, L., Arluke, A., & Michael, M. (2007). *The sacrifice: How scientific experiments transform animals and people*. West Lafayette, IN: Purdue University Press.

Kellert. S., & Felthous, A. (1985). Childhood cruelty toward animals among criminals and noncriminals. *Human Relations, 38*, 1113–29.

Macdonald, J. (1961). *The murderer and his victim*. Springfield, IL: Charles C. Thomas.

Piper, H. (2003). The linkage of animal abuse with interpersonal violence: A sheep in wolf's clothing? *Journal of Social Work, 3*, 157–74.

Arnold Arluke

■ Cruelty to Animals
Law Enforcement of Anticruelty Legislation

Special police departments devoted to enforcing animal cruelty laws strike many as a very modern concept, but they have nineteenth-century origins. Creating animal police forces followed the development of humane societies in Boston and New York. After George Angell founded the Massachusetts Society for the Prevention of Cruelty to Animals (MSPCA) and Henry Bergh the American Society for the Prevention of Cruelty to Animals (ASPCA) in 1867, they both successfully lobbied for anticruelty laws.

Enacted in 1868 and revised in 1909, the Massachusetts animal protection law primarily focused on the abuse of horses. Although somewhat antiquated today, the code still stands. To enforce this law, and its parallel in New York, the MSPCA and the ASPCA created small police departments within their organizations. Little is known about the nature of early animal police work other than what has been recorded in the annual reports of humane societies having such departments. For the most part, these brief records only note the numbers and kinds of cases prosecuted by officers. Humane agents, empowered as police officers, primarily investigated cruelty to horses, because the urban infrastructure required these animals to be well tended and healthy. One typical entry catalogued the ASPCA's work in New York, saying that agents carried out 768 prosecutions, of which 446 involved the mistreatment of horses for offenses such as beating, abandoning, starving, overloading, driving until they fell dead, and working sick, lame, or worn-out horses. Other prosecutions involved dog and cockfighting; rat baiting; feeding cows swill and garbage; keeping cows in filthy conditions; refusing to relieve cows with distended udders; cruelty to cattle, dogs, cats, and poultry; and maliciously killing, mutilating, and wounding animals with knives and other instruments. The only other information is the rare commentary about the work of humane law enforcement agents. In one case, the ASPCA report noted how "discouraging" it was for agents to be criticized for "overzealousness."

By the middle of the twentieth century, the makeup and organization of humane law enforcement departments in cities such as Boston and New York resembled their present-day form. The MSPCA's department is made up of sixteen staff members, including eleven investigative officers, a consulting veterinarian, two dispatchers, and a director and assistant director. Except for the dispatchers, all have been appointed as Special State Police Officers by the State of Massachusetts, although they are restricted to the enforcement of animal protection laws and regulations. They do, however, conduct investigations, obtain and execute search warrants, make arrests, and sign and prosecute complaints. Officers are assigned throughout the state to investigate whether individuals and, less often, organizations, have been cruel or neglectful. The bulk of their cases involve "everyday" animals—the strays, pets, vermin, and small-farm livestock—that are neglected or sometimes deliberately mistreated by individuals. They also visit and inspect stockyards, slaughterhouses, race tracks, pet shops, guard dog businesses, hearing ear dog businesses, horse stables that rent or board horses, kennels, and animal dealers licensed by the United States Department of Agriculture. During a typical year, MSPCA officers conduct approximately 5,000 investigations and 1,000 inspections, involving more than 150,000 animals. Because such complaints are also lodged with other organizations in the state, estimates of abuse complaints easily surpass 10,000 annually in Massachusetts and show evidence of steadily mounting over time. Of course, this increase may be due growing public sensitivity to animal welfare, greater visibility of humane law enforcement departments, or simply improved recordkeeping.

According to the MSPCA's official job description, the primary purpose of officers' work is "to prevent cruelty to animals, to relieve animal suffering, and to advance the welfare of animals whenever and wherever possible. Such purposes are to be achieved through the pursuit and implementation of a combination of activities, including, but not necessarily limited to, the enforcement of Massachusetts anticruelty and related laws, and the dissemination of animal protection/welfare related information." To do this work, prospective employees are expected to have a number of skills, the first of which is "humane sensitivity, with affinity for, and ability to empathize with animals and respond with compassion and objectivity."

When investigating cruelty complaints, rookie officers think of themselves as a brute force because they believe that they have legitimate authority to represent the interests of abused animals. They see themselves as a power for the helpless, a voice for the mute, representing and speaking for animals when their welfare or lives are in jeopardy. With more time on the job, this view changes. Although they are expected to represent the animal's "side" when investigating cruelty complaints, officers encounter a number of problems that make it difficult to do this. For the rookie officer fresh from training, these problems can be confusing and discouraging. Hired in part because of their humane sensitivity, this strong concern for animals plus their recent police training creates a number of expectations in them. Rookies expect to handle complaints against animals that violate the legal definition of cruelty as well as their own standards, to observe animals to ascertain the nature and extent of cruelty, to counsel "respondents" or "perpetrators" when necessary to improve the treatment of their animals, to prosecute those who commit egregious acts of cruelty or who do not comply with advice, and to be understood and respected as both police and humane officers. These expectations are quickly disappointed as rookies begin investigating complaints.

First, professional identity is a problem. Rookie officers experience a disparity between how they see themselves and how others see them. On the one hand, officers see themselves as professional law enforcers and animal protectors. As one officer said of the department's general job expectation: "They want you to be a humane officer, but have the authority or the presence of a police officer. It's hard to do both." On the other hand, one reason why it is "hard to do both" is that friends, family, strangers, and other professionals often are confused by this combination and either have no idea what humane officers do or relegate them to the level of "dogcatcher."

Second, officers must enforce a problematic law. Massachusetts, like other states, has an anticruelty code specifying that animals should not be deliberately mistreated. The law prohibits many types of abuse and neglect that threaten the safety and well-being of animals, including but not limited to beating, mutilating, or killing them as well as failing to provide them with proper food, drink, and protection from the weather. Those convicted of violating this law can be fined up to $1,000 and imprisoned for as long as one year, or both. Newer animal protection laws have classified cruelty as a felony, thereby increasing the maximum prison sentence to as long as five years.

Despite the existence of the law, officers find it difficult to enforce because of vague use of terms such as "neglect," "abuse," "proper care," "necessary veterinary care," and "suffering." Nor can officers fall back on more general cultural conceptions of suffering, because these, too, are vague and contested by different groups. This problem forces officers to interpret the meaning and application of the law on a case-by-case basis, a point made by Walter Kilroy, the former director of the MSPCA humane law enforcement department, who noted the "continuing absence of a widely accepted definition of cruelty to animals. Every activity that threatens the well-being of animals . . . must be challenged and overcome on a largely individual basis."

Third, there is a problem with evidence. The best witness to the abuse of humans is the victim; their testimony certainly facilitates, although does not guarantee, successful prosecution. Yet animals obviously cannot report or articulate their harm. Rookies must learn how to figure out whether an animal has been mistreated, relying on indirect evidence in order to "tell the story" of an act of abuse. Rookies discover that a large part of this indirect evidence comes from investigating humans. In fact, this human side of animal cruelty often becomes the deciding factor in handling and resolving complaints.

Finally, there is a problem with enforcement and prosecution. Rookies encounter very few clear-cut cases of animal cruelty that lead to prosecution and punishment. Instead, they encounter respondents whose behavior toward their animals does not violate the law but falls short of what officers would prefer to see. Without a technical violation of the cruelty law, officers feel that they have little, if any, authority to force respondents to improve their treatment of animals. When they meet respondents whose acts violate the law, officers see their advice ignored. Rather than giving up entirely at these times, rookies must learn how to get their message across to respondents and, if necessary, take them to court. This final option also can be particularly frustrating, especially for rookies as they encounter a judicial system that seems indifferent or hostile to the concerns of animals.

Most officers learn to cope with these problems by developing an attitude of "humane realism." With little legitimate authority to enforce the law, officers become humane educators who try to make abusers or others they meet on the job into responsible animal owners. With few victories in court, they discover alternative ways to feel effective in their fight against cruelty. And with public confusion about, or derision of, the role of humane law enforcement, they emphasize the police side of their work without forgetting their commitment to animal protection.

Further Resources

Alexander, L. (1963). *Fifty years in the doghouse: The adventures of William Ryan, Special Agent No. 1 of the ASPCA*. New York: G.P. Putnam's Sons.

Arluke, A. (2004). *Brute force: Animal police and the challenge of cruelty*. West Lafayette, IN: Purdue University Press.

Arnold Arluke

■ Culture, Religion, and Belief Systems
Animal Burial and Animal Cemeteries

The ritualized burial of animals has been practiced, at some point, in virtually every part of the world. In many societies, it was (and is) a means of honoring animals who had (and have) endeared themselves to their human families or served as symbols of virtue and worship. Beyond a simple sense of grief at the loss of a pet, 150 centuries of posthumous animal veneration speaks to humanity's ongoing kinship with the natural world at large. That such reverence is so universal attests to the most common of spiritual convictions: that we share a "next life" with other animals—just as we do this one.

Some of the earliest interments are of domestic dogs—such as the 7,000-year-old grave discovered in Sweden in which the dog was arranged in repose over the legs his master—suggesting that even then there was a belief that humans and animals were spiritual partners. In ancient Egypt, many animals were thought to be earthly representatives of the gods and were therefore treated as mascots or votive offerings. Fingertip-sized shrews to one-ton bulls were mummified and deposited in underground vaults, some of which could hold thousands of animals. Today, only a small number of these mummies survive, housed in museums around the world. Early archaeologists found these specimens in such large quantities that they deemed them to be more a nuisance than of any scholarly value. Entire vaults of mummified birds, dogs, and other creatures were hauled away for processing into garden fertilizer. One such shipment of mummified cats to England, in the late 1800s, weighed nineteen tons, representing about 180,000 specimens—of which only a single skull today survives in the British Museum.

Egyptian personal pets were entombed in private family vaults. Wall paintings in these crypts often show the household dog or cat typically reposed under a chair at the dinner table, with its name inscribed in hieroglyphics. "The limbs of Tamyt, one true of voice before the great god, shall not be weary," reads the sarcophagus inscription for one cat. In carvings, depictions of the gods as Tamyt's pallbearers suggest she enjoyed a status equal to humans in mortal life, at least in the eyes of her doting owner, who prayed for his pet's rebirth as an "imperishable star" in heaven. And when the royal hound Abitiu died in 2180 BCE, his grieving pharaoh ordered that fine linens, incense, and scented oil be used in the mummification process. He further decreed that Abitiu be laid to rest in a tomb of his own, so that the dog would become one of "the Blessed."

The Phoenicians, Greeks, and Romans expressed similar beliefs about honoring animals after death. In the Mediterranean port of Ashkelon (modern-day Israel), one thousand dogs were buried in a planned cemetery overlooking the sea, around 2500 BP. It is unclear if these were pets or part of a religious cult based on the popular belief that canines possessed curative powers. Regardless, each dog was carefully arranged in its own grave, reflecting some degree of human attachment.

Centuries of affiliation with pagan gods made many animal-loving people subject to persecution in the Christian era, beginning around 700 CE. Those who doted on companion animals risked being tried and executed as witches—along with their pets. In the centuries that followed, the only issue more controversial than giving an animal its own grave was the idea of burying humans and animals together. Clerics objected to the belief that animal and man might be on equal spiritual footing, and to express such an opinion was considered heresy. Even those with wealth and status often could not have their last request to be buried with their pets honored. On his deathbed, in 1786, the king of Prussia, Frederick the Great, stipulated that he wanted to be buried with his beloved Italian greyhounds, but his wish was ignored for two centuries. Finally, in 1991, his remains were returned to his Stuttgart estate and entombed in the garden amidst the graves of his dogs. By contrast, the German music composer Richard Wagner (1813–83) fared somewhat better—when his dog Russ died, it was smuggled into the composer's personal crypt, and then Wagner commissioned a stonemason to inscribe the words "Here Russ rests, and waits" on a discreet corner of the mausoleum.

It wasn't until the latter half of the nineteenth century, as industry and social reforms revolutionized the standard of living for the working masses, that companion animals again became accepted personal fixtures of mainstream life. For many people feeling estranged in this rapidly changing society, animals were the only "family" they could depend on to love them regardless of their social standing. The number of pet lovers swelled into the thousands in cities throughout Europe and America, and, in turn, closely held beliefs about animal spirituality began to morph into a social movement.

"Is not the dog an animal that we can exalt without any reservation?" asked Parisian George Harmois in an 1890 essay advocating a more dignified means of saying goodbye than putting one's pet in a weighted sack to be tossed into a river. Hence, the idea of establishing public pet cemeteries near major urban areas was welcome news to

The Hartsdale Pet Cemetery, as it appears today after more than a century of interments. ©Animal Image Photography.

The Hyde Park Dogs' Cemetery in London, as it appeared around 1900. ©Animal Image Photography.

A dog's grave marker in the Hyde Park Dogs' Cemetery. ©Animal Image Photography.

thousands of landless tenants. The accidental death of the royal dachshund Prince in 1888 led to the creation of the first public animal cemetery in London's Hyde Park, where it exists today—virtually unchanged by the last hundred years. North of New York City, in Westchester County, is the Hartsdale Pet Cemetery, the first of its kind in America. From its simple beginning in an apple orchard in 1896, it today holds several thousand burials, as does the Dog's Cemetery on the Asnieres islet just outside of Paris, which evolved from a popular picnic spot, in 1899, to a memorial park famous for its funerary artworks. Dressed in black, Victorian animal lovers conducted their own graveside services by playing music, reading Bible verses, eulogizing their pet's achievements, and burying them in satin-lined rosewood coffins. "Not one is forgotten before God," read more than a few animal epitaphs from this era.

With people keeping pets in record numbers in the twenty-first century, memorials for animals are continuing to evolve. Freeze-drying and traditional methods of taxidermy are among the more popular alternatives to burial or cremation, and there is one Utah-based company offering pseudo-Egyptian mummifications using modern chemical preservatives. There are even online virtual pet cemeteries that draw thousands of visitors who post pictures and personal stories. The current trend also indicates growing interest in joint interments for people and animals, with cemeteries such as Hartsdale permitting the scattering or burial of human remains in the graves of pets.

Now more than ever, animal memorialization is particularly germane, as we find ourselves increasingly dependent on animals not only for personal companionship and emotional health, but as psychological tethers to the natural world. As demonstrated by this history, we learn that a slow but inexorable revolution is taking place in humanity's concept of itself—that humankind is part of the larger community of creation, not separate or above it.

Further Resources

Haddon, C. (1991). *Faithful to the end: An illustrated anthology of dogs*. London: Headline.

Ikram, S. (2005). *Divine creatures: Animal mummies in ancient Egypt*. Cairo: The American University in Cairo.

Kete, K. (1994). *The beast in the boudoir: Petkeeping in nineteenth century Paris*. Berkeley: University of California Press.

Martin, E. Jr. (1997). *Dr. Johnson's apple orchard: The story of America's first pet cemetery*. Paducah, KY: Image Graphics Inc.

Rice, B. (1968). *The other end of the leash: The American way with pets*. Boston: Little, Brown and Co.

Ritvo, H. (1987). *The animal estate: The English and other creatures in the Victorian age*. Cambridge, MA: Harvard University Press.

Thurston, M. E. (1996). *The lost history of the canine race: Our 15,000-year love affair with dogs*. Kansas City: Andrews and McMeel.

Mary Thurston

■ Culture, Religion, and Belief Systems
Animal Immortality

Changing Attitudes and Beliefs

Christian teaching has long held that only humans possess an immortal soul, and it is commonly thought that this has had the effect of downgrading animal status. However, debates over the centuries reveal significant changes in thinking.

The word "soul" is ambiguous. When we say that a person "has no soul," it is commonly taken to mean that he or she lacks emotion, or is indifferent to moral questions, the arts, or religion. However, when we talk of "an immortal soul," we mean that the soul is everlasting and has a life beyond this. It is important to distinguish between them.

Debates surrounding the difference between human and animal souls have a long history, dating back to the time of the ancient Greeks. In the sixth century BCE, Pythagoras believed that animals possessed immortal souls, because, he thought, they incorporated reincarnated human souls. Plato (c. 428–348 BCE) held that the soul was separate from the living body, in order that it could survive the body's death. Socrates (c. 470–399 BCE) also identified the soul as distinct from the body. When asked the question "How shall we bury you?" he replied lightly, "However you please, if you can catch me and I do not get away from you." Later, Aristotle (384–322 BCE) considered the soul to have three parts: a nutritive part, which plants have; a sensitive part, which animals have; and a rational part, which is exclusive to man. This was taken to mean that only the rational, or thinking, part is immortal, and in animals this is lacking. Aristotle exalted reason at the expense of all other human attributes, a view that gave humans the highest status and was clearly detrimental to that of animals. The assertion that animals had no souls implied that they were inferior beings, and less entitled than humans to consideration. It became accepted that they were here for human use. Over the centuries, the views of these Greek philosophers were embedded in Western, Latin-speaking Christianity. Saint Augustine (354–430 CE) reinforced Aristotle's view, adding that we have no duties to beasts because they are disassociated from us by their irrationality. However, the modern academic, Richard Sorabji, says that the early Greek philosophers held more kindly views about animals, and that Christianity concentrated on only one half—the anti-animal half—of the far more evenly balanced ancient Greek debates (1993).

References to a future life, even for humans, are scarce in the Old Testament, and that of animals is hardly included. One exception, however, is in Ecclesiastes 3:19–20, which says: "For the fate of the sons of men and the fate of beasts is the same: as one dies, so dies the other. They all have the same breath . . . . All go to one place; all are from dust, and all turn to dust again. Then shall the dust return to the earth as it was; and the spirit shall return unto God who gave it." In the New Testament, the question of whether animals have immortality is neither put nor answered.

In the thirteenth century, the philosopher-theologian, Saint Thomas Aquinas (1225–74), citing Aristotle as his authority, taught that because animals lacked reason, they could have no immortal soul. Only *man* had been created a rational being, and he alone had a life beyond this. It could thus be argued that the doctrine that animals have no immortal soul is of pagan, rather than Christian, origin. The contemporary philosopher Mary Midgley has pointed out that the uniquely human possession of immortality is not at all easy to defend once it ceases to be universally taken for granted (1986). James Serpell suggests that St. Aquinas had rescued Christians from the otherwise-alarming prospect of encountering the vengeful spirits of their hapless animal victims somewhere in the hereafter (1996).

Charles Darwin (1809–82) added fuel to the debates. There was considerable reluctance to admit even physical, and still less *mental,* continuity between animals and mankind. The historian Keith Thomas says that the acceptance of evolution posed a sharp dilemma, for if men had evolved from animals, then either animals also had immortal souls or men did not (1984). To many people, this was totally unacceptable. However, traditional teaching on the matter has tended to become more liberal.

During a public audience in 1990, Pope John Paul II affirmed that animals, like men, were given the "breath of life" by God and stated that animals possess a soul. Although he did not address the issue of whether their souls are immortal, it has been generally taken to have softened traditional Catholic teaching. In fact, some theologians of the Franciscan order have, for many years, allowed (if only on a hypothetical basis) the idea of a survival of the souls of animals.

In his book, *Animals on the Agenda,* the contemporary theologian Andrew Linzey argues that whatever hope there may be for a future life for humans applies equally to animals and describes them both as "spirit-filled" creatures (1998). The Christian philosopher Roger Scruton says that people have become less sure of their status than was the case when they believed themselves to be the highest order of creation, alone blessed with an immortal soul (1996).

Those who have loved and enjoyed the companionship of pets are especially likely to question the idea that animals cannot have an afterlife. Citing the good qualities of their animals—such as their loyalty, patience, and sympathy—many pet owners see no reason why animals should not merit immortality. In his book *What Is Man?*, the American writer Mark Twain (1835–1910) remarked, "Heaven is by favour, if it were by merit your dog would go in and you would stay out."

The grounds upon which an afterlife has been "denied" to animals have predominantly been their lack of rationality and their lack of a moral sense. However, academics have now produced firm evidence of both intelligence and altruistic behavior in many animals. A recent three-generational survey carried out at the University of Southampton found that many people—especially those who are younger—no longer take it for granted that immortality is confined to humans.

See also

Culture, Religion, and Belief Systems—*Religion and Human-Animal Bonds*

Further Resources

Fidler, M. (2004). The question of animal immortality: Changing attitudes. *Anthrozoös, 17*(3), 259–66.

Hume, C. W. (1957). *The status of animals in the Christian religion.* Hertfordshire, UK: Universities Federation for Animal Welfare.

Linzey, A., & Yamamoto, D. (Eds.). (1998). *Animals on the agenda.* London: SCM Press Ltd.

Midgley, M. (1986). *Conflicts and inconsistencies over animal welfare.* Hertfordshire, UK: Universities Federation for Animal Welfare.

Scruton, R. (1996). *Animal rights and wrongs.* London: Demos.

Serpell, J. (1996). *In the company of animals.* New York: Cambridge University Press.

Sorabji, R. (1993). *Animal minds and human morals: The origins of the Western debate.* Ithaca, NY: Cornell University Press.

Thomas, K. (1984). *Man and the natural world.* London: Penguin Books.

de Waal, F. B. M. (1996). *Good natured: The origins of right and wrong in humans and other animals.* Cambridge, MA: Harvard University Press.

Margaret Fidler

■ Culture, Religion, and Belief Systems
Animal Mummies in Ancient Egypt

Not unlike modern humans, the ancient Egyptians had a variety of strikingly different, and often inconsistent, relationships with animals. They used them as help in the field, kept them as pets, worshipped certain of them as manifestations of deities, and ate their meat for food. They also mummified them.

There are four categories of animal mummies, each of which reveals a little more about human-animal relationships in ancient Egypt: pets, preserved so that they might be reunited with their owners after death; sacred animals, thought by the Egyptians to have been incarnations of gods and honored with an appropriately elaborate burial ceremony upon death; offerings, given in order to honor the dead (or, alternatively, some deity associated with the animal in question); and those preserved as food, for use by the dead in the afterlife. Archaeologists have also discovered a number of *fake* sacred and

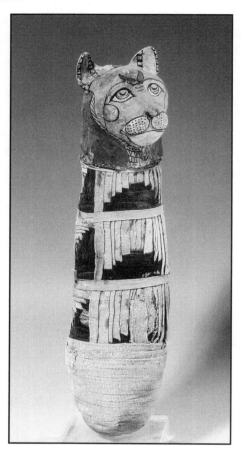

A mummified cat from Ancient Egypt, stuccoed and painted. Cats were seen as incarnations of Bastet and given an appropriately elaborate burial upon death. ©Art Resource, NY.

votive mummies (some of which may be found in the British Museum), most likely sold as genuine offerings to unsuspecting pilgrims.

Pets (including gazelles, monkeys, and hunting dogs) were often given elaborate burials—and, although we cannot rule out the possibility that when their owners died, the pets were killed in order to be buried with them, it seems likely that they were added to their respective tombs only after their own natural deaths. As I write this, pet-animal mummies from the Cairo museum are being X-rayed, in order to determine the causes of their deaths, which would in turn help to clarify the aforementioned question.

Examples of gods worshipped as beasts (which would invariably be turned into sacred mummies after their deaths) include Horus (the falcon or hawk), Anubis (the dog, fox, or jackal), Khephir (the scarab), Bastet (the cat), Sobek (the crocodile), Hathor (the cow), Atum (the eel), Khnum (the ram), and Apis (the bull). Mummified votive offerings include cats, shrews, snakes, dogs, raptors, and ibises, and typical examples of victual mummies would be cow ribs and geese.

While animal mummies have historically been neglected by most Egyptologists, the Animal Mummy Project at the Cairo museum (developed with the help of the American University in Cairo, and, in particular, Salima Ikram and Nasry Iskander) recently presented a praiseworthy series of studies on the different types of animal mummies, the methods of mummification, and the animal cemeteries located at sites throughout Egypt. These studies provide information not only about ancient Egyptian animal life (which includes a number of extinct species), but also about animal domestication, veterinary practices, human nutrition, and the role (symbolic or otherwise) of animals in the ancient Egyptians' religious practices. The studies have also inspired the book *The Cat Mummy*, by children's author Jacqueline Wilson. For recent updates on the Animal Mummy Project, and ways in which you can help a mummy (e.g., through adoption), visit http://www.animalmummies.com/.

Further Resources

Ikram, S. (Ed.). (2004). *Divine creatures: Animal mummies in ancient Egypt*. Cairo: American University in Cairo Press.

———. (2006). *Beloved beasts: Animal mummies from ancient Egypt*. Cairo: American University in Cairo Press.

Martin, G. T. (1981). *The sacred animal necropoli at north Saqqara*. London: Egypt Exploration Society.

Wilcox, C. (2002). *Animal mummies: Preserved through the ages*. Mankato, MN: Capstone Press.

Constantine Sandis

■ Culture, Religion, and Belief Systems
Animal Symbolism in Native American Cultures

There are approximately 500 American Indian nations inhabiting the northern hemisphere, and the culture of every Indian nation reflects the land on which they live. In the United States, there are the Woodland people (of the northeast and northwest), the Desert people, the Plains people, and the Coastal people. Although the nations of Native America can be characterized geographically, they each have their own creation stories, government, language, religion, and kinship systems.

In each nation, animals are used as a source of practical knowledge and spiritual insight. Animals have been used as clan symbols, to illustrate morality, and as inspiration in dress, music, and art. Animals have also been used to teach hunting and tracking skills and used in birth rituals; Ponka mothers, for instance, would save the umbilical cords of their newborns and fashion them into the forms of animals: turtles for boys, and lizards for girls. Animals are also often the central theme of American Indian dances; many of these dances have names to honor animals—such as the Eagle Dance and the Rabbit Dance. The Bear Dance is performed by the Utes and the Shoshones to ensure that no one goes hungry. Dance is also used by Native Americans to celebrate family events, such as the birth of a child; during such occasions, the dancers may take on the characteristics of various animal spirits, with the aid of elaborate costumes representing the spirit animal.

The Native people of the Pacific Northwest often construct totems to tell stories about individual clan members and tribal legends. Totems are elaborate pieces of art, carved from wood, that serve as a type of "institutional clan memory." A totem is also a symbol that each clan adopts. These symbols often represent animals, resulting in names such as "the Bear Clan" and "the Raven Clan."

Among the most powerful animal symbols in many Native cultures is the wolf. The wolf is considered a teacher and mentor. By traveling "the trail of the wolf," we learn the values of cooperation, fidelity, and loyalty. We also learn hunting skills. Rather than fearing the wolf or considering it a competitor, Native people considered its presence a sign of good hunting. Following a successful hunt, food would be presented to the wolf as a gesture of gratitude, and, over time, the wolf became what we call "man's best friend."

The animal considered by many Native American cultures to be the closest to humans is the bear. Like the bear, the human walks upon two legs, and, like the bear, the human has a wild, uncontrollable emotional dimension. This emotional dimension is considered so strong that many Native tribes do not directly discuss this dimension of their personality. The Blackfoot, for example, call the bear "unmentionable big animal," and the Haida call the bear the "dark thing."

In addition to the identification of the bear with the darker, uncontrollable dimension of our personality, the bear is considered a source of introspection; the bear is a solitary animal that spends half its life in a dark cave, thereby teaching humans to search within ourselves for knowledge. The bear awakens from hibernation a new animal, and so, too, are *we* transformed by introspection.

The animal most associated with Native people of the plains is the buffalo. The buffalo represented life to the Native people, because it was a source of food, clothing, and various practical objects (such as toys and cooking utensils). The buffalo also fulfilled an important spiritual need and was associated with wisdom and renewal. The skull of the buffalo continues to be used by Native shamans of the plains as a conduit to the spirit world and as an altar during the Sun Dance. The Sun Dance focuses on the buffalo as a way to honor the animal for its sacrifice and its contributions to the physical and spiritual well-being of the Native people.

Another important animal for the Native people is the eagle. The eagle represents many of the best human traits, including bravery, foresight, and strength. Moreover, because the eagle flies close to the sun, it is seen as a facilitator between the physical and spiritual worlds. The eagle feather has medicinal properties; its touch is known to heal.

Animal symbolism plays an important role in creation stories. One such story is told by the Makah of Washington, to account for how animals and birds were placed on Earth: In order to get the planet ready for the first people, two brothers, known as "Two-Men-Who-Changed-Things," summoned all entities. Some of these, the brothers changed into animals; others, they changed into birds. One entity was fond of stealing from those who could hunt and fish. The brothers shortened its arms and bound its legs so that only its feet could move. Once bound, the brothers threw the entity into the sea, and it became a seal that had to catch its own food.

A Creek story relates how fire was brought to the people. In the beginning, only weasels possessed fire, and they were not about to share it. The people summoned a council to discuss how to get fire from the weasels. The problem was complicated, because the weasels lived on an island and could only be reached by skilled swimmers. After consideration, the rabbit volunteered to go because of his dancing and swimming skills. In preparation, he covered the top of his head with pine tar. Rabbit swam to the island, joined the weasels in dance, and, at the appropriate moment, put his head near to the fire and ignited the pine tar. Rabbit swam back to the people, and the people had fire.

Animal symbolism is woven into the fabric of Native life. Animals are not seen as distinct from humans; they serve to teach us about the most important aspects of life and to inspire us to be greater than we are.

Further Resources

Caduto, M., & Bruchac, J. (1991). *Keepers of the animals*. Golden, CO: Fulcrum Publishing.
Kopper, P. (1986). *North American Indians*. Washington, DC: Smithsonian Books.
Olson, D. L. (1995). *Shared spirits: Wildlife and Native Americans*. Minnetonka, MN: North Word Press.

Wilma Bold Warrior and Charles I. Abramson

■ Culture, Religion, and Belief Systems
Animalism and Personal Identity

The problem of personal identity is one of the oldest and most bewitching in all of philosophy. The challenge, in short, is to persuasively articulate the conditions under which beings like you and me persist through time and change. In answer to this challenge, the view known as "animalism" maintains that, since we are essentially and most fundamentally human animals, our persistence conditions are no different from those of other biological organisms.

The Problem of Personal Identity

In order to appreciate the significance and implications of the animalist answer to the problem of personal identity, a better sense of the problem is required. We begin with the uncontroversial observation that you existed ten years ago—that something that existed ten years ago is identical to you today. This observation is true despite the fact that today's you and yesteryear's you bear little resemblance to one another. Today's you is made up of different physical constituents, possesses different capacities and tendencies, holds different beliefs and desires, and so on. Yet, despite these differences, an uninterrupted line of continuity binds yesteryear's you with today's you, and it is this continuity that grounds the fact that yesteryear's you and today's you are one and the same thing. To characterize and explain the nature of this continuity is to answer the problem of personal identity.

The complexity of this task emerges from two related sources. First, any adequate account of personal identity must accommodate not only the standard and uncontroversial cases (like yours), but various nonstandard cases as well. Any account that fails to attend to these more unusual cases risks charges of parochialism, of failing to limn the boundaries of the concepts in play, for our aim is not only to describe the criteria wherein satisfaction is *sufficient* for persons to persist through time and change, but also to isolate those criteria wherein satisfaction is *necessary* for such persistence—to pinpoint the conditions which a person who exists at one point in time *must* satisfy in order to be identical to a person who exists at another time. To isolate these necessary conditions would enable us to differentiate those changes that persons can survive from those that they cannot.

Not for want of trying, philosophers have found this to be a difficult task. Indeed—and here is the second complicating factor—several of the most prominent proposals are both independently plausible and mutually inconsistent. (In addition to the three discussed below, these views include the so-called "brain criterion of personal identity," as well as various dualist theories. Also, Derek Parfit has challenged the evaluative significance of facts about personal identity on the grounds that, rather than strict identity, one's survival in *all important respects* is what matters.) Consequently, although no consensus has yet been reached, the debate concerning the problem of personal identity remains vigorous, with philosophers developing ever-more-nuanced views—views that bring to bear not only our pre-philosophical intuitions about fantastical thought experiments, but also (increasingly) empirical data.

Three Criteria of Personal Identity

According to the "bodily criterion of personal identity," the persistence conditions for persons are bodily in character. On this view, a person (P) existing at one time (t) is identical to something (S) existing at another time (t'), just in case P's body at t is physically continuous with S's body at t'. (Those who have advocated some version of the bodily criterion include A. J. Ayer, Judith Jarvis Thomson, and Bernard Williams.) The appeal of this view is not difficult to appreciate. After all, one need not be a philosopher to observe that wherever you are, your body is there too. Likewise, notwithstanding religious commitments to the afterlife, who would deny that you would cease to exist if your body were obliterated?

But the bodily criterion faces a host of objections, perhaps the most compelling of which derives from the strong intuition that a person could come apart from the particular body (or the particular animal) with which she happens to be associated. Thus, in a famous passage from his *An Essay Concerning Human Understanding*, John Locke invites his reader to agree that

> should the Soul of a Prince, carrying with it the consciousness of the Prince's past Life, enter and inform the Body of a Cobler as soon as deserted by his own Soul, every one sees, he would be the same Person with the Prince, accountable only for the Prince's Actions: For who would say it was the same Man? (1975)

Suitably generalized and updated, Locke's body-transfer thought experiment seems to recommend two important conclusions. The first is that it is one thing to be the same person over time, and quite another to be the same body (or the same animal). Today's you could be identical to yesteryear's you, even if the body (or the animal) associated with today's you was once associated with a completely different person. There seems to be nothing incoherent in the thought that you could wake up in your neighbor's body. (Indeed, many a science-fiction narrative has relied on the coherence of just this intuition!)

The second conclusion suggested by Locke's thought experiment is that one's identity over time necessarily involves the identity of one's consciousness. This suggestion provides the basis for a family of views generally labeled the "psychological criterion of personal identity." The basic idea here is that a person (P) existing at one time (t) is identical to something (S) existing at another time (t'), just in case P at t is psychologically continuous with S at t'. (A common, though not uncontroversial, way of characterizing this continuity is to say that P at t is psychologically continuous with S at t' just in case P and S are connected by a series of non-branching, intermediary person-stages, $PS_1 \ldots PS_n$, such that each PS_{i+1} contains apparent memories of events which occurred to PS_i.) And while substantial debates concerning the nature, strength, and relevance (e.g., for warranted accountability) of this psychological continuity have led to various refinements of the psychological criterion, in one form or another, this view has been the dominant account of personal identity for the last three centuries. Any representative sample of the many philosophers who have advocated some version of the psychological criterion would include (in addition to Locke) Mark Johnston, Carol Rovane, and Sydney Shoemaker.

A relative newcomer to the personal identity debate, the "animalist criterion of personal identity" opposes the psychological criterion by denying the Lockean distinction between the persistence of persons and the persons of human animals. According to the animalist, each of us is essentially—and most fundamentally—a human animal; we could not exist except as human animals, and the conditions of our persistence derive from our status as human animals. That is, a human person (P) existing at one time (t) is identical to something (S) existing at another time (t'), just in case P at t is biologically continuous with S at t'. (As it is with psychological continuity, the characterization of biological continuity is problematic. One way of spelling it out would be to say that that P at t is biologically continuous with S at t' just in case P and S are connected by a series of non-branching, intermediary person-stages, $PS_1 \ldots PS_n$, such that each PS_{i+1} is animated by life processes that naturally continue those life processes which animated PS_i.) Among those credited with developing this view are Michael Ayers, William Carter, Eric Olson, Paul Snowdon, Peter van Inwagen, and David Wiggins.

Debating Animalism Positive arguments for animalism are typically presented hand in hand with critical arguments against the psychological criterion and its associated person-animal distinction.

First, consider an animalist objection to the psychological criterion. In order for today's you to be psychologically continuous with yesteryear's you, both today's you and yesteryear's you must possess various psychological capacities (e.g., the capacity to remember, the capacity for self-consciousness). Yet these capacities are notably absent in two entities to which, intuitively, we take ourselves to be actually or possibly identical, *viz.* an unborn fetus and a patient in a persistent vegetative state. If each of us was once an unborn fetus, and if each of us might one day end up in a persistent vegetative state, then that strongly suggests that our persistence conditions need not be psychological, but rather are biological.

Second, by arguing as follows, the animalist may rely on empirical considerations to validate the proposed rejection of the person-animal distinction: If you are not an animal, then presumably nor are your parents animals. But then, nor are your parents' parents, nor their parents' parents, and so on, as far back as your ancestry extends. Taking the phrase "as far back as your ancestry extends" seriously, this suggests that denying your identity to an animal entails the rejection of evolutionary theory, because denying your identity to an animal is tantamount to denying that your distant ancestry includes entities that were animals. But the rejection of evolutionary theory is too high of a price to pay. Therefore, we should reject the assumption that you are not an animal.

Lastly, anyone who denies that each of us is identical to a particular human animal must face what has come to be known as the "too many minds argument" (or the "thinking animal argument") for animalism. According to this argument, even one who denies that the animal presently sitting in your chair is you must grant that there is an animal presently sitting in your chair. Yet it is increasingly implausible to deny that animals of many types think and perceive; and if any do, surely the type of animal sitting in your chair does. Moreover, you are sitting in your chair, and clearly you are thinking and perceiving. But if you were not identical to this animal, then there would be two mental lives simultaneously running in parallel: the thoughts and perceptions had by you and the qualitatively identical thoughts and perceptions had by the animal. But this is absurd. Therefore, the animal presently sitting in your chair is you.

In reply to the preceding arguments, critics of the animalist criterion have attempted to maintain the distinction between the persistence conditions of persons and the persistence conditions of human animals by deploying a fine distinction between the "is" of identity and the "is" of constitution. On their view, although we are animals in the sense of being non-identically constituted by animals, it is not the case that we are identical to animals.

While the debate between advocates of the animalist and the psychological criteria of personal identity continues to thrive, little consensus has been reached. Both the animalist position and objections to it remain under active development, with avenues of inquiry into such fields as the philosophy of biology, animal cognition, and bioethics only just beginning to open.

See also

Sentience and Cognition

Further Resources

Baker, L. R. (2000). *Persons and bodies: A constitution view.* Cambridge: Cambridge University Press.

Locke, J. (1975). *An essay concerning human understanding.* P. Nidditch (Ed.). Oxford: Clarendon Press.

Noonan, H. (2003). *Personal identity.* London: Routledge.

Olson, E. (1997). *The human animal: Personal identity without psychology.* Oxford: Oxford University Press.

———. (2002, Fall). *Personal identity. Stanford Encyclopedia of Philosophy.* http://plato.stanford.edu/archives/fall2002/entries/identity-personal/

Parfit, D. (1984). *Reasons and persons.* Oxford: Oxford University Press.

Shoemaker, S., & Swinburne, R. (1984). *Personal identity.* Oxford: Blackwell.

Snowdon, P. (1990). Persons, animals, and ourselves. In C. Gill (Ed.), *The person and the human mind* (pp. 83–107). Oxford: Oxford University Press.

Wiggins, D. (2001). *Sameness and substance renewed.* Cambridge: Cambridge University Press.

Williams, B. (1973). *Problems of the self.* Cambridge: Cambridge University Press.

Stephan Blatti

■ Culture, Religion, and Belief Systems
Animals as Sources of Medicinal Wisdom in Traditional Societies

In recent years a revival has taken place in the search for new medicines from plants to cure a variety of health problems that have emerged because of our changing lifestyle and the growing resistance to modern pharmaceuticals by once-controllable diseases caused by parasites and pathogens. In this search for new medicines, ancient sources have yielded some surprising clues. A common approach has been to search the ethnographic database for the traditional uses of plants as medicine (ethnomedicine) followed up by scientific validation of their reported medicinal properties (ethnopharmacology). A new source has emerged from the study of what animals do when sick ("animal self-medication" or zoopharmacognosy) and is already revealing some ancient sources of medicinal wisdom of potential use in the modern world. Primatologists, including myself, have documented cases of animals using the same medicinal plants as humans for the relief of similar symptoms. Synthesizing these different areas of research helps bring to light a unique relationship in which humans have long looked to other animals for medicinal wisdom in their daily and spiritual lives.

Drawing from a wide variety of sources, it is clear that people have long looked to animals for sources of herbal medicine (Table 1). Numerous independently reported records of the possible use of plants as medicine by wild animals exist and are still being used by humans today. For example, the Navajos of the southwestern United States credit the brown bear for their obtaining knowledge of "osha" (*Ligusticum porteri*), a plant with demonstrated antifungal, antiviral, and antibacterial properties. The active ingredients of this plant, lactone glycoside and saponin, extracted from its roots, can be bought over the counter at health-food stores in North America for the treatment of colds and sore throats.

"Iboga" (*Tabernanth iboga*), as it is known in parts of Gabon in western Africa, contains several indole alkaloids. It is used as a powerful stimulant and aphrodisiac in secret religious societies in Gabon. It is speculated that because of the widespread reports from local people of bush pigs, porcupines, and gorillas going into wild frenzies after digging up and ingesting the roots, they probably learned about these peculiar properties of the plant from watching the animals' behavior. The most active ingredient, found in the root, is called "ibogaine" and is shown to affect both the central nervous system and the cardiovascular system. Two other active known compounds in the plant are tabernanthine and iboluteine. The stimulating effects are similar to caffeine. The active component of this plant has been under clinical trial as an alternative to methadone, in order to help treat drug abusers.

The sloth bear and local people of central India are noted to become intoxicated from eating the fermented "madhuca" flowers. Reindeer and the Lapp reindeer herdsmen of Norway consume the same mushrooms known for their intoxicating effects.

One version of the discovery of coffee says that the observations by a shepherd in the Ethiopian Highlands of his goats becoming more alert and active after grazing on the berries of wild coffee provided the clue for people to exploit the plant as a stimulant. Coffee was used by monks in a nearby monastery to stay awake during their all-night ceremonies. Dr. Jaquinto, the trusted physician to Queen Anne (wife of James I) in seventeenth-century England, is said to have made systematic observations of domestic sheep foraging in the marshes of Essex that led to his discovery of a treatment for consumption (tuberculosis). Such accounts of animals' use of medicine are numerous in

Table 1. Some Anecdotal Evidence for Self-Medication in Animals

Species	Plant (Chemical Evidence)
Malay elephant	*Entada schefferi* (LEGUMINOSAE): for stamina before long walk, possible pain killer?
African elephant	*Boraginaceae* sp.: induce labor; used by Kenyan ethnic group to induce labor and abortion. Similar story related to Huffman about observations made in Tanzania
Indian bison	*Holarrhena antidysenterica* (APOCYNACEAE): bark regularly consumed. Species name suggests antidysentaric action.
Wild Indian boar	*Boerhavia diffusa* (NYCTAGINACEAE): called pig weed. Its roots are selectively eaten by boar, and it is a traditional Indian antihelmintic.
Pigs	*Punicum granatum* (PUNICACEAE) pomegranate: root sought after by pigs in Mexico. Alkaloid in roots toxic to tapeworms.
Indian tigers, wild dogs, bears, civets, jackals	*Careya arborea* (BARRINGTONAEACEAE), *Dalbergia latifolia* (LEGUMINOSAE), etc.: fruits of various species eaten by large carnivores. Possibly helps in elimination of parasites ingested along with contents of intestines of herbivore prey
South American wolf	*Solanum lycocarpon* (SOLANACEAE): rotting fruit said to be eaten to cure stomach or intestinal upset.
Asiatic two-horned rhinoceros	*Ceriops candoleana* (RHIZOPHORACEAE): tannin rich bark eaten in large amounts enough to turn urine bright orange. Possible use in control of bladder and urinary tract parasites.
Black howler monkey, Spider monkey	Indigenous peoples living in primate habitats of the Neotropics claim that some monkey species are parasite-free because of the plants they eat.

literature and are corroborated by similar stories from other ethnic groups, living far away, who use the same or related species for similar purposes.

Scientists, it seems, have just recently begun to rediscover this important relationship between humans and animals and are attempting to critically evaluate—and, in some cases, actively validate—what many long considered folklore or "just so stories."

"Asiri ya dawa"—Tongwe Stories on the Animal Origins of Medicine

This essay focuses on information conveyed firsthand to the author in Tanzania by a long-term collaborator in chimpanzee self-medication research, Mohamedi Seifu Kalunde. A member of the Tongwe ethnic group in Western Tanzania, Mohamedi is a senior game officer in the Mahale Mountains National Park and a practicing traditional healer. He has assisted the Kyoto University chimpanzee research project since it began in the early 1960s. The following information provides vivid examples of the status of animals in the daily lives of people still living close to nature in their traditional areas. Mohamedi obtained these stories from his mother, Joha Kasante; his grandfather, Kalunde; and his uncle, Musalangi, only weeks before Musalangi's death in mid-June

2000. Some of these stories were told to Musalangi by his father, Kabingo, undoubtedly passed to *him* from along this long family line of traditional healers.

It is not the intention of this essay to in any way validate the effectiveness of the treatments described below, although it can be argued that some of these medicines have undergone many years of *in situ* drug trials and have thus proven their utility. It is the purpose of this essay to describe the level of importance some traditional societies put on animal behavior as a source of medicinal wisdom.

Animals in Dreams, and the Selection of Medicinal Plants by the Sick

One day, a sick man lay in his bed trying to rest. He was extremely tired from fighting the illness, and was in a state of semiconsciousness. Drifting in and out of dreams, he saw a vision of a plant in the forest that he had often seen before but did not know of any useful purpose of it for humans. Suddenly, a voice told him that if he used this plant it would cure him. The sick man was startled by the voice and unsure if he should take its advice. The voice came again, encouraging him: "Just take it, why are you afraid? It will make you better." Coming in and out of this dreamy state, the man was not sure whether he had been dreaming or not, but was scared and still not prepared to take his chances with the plant. He drifted off to sleep again, and this time he saw a sick animal appear and eat the plant. The voice came to him again, saying "Look at that, the sick animal knows to take this plant to recover from its illness." After awakening, he became more confident about this advice and went out to look for the plant. The story goes that he prepared the plant and recovered just as the voice in his dream had assured him he would.

In many societies, traditional healers are known to actively seek out solutions to cure difficult illnesses through visions or dreams. Indeed, Mohamedi says that his mother, too, has sought out and obtained successful cures, for difficult-to-treat patients, through dreams.

"Likibanga" and "Kaselenje"—Wild Pig Medicines

A long time ago, it is said, the people of a Tongwe village were afflicted with an outbreak of cholera. Many people were suffering, and some had already died. One day, a hunter went into the forest looking for something with which to feed his family. He came across a bush pig and was about to spear it when he noticed that it appeared to be sick. It was weak and could move only very slowly. The hunter could not kill and eat a sick animal, so instead he continued to follow it for awhile in order to see what the problem was. He realized that the bush pig was suffering from diarrhea, and so he thought that if he observed its habits for awhile, he might learn what the bush pigs took for that illness.

He left the sick bush pig in the evening, but he returned early the next morning and continued to follow it again. That day, he observed the pig digging up and chewing the roots of two plants known to him as "likibanga" and "kaselenje." Curious, the hunter tasted these plants and found them both to be quite bitter. Keen on finding out whether these plants would help the sick pig—and thus possibly help his family, too—he followed the bush pig for a few more days until it recovered from its sickness. Finding out what he had wanted to know, the hunter took these plants back to his village. The roots and bark were crushed and put in water, to drink or to use as an enema.

The medicine worked on the people of his village, and, according to Mohamedi, likibanga and kaselenje are, to this day, still used in rural villages of Tongwe for the treatment of cholera-like symptoms.

"Munyonga nTembo"—Elephant Medicine

Munyonga nTembo is a medicinal plant used by the WaTongwe as a treatment for stomach problems. The origin of the name of this plant comes from the Tongwe verb "kunyonga" (to twist and pull off). Elephants ("tembo" in both Swahili and the Tongwe language) twist and pull off the leaves of this plant before ingesting it. One day, Kalunde observed an elephant suffering from an upset stomach. The elephant put a few bunches of the leaves of this plant into its mouth and chewed them up a bit without swallowing them. The elephant then took in some water with its trunk and transferred it into its mouth. After holding the water in its mouth together with the leaves for awhile, it spit out the leaves and swallowed the water.

For the Tongwe, the leaves are crushed and placed in water. The resulting solution is drunk, or the crushed leaves are used as a suppository for an upset stomach.

"Mulengele"—Porcupine Medicine

Some events are still fresh in the memories of living people and demonstrate that this unique relationship between humans and animals continues in practice. Mohamedi's grandfather, the late Kalunde, was a traditional healer and a skilled hunter. He discovered a treatment still widely used today for dysentery by watching a sick porcupine ingest the roots and recover from symptoms suggestive of the disease. Prior to this time, the villagers knew the plant called "mulengele," but regarded it as being highly toxic and did not use it for medicine.

One day, while Kalunde was out checking his snares, he came across a porcupine caught in one of them. He quickly killed the porcupine, but discovered afterward that it had been a mother—her young were hiding in the bushes nearby. Kalunde took the babies home to look after them, putting them inside a small enclosure so that they would not stray off alone. The young grew ill with passing time, displaying, among other symptoms, bloody stools—the same symptom being shown by the members of his village that Kalunde was desperately trying to cure. One of the sickly porcupines escaped and wandered off into the forest. Kalunde followed it to see what it would do and observed it dig up and chew on the roots of the mulengele found at the edge of the village. As far as he knew, it had never been used as medicine before, but he was curious as to why the porcupine would eat this toxic bush. He took the young porcupine back to the enclosure. In a few days, its symptoms had disappeared, convincing Kalunde that he should try the roots on his patients in the village. He told them what he had seen the porcupine do and that he wanted to give it a try. People were at first reluctant, because they knew of the plant's poisonous qualities. Desperate to get well, some of the sick said that they would take this new treatment if Kalunde first showed that it would not kill them instead. He prepared a concoction of crushed root in water and drank some of the bitter liquid. Satisfied that the dosage was safe, Kalunde put the medicine to work. As he had hoped, many people inflicted with dysentery soon began to recover.

From that time on, the plant has been an important medicine of the Tongwe in western Tanzania. Few young people today who work at the research site in Mahale National Park are aware of the history behind the discovery of mulengele, but they all know of its medicinal value and have been treated with it at one time or another. Mohamedi has experimented with the plant on various illnesses and has found it to be quite effective against sexually transmitted diseases, suggesting that the effectiveness of this plant lies in its antibiotic properties. In further support of this, Mohamedi has heard that a fellow healer in the Mpanda region, an area further inland from Lake Tanganyika, is using the root of mulengele to treat various secondary infections in AIDS patients. While there is no published medicinal information on this species, another closely related species of the same genus is known for its antibiotic properties. We are now working together with colleagues in Tanzania and France to verify the antibiotic activity of mulengele.

Over the years, Mohamedi has patiently shared with me his knowledge of the medicinal, and other practical, uses of various plants that we encounter in the forest of Mahale, whether they are eaten by chimpanzees or not. It was an exhilarating and humbling revelation when Mohamedi confided in me one day about the WaTongwe legends regarding the animal origin of some of their important medicines. The historically recent examples, such as mulengele, are based on living memories of those involved and suggest that the discovery of new medicines from clues gained by observing animal behavior has continued into the recent past.

"Mjonso," "Mhefu," and Other Chimpanzee Medicines

In the course of my research at Mahale, Mohamedi and I have come across numerous sick chimpanzees that we followed closely over the course of their illness and recovery after the consumption of one of the most widely used traditional African medicinal plants, *Vernonia amygdalina*. It is known locally in the Mahale region as "mjonso" and is extremely bitter tasting—definitely not the preferred taste of chimpanzee foods. Given its wide use across Africa, the medicinal properties of this species have been studied in laboratories around the world since the 1960s. Prior to our work on chimpanzees, however, the main active ingredients reported for this plant were isolated from the bark, leaves, and roots of the plant—those parts used in traditional medicine. Those compounds described come from a group of chemicals known as the sesquiterpene lactones and have demonstrated anti-tumor, antibacterial, anti-parasitic, and antibiotic properties. The plant has long been used by humans across Africa for—among other things—treatment for a wide range of parasite infections and related symptoms of gastrointestinal upset (Table 2). Japanese collaborators in this research—Koichi Koshimizu, Hajime Ohigashi, and others—in their laboratory at Kyoto University were quite surprised when they isolated active ingredients, from the young shoots of this species, that had never before been described. They named this group of compounds the "steroid glucosides." In total, nine new compounds belonging to this new group of chemicals were discovered. Further laboratory studies of these compounds by our group revealed significant pharmacological activity against four major parasite-caused illness of the tropics: falciparum malaria, leishmania, amoebic dysentery, and schistosomiasis (not to mention an important tumor-growth inhibitor, thanks to the observations of sick chimpanzee behavior).

In 2003 I was able to witness firsthand the acquisition of a new medicine by Mohamedi as a consequence of our observations of the use of a plant by sick chimpanzees suffering from parasite infections. The plant, "mhefu" in Tongwe and *Trema orientalis* in Latin, is a common pioneer tree species found in recently cleared forests or fields. It is one of the plants from which rough leaves are swallowed whole by chimpanzees for the expulsion of the same parasites that cause the illness relieved after the use of mjonso by chimpanzees at Mahale. For chimpanzees, swallowing the rough leaves of this plant on an empty stomach, without chewing them, induces temporary diarrhea that helps to physically purge nematodes from the chimpanzees' intestinal tracts. Mohamedi, on the other hand, prepared a water extract from the crushed leaves of mhefu and tried them to see how they would work to stop diarrhea. He found the extract to be quite effective, and Mohamedi and his mother Joha Kasante now use it on their patients. Mohamedi's mother was at first skeptical that it would be of any medicinal value. She knew the plant well, but considered it only as a source of firewood and building material.

Table 2.   Ethnomedicinal Uses of *Vernonia amygdalina* in Africa

Application	Plant Part Used	Region Used/Comments
General intestinal upsets:		
enteritis	root, seeds	Nigeria
constipation	leaves, sap	Nigeria, Tanzania, Ethiopia: as a laxative
diarrhea	stem, root-bark, leaves	W. Africa, Zaire
stomach upset	stem, root-bark, leaves	Angola, Ethiopia
Parasitosis:		
schistosomiasis	root, bark, fruit	Zimbabwe, Mozambique, Nigeria: sometimes mixed with *Vigna sinensis*
malaria	root, stem-bark, leaves	E. Africa, Angola, Guinea, Nigeria, Ethiopia: a quinine substitute
trematode infection	root, leaves	E. Africa: treatment for children used as a suppository
amoebic dysentery	root-bark	S. Africa
ringworm	leaves	Nigeria: ringworm and other unidentified epidermal infections
unspecified	leaves	Nigeria: prophylactic treatment for nursing infants, passed through mother's milk
	root, seeds	Nigeria: worms
	leaves	W. Africa: crushed in water and given to horses as a vermifuge, livestock fodder supplement for treating worms
	leaves	Ghana: purgative
Tonic food	leaves	Cameroon, Nigeria: boiled or soaked in cold water prepared as soup or as a vegetable fried with meat; "n'dole," "fatefate," "mayemaye," leaves sold in markets and cultivated in home gardens
Other ailments:		
amenorrhoea	root	Zimbabwe
coughing	leaf	Ghana, Nigeria, Tanzania
diabetes	all bitter parts	Nigeria
fever	leaves	Tanzania, Kenya, Uganda, Congo-Kishasa: leaves squeezed and juice taken
gonorrhea	roots	Ivory Coast: taken with *Rauwolfa vomitoria*

(Continues)

Table 2. (Continued)

Application	Plant Part Used	Region Used/Comments
'heart weakness'	root	W. Africa: vernonine is a cardiotonic glycoside comparable to digitalin
lack of appetite	leaf	W. Africa: leaves soaked in cold water to remove bitter and then boiled in soup
pneumonia	leaf	Ivory Coast: taken with *Argemone maxicana* or used in a bath
rheumatism	stem, root-bark	Nigeria
scurvy	leaves	Sierra Leone, Nigeria, W. Cameroon: leaves sold in markets and cultivated in home gardens
General hygiene:		
dentrifice	twig, stick	Nigeria: chew stick for cleansing and dental caries
disinfectant	not given	Ethiopia
soap	stems	Uganda

Adapted from Huffman (2003).

And so, the relationship between humans and animals continues. It will continue for as long as people are curious about animal behavior and believe that they have something to teach us.

These examples, from the Tongwe ethnopharmacopeia and recent primatological research, point to practical uses for plants that people have long attributed their knowledge of to their relationship with animals in their daily and spiritual lives. This is a relationship in which people of traditional societies look to animals as a source of medicinal wisdom. I have spent much of the last two decades on a quest to understand what animals do when they are sick and what they can teach us about curing ourselves. For me, the study of animal self-medication is not just about finding solutions for practical problems; it is also about attaining knowledge for knowledge's sake, a means to a richer understanding and appreciation of the world around us. In the process, I have come to appreciate the wisdom of traditional cultures living much closer to the "natural" world than most of us do today. It is also a humbling rediscovery of an important relationship between humans and animals and—indeed—of the magic of studying animal behavior.

Further References

Balick, M. J., & Cox, P. A. (1996). *Plants, people, and culture. The science of ethnobotany*. New York: Scientific American Library.

Berry J. P., McFarren, M. A., & Rodriguez, E. (1995). Zoopharmacognosy: A biorational strategy for phytochemical prospecting. In D. L. Gustine & H. E. Flores (Eds.), *Phytochemicals and Health* (pp. 165–78). Rockville, MD: American Society of Plant Physiologists.

Huffman, M. A. (2002). Animal origins of herbal medicine. In J. Fleurentin, J.-M. Pelt, & G. Mazars (Eds.), *From the sources of knowledge to the medicines of the future* (pp. 31–42). Paris: IRD Editions.

———. (2003). Animal self-medication and ethnomedicine: Exploration and exploitation of the medicinal properties of plants. *Proceeding of the Nutritional Society, 62*, 371–81.

Huffman, M. A., Koshimizu, K., & Ohigashi, H. (1996). Ethnobotany and zoopharmacognosy of *Vernonia amygdalina*, a medicinal plant used by humans and chimpanzees. In P. D. S. Caligari & D. J. N. Hind (Eds.), *Compositae: Biology & utilization* (pp. 351–60), Vol. 2. Kew: The Royal Botanical Gardens.

Huffman, M. A., Ohigashi, H., Kawanaka, M., Page, J. E., Kirby, G. C., Gasquet, M., Murakami, A., & Koshimizu, K. (1998). African great ape self-medication: A new paradigm for treating parasite disease with natural medicines? In Y. Ebizuka (Ed.), *Towards natural medicine research in the 21st century* (pp. 113–23). Amsterdam: Elsevier Science B.V.

Koshimizu, K., Ohigashi, H., Huffman, M. A., Nishida, T., & Takasaki, H. (1993). Physiological activities and the active constituents of potentially medicinal plants used by wild chimpanzees of the Mahale Mountains, Tanzania. *International Journal of Primatology, 14*(2), 345–56.

Krief, S., Huffman, M. A., Sévenet, T., Guillot, J., Hladik, C.-M., Grellier, P., Loiseau, M., & Wrangham, R. W. (2006). Bioactive properties of plant species ingested by chimpanzees (*Pan troglodytes schweinfurthii*) in the Kibale National Park, Uganda. *American Journal of Primatology, 68*, 51–71.

Krief, S., Martin, M.-T., Grelier, P., Kasenene, J., & Sevenet, T. (2004). Novel antimalarial compounds isolated in a survey of self-medicative behavior of wild chimpanzees in Uganda. *Antimicrobial Agents and Chemotherapy, 48*(8), 3196–99.

Ohigashi, H., Huffman, M. A., Izutsu, F., Koshimizu, K., Kawanaka, M., Sugiyama, H., Kirby, G. C., Warhurst, D. C., Allen, D., Wright, C. W., Phillipson, J. D., Timmon-David, P., Delmas, F., Elias, F. R., & Balansard, G. (1994). Toward the chemical ecology of medicinal plant use in chimpanzees: The case of *Vernonia amygdalina*, a plant used by wild chimpanzees possibly for parasite-related diseases. *Journal of Chemical Ecology, 20*(3), 541–53.

Michael A. Huffman

■ Culture, Religion, and Belief Systems
Animals in Nazi Germany

To the extent that one can even speak of a Nazi ideology, it was a highly unsystematic blend of several strands—including anti-Semitism, racialist theory, Nietzschean philosophy, neo-paganism, Bolshevism, conservatism, and agrarianism. It should, therefore, not come as a surprise that the attitudes of the Nazis toward animals were complex and often contradictory. On the one hand, the animal-protection laws enacted by the Nazis in 1933 (and expanded in 1938) were the strictest in the world at the time. At the same time, animals were often tormented in training exercises designed to toughen up members of the SS and Hitler Youth.

The issue that received the most attention from the Nazi Movement was animal experimentation. The rhetoric of the Nazis regularly condemned "vivisection" as a symbol of the sort of rationalism that they associated with Jews, which they then contrasted with the organic thinking of the Aryan races. In August 1933, Hitler's deputy, Hermann Göring, announced that he would send anyone who engaged in vivisection to a concentration camp. His speech is probably the origin of the popular misconception that the Nazis completely banned animal experimentation, though Göring did made a few exceptions for prestigious research institutes. The Law on Animal Protection, first enacted in 1933 and updated in 1938, allowed animal experimentation only with explicit permission of the Ministry of the Interior, and then subject to a long list of restrictions.

Though falling far short of an unconditional prohibition, the Law on Animal Protection did considerably reduce the number of experiments performed on animals in Germany. Prior to 1933, half of the articles in German medical journals featured animal experiments, but the percentage was reduced to less than a tenth after the Law on Animal Protection was enacted. The combination of restrictions on animal experimentation with grisly experiments on unwilling human subjects during the Third Reich prompted the Nuremburg Tribunal to mandate that all experiments on human beings be performed first on animals. In fact, the Nazis did almost always precede experiments on human beings with tests on animals, though far more out of reasons of economy than humanitarianism.

Animals also figured prominently in the theoretical works of the Nazis. Konrad Z. Lorenz, who was to share the Nobel Prize in 1971, was a member not only of the Nazi Party but also of its Office of Race Policy. He compared the effects of domestication in human beings with those of civilization in human beings, concluding that both lead to genetic decline because they interfere with natural selection. He believed, however, that this degeneration could be prevented by instituting strict eugenic measures.

The combination of concern for animals and cruelty to human beings often found among the Nazis does remain a haunting paradox, even if it is very far from being a universal pattern. In an entry in his private diary dated December 28, 1939, Nazi Minister of Propaganda Joseph Goebbels wrote that Hitler "has little regard for homo sapiens. Man should not feel so superior to animals . . . . Man believes that he alone has intelligence, a soul, and the power of speech. Have not the animals these things?" (1983) The Nazis consistently blurred the boundary between human beings and animals, which helped them to extend practices usually reserved for animals—such as forced sterilization, mass killing, and experimentation without consent—to human beings. What remains highly debatable is whether such treatment of human beings by the Nazis should be attributed more to that blurring of boundaries, to our customary practices with animals, or to some elusive combination of the two.

Despite the rhetorical use by many parties in debates on animal welfare, the record of Nazi Germany in the treatment of animals has no unequivocal moral or political lesson for us today. It should, for example, be almost too obvious to merit stating that the sometimes-disputed vegetarianism of Hitler tells us absolutely nothing about whether or not it is "right" to eat meat. Nevertheless, this example is regularly invoked by critics of vegetarianism and vehemently denied by defenders of it—as though the practice of vegetarianism was on trial.

One of the most abiding cultural legacies of the Nazi period has been to cast new doubt on the expectation of human progress, which had been widely accepted in the nineteenth century but already shaken by the brutalities of World War I. In a similar way, the experience of Nazi Germany places in question "expanding circle" theories of rights, according to which ethical consideration has been expanded to ever more creatures throughout history. Variants of this idea have been propounded by many thinkers, including Aldo Leopold, Tom Regan, and Peter Singer. It is not easy, in such a perspective, to explain how a regime that was so notorious for ethnic and racial prejudice should be responsible for breaking new ground in the protection of other creatures.

Further Resources

Föger, B., & Taschwer, K. (2001). *Die andere seite des spiegels: Konrad Lorenz und der National-sozialismus*. Vienna: Czernin Verlag.

Jütte, D. (2001, December 12). Tierschutz und Nationalsozialismus—Eine unheilvolle Verbindung. *Frankfurter Allgemeine Zeitung*, N3.

Sax, B. (2000). *Animals in the Third Reich: Pets, scapegoats and the Holocaust*. New York: Continuum.

Taylor, F. (Ed.). (1983). *The Goebbels diaries: 1939–1941*. New York: G. P. Putnam's Sons.

Boria Sax

■ Culture, Religion, and Belief Systems
Animals in the Hebrew Bible

According to a biblical tradition, when creation was completed, God instructed mankind: "Be fruitful, and multiply, and replenish the earth, and subdue it: and have dominion over the fish of the sea, and over the fowl of the air, and over every living thing that moves upon the earth" (Gen. 1:28). The human-animal relationship was established in this dictum instructing humans to be in charge but not exploitative. Creation was to be used for the benefit of mankind, but it was *not* to be exploited.

The areas of human-animal interaction in biblical times were multiple, but they can be divided into two major spheres: secular and cultic. However, because of the nature of life in those times, the boundaries between these areas were not exclusively defined. This is to say that, at times, one would encroach into the territory of the other; the religious and the profane could—and did—intermingle.

Animals for Consumption and Labor

When it comes to the secular sphere, there were two main areas where humans and animals interacted: diet and labor. Some animals were components of the diet, and others were used for accomplishing tasks. There were also animals that had a role in both areas at different parts of their lives. When animals relate to the diet, the biblical rules are very strict. Certain creatures can be eaten while others are forbidden. When it applies to mammals, the Old Testament specifies:

> Any animal that has divided hoofs and is cleft-footed and chews the cud—such you may eat. But among those that chew the cud or have divided hoofs, you shall not eat the following: the camel, for even though it chews the cud, it does not have divided hoofs; it is unclean for you. The rock badger, for even though it chews the cud, it does not have divided hoofs; it is unclean for you. The hare, for even though it chews the cud, it does not have divided hoofs; it is unclean for you. The pig, for even though it has divided hoofs and is cleft-footed, it does not chew the cud; it is unclean for you. Of their flesh you shall not eat, and their carcasses you shall not touch; they are unclean for you. (Lev. 11:3–8)

The rules of dietary cleanliness pertain also to wild animals: "the deer, the gazelle, the roebuck, the wild goat, the ibex, the antelope, and the mountain-sheep" (Deut. 5) can be eaten. Which water fauna can be eaten is also strictly defined:

> These you may eat, of all that are in the waters. Everything in the waters that has fins and scales, whether in the seas or in the streams—such you may eat. But anything in the seas or the streams that does not have fins and scales, of the swarming creatures in the waters and among all the other living creatures that are in the waters—they are detestable to you and detestable they shall remain. Of their flesh you shall not eat, and their carcasses you shall regard as detestable. Everything in the waters that does not have fins and scales is detestable to you. (Lev. 9–12)

The list of birds includes, by name, all those that should not be eaten (Lev. 11:13–19); unfortunately, several on the list cannot be securely identified. However, the common characteristic of these birds is that they feed on prey and carrion. The list does not include or define birds that can be eaten, but from other references it is known that doves, turtledoves, quail, and rock partridges are permitted. Rodents and creeping creatures (Lev. 11:29–30) are strictly forbidden, as well as insects (Lev. 11:23), with the exception of certain kinds of grasshoppers (Lev. 11:21–22). Any carcass cannot be consumed, and whoever (or whatever) touches it is unclean and needs to go through purification procedures.

Domestic mammals that can be eaten are ruminants, and they include sheep, goats, and cattle. These animals were raised primarily for their by-products, that they could provide while alive, including milk, dung, wool, and hair—and upon their death, meat, skin, horns, and bones.

Cattle, together with equids (e.g., donkeys, mules), were used as draft and pack animals. They provided traction when hitched to plows and wheeled carts or wagons; they could haul, on their backs, heavy loads in places where the terrain did not permit the use of vehicles. They were also harnessed and saddled for riding, and thus used for human transportation. Horses were considered prestigious animals and were used only for military purposes—either as riding animals in the cavalry or hitched to a spoked-wheel chariot. Although the camel is closely identified with the modern Near East, it did not appear on the scene until relatively late (1200–1000 BCE) and was used both for transport and for military purposes (camelry). The use of camels in this period is recorded in the book of Judges (7:12), where the Amalekites' attack upon the Israelites is described. Also, the Arabian tribes were known to use camels in their military encounters with the

Assyrians. The Persians used camels as pack animals when, circa 525 BCE, they invaded Egypt through the Sinai desert.

Domestic animals were housed in sheds, folds, and other sheltering structures. When in small numbers, they were kept on the ground floor of the house, while the family resided either in other rooms on the same level or on the second floor. According to available textual and archaeological information, private individuals did not own horses; these animals were traded and stabled by the central government. Biblical tradition maintains that King Solomon was engaged in large-scale horse trading (1 Kings 10:28–29; 2 Chron. 1:17). Ancient documents report that the Israelite king Ahab (c. 871–852 BCE) fielded large numbers of chariots and horses when he supported a rebellious coalition against Assyria. Certain structures referred to, on the basis of their floor plan and construction method, as "tripartite," or pillared, buildings, which were uncovered at several sites in layers dated to the period of the Israelite monarchy, were interpreted as stables. However, a large number of scholars think that most of the structures were used for the storage of commodities by the central government. It is possible that a small number, especially those discovered at Megiddo, were used as stables. Furthermore, the original dating of the Megiddo buildings to the time of Solomon (tenth century BCE) has been revised to the ninth century, possibly to Ahab's time.

When discussing domestic animals in biblical times, there are three more that need mentioning: the pig, the dog, and the cat. Although the pig was forbidden to the Israelites for consumption, it was a favorite of the Philistines during the first century or two of their arrival and settlement on the coast of Palestine (1200–1000 BCE). When the Philistines became acculturated, one of the customs they abandoned was the consumption of pork. The adherence of the Israelites to the restriction on pork consumption is attested to in the zooarchaeological record, though the absence of pig bones at a site should not necessarily be interpreted as that site being Israelite.

One of the common animals in the Israelite farmyard that was not associated directly with food production was the dog, which earned both respect and contempt. Domestic dogs were used for hunting, herding (Isa. 56:11), guarding (Isa. 56:10), and recreation. However, unmanaged dogs ("pariah") were well known in the biblical world and aroused ill feelings. Although not mentioned in the biblical text, the cat must have been part of the Israelite domestic menagerie, simply because it is known to have existed in the surrounding cultures of Egypt and Mesopotamia. Its ability to exterminate mice must have made the cat attractive also to the Israelites. But because of similarities, it is hard to distinguish between wild and domestic skeletal cat remains.

Many wild animals, such as ungulates (e.g., deer, gazelle), certain birds, fish, and insects, were also part of the diet. Other contributions to the diet by these animals were their by-products that enriched the diet, as, for example, bird eggs (Isa. 10:14). Even insects contributed to the diet; one such contribution was bee honey. Interestingly, not all fish bones recovered from sites identified as Israelites' belong to the approved species; some of them originated in the Nile in Egypt. This illustrates that not all biblical dietary rules were strictly observed.

A special contribution to the economy was made by the use of certain mollusks for dyeing yarn and fabric. Three kinds of murex (*Murex brandaris, M. trunculus, Purpura haemostoma*) were available near the coasts of the Mediterranean. The Phoenicians became experts in the use of these mollusks for the production of purple dye, which was considered a luxury item and was coveted by royalty and the nobility. Purple-dyed garments, which are mentioned in the Bible several times, were items of trade, tribute, and booty.

Many references in the Hebrew Bible show that ecological concerns were important to the ancient Israelites. Some of these concerns are expressed in the narratives, laws, parables, and other literary genres. One concern was for the extinction of certain species. The understanding of this principle is well illustrated in the injunction, "When you come upon a bird's nest by the road, in a tree or on the ground, with fledglings or eggs in it and the mother bird on the nest, do not take both mother and young. Let the mother go free, and take only the young." (Deut. 22:6–7). Although humankind is allowed to eat animals (Gen 9:2–3), it is not permitted to destroy them completely. This applies not only to wildlife; domestic animals can also benefit from wise management, as stated in Leviticus 22:28, which says: "You must not slaughter a cow or a sheep at the same time as its young." Wise management of resources is the background for the story of Abraham and Lot and their ever-increasing herds (Genesis 13). According to the narrative, Abraham appears to be aware of the results of overgrazing and is attuned to the principle of land-carrying capacity. For the Israelites who were engaged in herding, this story might have served as a lesson or an illustration of a principle with which they were well familiar.

One other concept that the ancients were cognizant of, to a certain degree, is the area of genetics. Although they did not see it as modern scientists do, they knew how to create hybrids—like mules—and tried their hand in improving the breeds by "genetic engineering," as the story of Jacob and his father-in-law, Laban (Gen. 30), shows. Here, Jacob was trying to manipulate, in his favor, the type of sheep making up the herd.

The Israelites knew that in order to get the most out of their animals, they had to take care of them in a way that was reminiscent of taking care of family members: "Six days you shall do your work, but on the seventh day you shall rest, so that your ox and your donkey may have relief, and your homeborn slave and the resident alien may be refreshed" (Exod. 23:12). Care for animals was supposed to be extended even outside the household: "When you see the donkey of one who hates you lying under its burden and you would hold back from setting it free, you must help to set it free" (Exod. 23:5).

Animal Sacrifice

The cult was a second sphere in which humans and animals interacted, through the practice of sacrifice. Sacrificing an animal was seen as a way to thank the deity for bestowing its favors upon the individual and the community. There were different types of sacrifices, and each had certain prescriptions. The underlying rule was that a sacrificial animal had to be "without blemish." Animals from the herd had to be of a certain age; this rule suggests that sacrifice was one way of culling the herd and maintaining its proper size. When an individual could not afford a four-legged animal for a sacrifice, a bird (or two) was a good substitute.

Sacrificing animals was an old Near Eastern tradition, and the Bible demonstrates it. Several narratives portraying events that took place in the dawn of history include references to sacrifices. The earliest sacrifice in the biblical tradition, which was offered by the brothers Cain and Abel, included "fruit of the ground" and "firstlings." The deity accepted the latter, and this led to jealousy and the first murder (Gen. 3). Another sacrifice that took place in early times was the one offered by Noah after the flood (Gen. 8:20–21). These traditions suggest that the Israelites were aware of the fact that sacrificing animals was a very old tradition, practiced even before the formation of the Israelite entity.

Sacrificing was carried out throughout Israelite history and was practiced during the formative period, through the process of settling the land, to the monarchical period. Sacrificing animals took place on temporary altars, in well-established cult centers, and in central temples. Sacrifices were carried out in family circles and under the auspices of the central government. They were a way of offering gratitude for past benevolence and future security. Evidence of sacrifices is provided by the zooarchaeological record that has been produced through excavations. That the deity was seen as closely connected to the animal world is evident by the fact that animals were believed to serve as its footrest. Near Eastern artistic traditions show the gods appearing on the backs of animals (e.g., lions and bulls). According to biblical traditions, portrayed in the Jerusalem Temple, God's footrest was the cherubim; in the temples of the north (Dan, Bethel), he was portrayed as riding on the back of a bull.

Animals as Symbols

Animals were very much appreciated for whatever they could contribute, and one way that this was manifested was in the role that they were assigned in the literature. Animal images appear in all literary genres contained in the Bible. Besides appearing in narratives and cultic laws, where they were depicted in their practical roles, animals were used symbolically and as metaphors in proverbs, parables, and poetry (just to name a few). They were used as omens, as in the Joseph story (Gen. 41:17–32), when fat and lean cows appear in a dream. Upon interpretation, it was revealed that the cows represented years of plenty and years of drought. Job asserted his right to complain, by using a proverb comparing his situation to that of a braying donkey and lowing ox (Job 6:5). The book of Proverbs utilizes animal behavior to instruct in areas related to human behavior: "Go to the ant, you lazybones; consider its ways, and be wise. Without having any chief or officer or ruler, it prepares its food in summer, and gathers its sustenance in harvest" (Prov. 6:6–8). Other animals used in proverbs include dogs, badgers, locusts, snakes, and more. Certain animals are used as symbols of beauty: "Let your fountain be blessed, and rejoice in the wife of your youth, a lovely deer, a graceful doe. May her breasts satisfy you at all times; may you be intoxicated always by her love" (Prov. 5:18–19). In several other instances, the gazelle and the deer appear as metaphors for beauty (e.g., Song of Sol. 3:5; Song of Sol. 4:5; 2 Sam. 2:18), and the pig appears as a symbol for ugliness and distaste (Prov. 11:22).

The sheep, being a common animal, is used symbolically in various ways. A well-known parable is the one in which the prophet Nathan relates a story to King David about a rich man who robbed the one little lamb owned by a poor man. This parable was told in reaction to David's affair with Bathsheba and the killing of her husband Uriah (2 Sam. 12:1–7).

The Hebrew Bible uses animal images as symbols for different entities and attributes. Domestic animals are used to convey beauty (Song of Sol. 4:1–2) and loyalty (Isa. 1:3). In many instances, the image of sheep symbolizes the people of Israel, while YHWH, as their shepherd, is the leader (e.g., Pss. 23; 74:1). Members of the bovid family symbolize strength (Isa. 10:13) and youthful vigor (Mal. 4:2; Ps. 29:6). Many times, YHWH himself is likened to a powerful bull.

Dogs play a mixed role; mostly they are depicted as pariahs (1 Sam. 17:43), but sometimes they convey loyalty (Eccles. 9:4) and humility (2 Kings 8:13).

Wild animals and birds provide much of the symbolic and metaphoric images. The fierce lion is a symbol of strength, retribution, destruction, and vengeance (Joel 1:6). The roar of the lion is like YHWH's speech (Amos 3:8). In the blessing of Jacob to his sons, Judah is likened to a lion (Gen. 49:9) and Dan to a snake (Gen. 49:17). The image of the lion is used in the blessing of Moses to the tribes when describing the attributes of Gad (Deut. 33:20) and Dan (Deut. 33:22). There are many other instances where the image of the lion plays a role. A parable in Ezekiel 19:1–9 uses the lion as a metaphor for Judah (with the "cubs" being kings of the House of David). The lion serves also as the subject of Samson's riddle (Judg. 14:14–18); his question was: "Out of the eater came something to eat. Out of the strong came something sweet." The response given by the Philistines was: "What is sweeter than honey? What is stronger than a lion?" And, because they used Samson's wife to get the answer, he countered them with another animal metaphor: "If you had not plowed with my heifer, you would not have found out my riddle."

The eagle is the most respected bird in biblical imagery. Its speed (Jer. 4:13), size (Jer. 49:22; 17:3), and place of habitat (Job 39:27; Jer. 49:16) made it an agent of rescue (Exod. 19:4; Deut. 32:11) and shelter (Ps. 91:3–4). All these attributes allowed Ezekiel to use the eagle's image in reference to both King Nebuchadnezzar (Ezek. 17:2–6) and Pharaoh Psammetichus II (Ezek. 17:7). Both the eagle and the lion were symbols of strength and nobility, as expressed in David's lament over the death of Saul and Jonathan: "they were swifter than eagles, they were stronger than lions" (2 Sam. 1:23).

Insects were also used in biblical references to provide images of attributes and entities. The ant is a model for diligence (Prov. 6:6–8), and the spider's web is likened to something that should not be trusted (Job 8:14). The image of the locust is invoked to portray enemy hordes (e.g., Judg. 6:5) and becomes a symbol of devastation (Joel 1:4; 1 Kings 8:37; Amos 4:9). The bee serves as a symbol for the army of Assyria, and the fly symbolizes the Egyptian troops (Isa. 7:18–19). The biblical wasp (Exod. 23:28; Deut. 7:20; Josh. 24:12) might have been used as a metaphor for Egypt.

In the Bible, there are times when animals are in possession of a distinct knowledge that humans lack—such as the knowledge of God and his accomplishments. Job, in his response to his friends, invokes appealing to the animals for such instruction (Job 12:7–10). This is very reminiscent of Isaiah's exhortation, "The ox knows its owner, and the donkey its master's crib; but Israel does not know, my people do not understand" (Isa. 1:3).

Some animals had the gift of speech. The snake in the Garden of Eden conversed with Eve and convinced her to eat from the Tree of Knowledge (Gen. 3:1–5). Balaam, the seer, had a she-ass who opened her mouth and protested after he mistreated her (Num. 22:28–30). But not all animals appearing in the Bible were of this world; some of them were mythical animals. The cherubim in the Jerusalem Temple depicted imaginary animals (see Ezek. 1:5–14; 10:14). Other mythological animals, including Rahab, Yamm, Tannin, and Leviathan (Isa. 51:9; Ps. 89:11; Job 9:13; Job 26:12; Isa. 27:1)—all of which seem to have existed in the sea—threatened world order. The "great fish" that swallowed Jonah (Jon. 1:17) can be also considered a fantastic animal. Satyrs, or "goat-demons," are also mentioned as disturbing the peace (Lev. 17:7; Isa. 13:21; Isa. 34:14; 2 Chron. 11:15).

There is no question that the human-animal relationship in biblical times was long and deep, and it was manifested in the leading role assigned to it in the biblical text, whether in the narratives, poetry, or other genres. There is hardly an aspect of life covered in the Hebrew scriptures that has no connection to the animal kingdom.

See also

Culture, Religion, and Belief Systems—*Christian Theology, Ethics, and Animals*
Culture, Religion, and Belief Systems—*Judaism and Animals*

Further Resources

Borowski, O. (1998). *Every living thing: Daily use of animals in ancient Israel.* Walnut Creek, CA: AltaMira.

———. (2002a). Animals in the literature of Syria-Palestine. In B. J. Collins (Ed.), *A history of the animal world in the ancient Near East* (pp. 289–306). Leiden/Boston/Cologne: E. J. Brill.

———. (2002b). Animals in the religion of Syria-Palestine. In B. J. Collins (Ed.), *A history of the animal world in the ancient Near East* (pp. 405–24). Leiden/Boston/Cologne: E. J. Brill.

Hesse, B., & Wapnish, P. (1985). *Animal bone archaeology*. Washington DC: Taraxacum.

King, P. J., & Stager, L. E. (2001). *Life in biblical Israel*. Louisville, KY: Westminster John Knox Press.

Matthews, V. H. (1991). *Manners and customs in the Bible*. Peabody, MA: Hendrickson Publishers.

Oded Borowski

■ Culture, Religion, and Belief Systems
Animism

Animism is a term used by anthropologists to label two distinct phenomena. It was originally used by Edward Tylor (the first professor of anthropology in Britain), in 1871, to refer to his theory about the nature of religion. Tylor's theory was challenged—and eventually rejected—by anthropologists and other academics interested in religions, and "animism" in Tylor's sense became obsolete. More recently, "animism" has been used in a quite different way in discussions of the worldviews and lifeways of various contemporary indigenous peoples. In both uses, "animism" is implicated in views about the nature of animals and about human relationships with animals.

Edward Tylor's most significant publication of his argument was in a two-volume work called *Primitive Religion* (1871). Tylor employed a word derived from ancient Greek, *anima* (commonly translated as "soul"), to encapsulate his theory. In this, he argued that all religions can be defined by one central characteristic: a "belief in spiritual beings." In other words, Tylor thought that religion was, by definition, a mistake in which dreams were confused as evidence for the reality of "souls" or "spirits." If someone dreamt about a dead relative, they might believe that they had been visited by a "spirit" living after death. If someone dreamt of visiting a distant place, they might believe in the existence of a "soul" that can detach itself from the body and travel independently. Tylor rejected the reality of such metaphysical or non-empirical beings and entities. He argued that all religions, beginning with such simple mistaken interpretations of dreams and evolving into more complex beliefs about universal deities, were about to disappear in the face of scientific rationality. Evidence about empirical reality would render metaphysical beliefs obsolete. Indeed, he argued that it was the task of academics, especially of the newly created discipline of anthropology, to correct mistaken beliefs and aid people in becoming properly rational.

Animals play various roles in Tylor's theory. He claimed that "primitive people" believed animals are like humans in possessing souls, and they believed that these souls were able to migrate between humans, animals, plants, rocks, the dead, and deities. Thus, both individual identity and the distinction between humans and animals were uncertain in many religions, including among "civilized theologians" who thought that dogs would go to heaven when they died. Not only does Tylor's animism involve the attribution of considerable degrees of likeness between humans and other soul-possessing beings, but Tylor claims that religion began as an animal-like mistaken interpretation of involuntary bodily and mental processes as intentional and communicative acts. Just as dogs respond to unexpected movements as potentially threatening, so did "primitive people"—and, ultimately, *all* religious people—attribute life to inanimate objects (such as rocks), impersonal beings (such as animals), or personified ideas (such as deities of love or justice). In this part of his argument, Tylor attributes "animal-likeness" to the humans he uses as evidence for his theory, even as he accuses them of failing to distinguish between humans and animals.

Tylor's theory of animism and religion provoked considerable disagreement and was eventually rejected as both inadequate and colonialist. However, the term "animism" had begun to be used in a quite different way from that envisaged by Tylor. Instead of defining the nature of religion, it identified a particular kind of religion distinct from monotheism and polytheism. In animistic religions, people are less interested in deities than in a wider community of significant living beings. They may, for example, make offerings and engage religious specialists, some called "shamans," to communicate with animals, plants, rocks, and other beings that are experienced as being responsive to these human efforts. This new use of the term "animism" is becoming increasingly important in discussions of a variety of contemporary indigenous religious cultures.

Unlike Tylor's theory, the new animism is rooted in indigenous knowledges and practices. Research by Irving Hallowell (1960) among the Ojibwe of Beren's River in southern central Canada provides a classic example of the new approach. The Ojibwe language recognizes a distinction between living persons and inanimate objects, but it includes animals, rocks, and clouds under the category "persons." Hallowell writes about human and other-than-human persons to convey this understanding of the world as a community of persons, most of whom are other than human. The point might equally well be made by talking about bears and other-than-bear persons. "Persons" is an overarching category, an umbrella term, beneath which are different groups of persons—for example, humans, bears, deer, rocks, clouds. It is not that the Ojibwe project human likeness onto animals, but that they experience personhood in a wide range of beings. The animists identified in this new approach ask different questions from those of Tylor's theory: not "Does this animal have a soul?" or "Is this rock alive?" but "How do I communicate respectfully with this animal?" and "Am I related to this rock?"

The new animism is about cultural engagements between persons, most of which are other than human, and all of which are related in some way. Two other words illustrate aspects of animist relationships. The Ojibwe word "totem" refers to clans in which humans and particular animals are related and have specific responsibilities for the well-being of other clan members. Within the larger animist quest for appropriate relationships between persons, totemism identifies more intimate relationships between humans and significant animals. Similarly, "shamanism" occurs within animism, because persons sometimes act insultingly toward others (e.g., hunters may offend animals by inappropriate behavior), and shamans are called upon to offer mediation and restore respect.

Further Resources

Bird-David, N. (1999). "Animism" revisited: Personhood, environment, and relational epistemology. *Current Anthropology, 40*, S67–S91. Reprinted in G. Harvey (Ed.), *Readings in indigenous religions* (pp. 73–105) London: Continuum.

Hallowell, A. I. (1960). Ojibwa ontology, behavior, and world view. In S. Diamond (Ed.), *Culture in history: Essays in honor of Paul Radin* (pp. 19–52). New York: Columbia University Press. Reprinted in G. Harvey (Ed.), *Readings in indigenous religions* (pp. 18–49). London: Continuum.

Harvey, G. (2005). *Animism: Respecting the living world*. London: C. Hurst & Co. / New York: Columbia University Press / Adelaide: Wakefield Press.

Howell, S. (1996). Nature in culture or culture in nature? Chewong ideas of "humans" and other species. In P. Descola & G. Pálsson (Eds.), *Nature and society: Anthropological perspectives* (pp. 127–44). London: Routledge.

Tylor, E. (1871). *Primitive culture*. London: John Murray.

Graham Harvey

■ Culture, Religion, and Belief Systems
Birds as Symbols in Human Culture

All human cultures relate intangible ideas to aspects of our physical environment in some symbolic way. Perhaps we allow our preoccupying self-perception to overflow into the rest of the world. Perhaps we symbolize in order to become more familiar with mysterious and complex things in our experience. Among the many objects in nature that are suitable for our symbolic use, the bird enjoys a privileged position because it is conspicuous, humanlike in many ways, and behaviorally interesting. Most birds are about in the daytime and tolerant of humans; they tend to be socially monogamous and provide parental care; they can be brightly colored and energetic; many of them sing loudly, distinctively, and beautifully; their activity tends to be obvious and provides easy insights to their lives; and among them are superb navigators, migrators, and builders.

Above all, however, birds are excellent objects for our symbolic exploitation because they *fly*. Probably the most universal symbol we have applied to birds derives from their ability to roam the three-dimensional world above us, the place we reverently call "the heavens," the home of celestial bodies, deities, and the best of the dead. Any small bird in most cultures can be conceived as a soul, or even as the particular soul of a dead person. Consistent with their nearness to the source of divinity, as well as their mobility and vocal proficiency, birds also tend to be viewed as communicators of arcane knowledge. They can be good or bad omens, advisors of the great and pious, and even tattlers on the unfaithful. The word *augury* literally means "bird talking," and *auspices* means "bird viewing"; thus an auspicious enterprise is one for which the birds have indicated a favorable future. Also, because of their flight, we tend to associate birds with freedom. Freedom can be represented by rulers as a dangerous—even *perilous*—thing, but in our own culture it is sweet liberty to be "free as a bird." Likewise, the caged bird is regularly a symbol of either loving protection or (much more commonly today) cruel confinement.

The ways in which bird symbols have infused cultures are as various as the birds themselves. Following are four brief examples from research performed in recent decades.

Birds as Omens for the Nage of Indonesia

The Nage people of the island of Flores, in Indonesia, recognize seventy-nine kinds of birds, fifty-five of which have symbolic value. The most striking example, and apparently most important to the Nage, is represented by the concept *po*, which is associated with malevolent spirits. *Po* as a noun refers mainly to any owl, but there are other birds, such as diurnal raptors, that can *po*—or, that can make a sound that reveals the bird to be indwelled by an evil spirit. These birds are the central members of a broader category of "witch birds" (*polo*) that includes any flesh-eating or scavenging bird, including the crow and the Wallacean drongo (*Dicrurus densus*). These, in turn, are a subset of the category of "spiritual birds," whose other main category besides *polo* is *mae,* birds that are associated with the souls of dead people. Among these, the black-faced cuckoo-shrike (*Coracina novaehollandiae*) makes a call that is considered to be a dead person summoning a relative (and is often seen as portentous). The only spiritual bird that is not ominous is the bare-throated whistler (*Pachycephala nudigula*), because it is the embodiment of the harmless dead: those who have died in infancy or who were aborted as fetuses. Finally, there are a few ominous birds that are nevertheless not associated with an evil spirit or dead soul. There is no spiritual category for the Nage that is not associated with some sort of bird. The primarily negative connotation of their symbolism is remarkable. As one resident told a researcher: "Many birds provide signs, and many of these are bad signs."

Bird Killing and Display as a Rite of Passage for the Maasai of Kenya

Maasai youths kill, stuff, and mount small, colorful birds on hats for competitive display, symbolically resurrecting them. This practice is probably unique in Africa. Given the Maasai's exclusion of birds from their diet, and their consideration of birds as links between humans and the divine, this rite of passage might seem difficult to reconcile. The Maasai place birds into two broad, spiritual categories. The first is comprised of birds of character, which are either auspicious or inauspicious; these, they do not kill. The other category includes anonymous birds that are considered decorative, and these they *do* kill. Only the woodpecker is a bird revered yet decorative. An apparent exception is the ostrich, whose plumes are second only to the lion's mane as the most prized decoration for a warrior; but, in fact, the Maasai consider the ostrich to be a mammal posing as a bird.

Cranes as a Subject of Art in Neolithic Anatolia

Cranes are represented more than any other bird in Neolithic (c. 7000–6000 BCE) Turkey. Bones have been found in midden heaps, indicating that they were killed and eaten by people. Wings with feathers attached have also been found, suggesting that here, as elsewhere, there was a "crane dance," an attempt to imitate this bird's beautiful courtship ritual. According to this evidence and the uses of cranes in paintings, the ancient Anatolians probably shared many other cultures' association of this bird with life and rebirth, possibly in opposition to the vulture as the symbol of death. The importance of cranes in human culture is likely due to their similarities to humans: their height, their long lifespan, their monogamous social system, their commonality and gregariousness, and, perhaps most importantly, the fact that they dance.

Bird Images as Indications of Character and Spirituality in Medieval Christian Art

Jerome was an early Christian father of the church, known especially for his translation of the Bible into Latin and his penitential stay in the wilderness, where he reported being accompanied daily by animals. Over 1,000 artistic representations of

Jerome depict animals in them—including thirty-three identifiable kinds of birds. In most cases, rich symbolic meanings have been deciphered. One example is *Saint Jerome in His Study* by Lucas Cranach the Elder, painted in 1526. This painting contains three birds, along with five mammals. The African gray parrot (*Psittacus erithacus*) that watches Jerome translate is the best avian talker, and thus is a frequent symbol of words, such as those spoken by an angel to Mary, announcing her motherhood of Jesus. Here, the parrot draws attention to Jerome's translating work, and perhaps even suggests divine help in the process. A pair of pheasants with young was a common representation of God's providence, not only because of the parental care depicted, but also because of a Teutonic Pleiades legend. Jesus, upon being given a loaf of bread by a woman and her daughters, set them in the sky as stars for their charity. The Pleiades were commonly depicted as a hen and chicks in Christian art. The pair of gray partridges (*Perdix cinerea*) are peculiar here because they were considered to be satanic creatures—including by Jerome himself—due to an old belief that the bird was oversexed and an egg-stealer to boot. Partridges were not *wholly* evil by the time of the painting, however, thanks to Leonardo da Vinci, who proposed that they could symbolize the eventual triumph of truth, because the chicks hatching from stolen eggs go

back to their real mothers (or so he thought). Taking the floor scene together, the lion, which symbolizes Christ, is keeping watch over the provision of the Bible to the people, symbolized by the pheasants. Part of this divine vigilance involves keeping away the evil partridges, whatever they might be—heresies, or else whatever sins Jerome sought to purge by his wilderness experience. Thus, the birds in this painting illustrate the spiritual significance and drama underlying Jerome's work.

Table 1 provides a taste of some of the frequently cited symbolic uses of birds. Distinct conceptions are separated by semicolons. Most of the representations originated in the civilizations of China, India, the Near East, and Europe (thus the birds are Eurasian). Sometimes, the symbolic values for a given species are widespread, because they are an imaginative extension of a feature of the bird; others stem from an interaction between the bird and the local geography or culture, and are thus parochial or idiosyncratic. Bird symbols are ubiquitous in politics and popular culture, but the table focuses on moral and religious symbolisms, which are, arguably, the roots from which

Saint Jerome in His Study, *by Lucas Cranach the Elder, painted in 1526. ©Sarasota, FL: John and Mable Ringling Museum of Art.*

Table 1. Birds Commonly Associated Symbolically with Concepts and Religious Figures

Bird	Associated Deities and Saints	Symbolic Values
albatross		Good fortune (esp. at sea)
bittern		Desolation, impending calamity
blackbird (European)		Poet, lover, song; solitariness; death; sensual passion
bunting		Vitality in winter (snow bunting); humble status
buzzard		Laziness, clumsiness, idiocy, lasciviousness
chicken: cock/rooster	Hermes/Mercury; St. Peter	Light, healing, resurrection; vigilance, overbearing insolence; pluckiness, priggishness (the Bantam); male sexuality, salaciousness
chicken: hen	Christian God or Church; St. Pharaildis	Maternal love, protection, fecundity, health, domesticity
chough		Detection of marital infidelity; marriage; thievery, lasciviousness
coot		Intelligence, prudence; baldness
cormorant		Voracious appetite, insatiability
crane	Apollo; St. Vincent	Sun, dance; longevity, happiness, enlightenment, vigilance, foresight
crow	Indra (India)	Contention, discord, strife; evil; longevity
cuckoo		Cuckoldry, avarice, jealousy; foolishness
dove	Yama (India); Aphrodite/Venus; Jehovah, Noah, Holy Spirit; Sts. Ambrose, Basil, Bridget of Sweden, Catherine, Catherine of Siena, Cunibert, David, Dunstan, Gregory the Great, Hilary of Arles, John Columbini, Lo, Louis,	Fertility, soul; faithfulness; peace, love, innocence, harmlessness, sorrow

(Continues)

Bird	Associated figures	Meanings
duck	Medard, Oswald, Peter of Alcantara, Peter Celestin, Sampson, Thomas Aquinas	Happiness, faithful married love; deceit
eagle	Zeus/Jupiter; Jesus Christ; Vishnu (India); Odin (Norse); supreme being, sky or sun spirit (many Near Eastern and New World aboriginal cultures); Sts. Augustine, Gregory the Great, John the Evangelist, Prisca	Divinity, majesty, inspiration
falcon/hawk	Horus, Montu, Sokar (Egypt); Sts. Bavo, Edward, Julian Hospitator, Otto (on cottage)	Nobility, victory; rapacity
gallinule, purple		Chastity
goldfinch	Mary and infant Jesus	Soul, salvation, light; fertility, sexual pleasure
goose	Ra, Geb (Egypt); Brahma (India); Ortiki (Siberia); St. Martin, Pharaildis (gosling)	Soul, sun; vigilance; sexual union; conceit, folly, old age
gull		Soul; gullibility
heron	Ra, Osiris (Egypt);	Cowardice; wisdom
hoopoe		Gratitude, filial piety; filth
ibis	Thoth (Egypt); Hermes/Mercury;	Heart; morning
jackdaw		Thieving, hoarding; vain assumption, empty conceit
jay		Senseless chatter
kestrel		Baseness; low social class
kingfisher (halcyon)	Tethys	Dawn, calm seas, respite; magnanimity
kite	Devil	Ruthlessness, rapacity, thievery

Table 1. (Continued)

Bird	Associated Deities and Saints	Symbolic Values
lapwing		Insincerity, treachery, deceit
lark		Soul, joy, gratitude, day, the awakening of lovers
magpie		Garrulity, verbosity, pilfering; struggle between good and evil
nightingale		Love, sexual emotions; heartbreak, grief; contentiousness
ostrich	Ma'at (Egypt)	Endurance; negligence, stupidity
owl	Pallas Athene/Minerva	Wisdom (little owl); death, doom, evil
parrot	Kama (India)	Mocking verbosity; love; poetry
partridge		Lust, excessive fecundity; trickery
peacock	Hera/Juno	Pride, vainglory; immortality
pelican	Jesus Christ	Maternal selfless love
pheasant		Authority; redemption, security; pleasure
pigeon		Cowardice, stupidity
quail		Reproduction; cowardice
raven	Elijah; Odin (Norse); Sts. Benedict, Erasmus, and Paul the Hermit (bringing food), Ida and Oswald (with ring in beak)	Prophecy, ill-luck; longevity; slaughter, death
robin (European)		Confiding trust, charity, Christmas
sparrow	Aphrodite/Venus	Lasciviousness; mankind

starling		Lechery, garrulousness
stork		Procreation; loyalty, filial piety
swallow (barn)		Spring, resurrection; gracefulness; omen of catastrophe; female Sexuality; maternal care despite poverty
swan	Aphrodite/Venus; Sts. Cuthbert, Hugh of Grenoble, Kentigern	Music, grace, serenity, innocence
thrush		Melody, song
tit		Fertility; gossip, carelessness
turkey		Official insolence
turtle-dove	St. Mary Magdalene	Marital fidelity
vulture	Nekhbet, Mut (Egypt); Ares/Mars	Rapine, filth, voraciousness, death; maternity, purity; penalty for sensuality
wall creeper	Mary (esp. regarding Jesus)	Impending death
woodpecker	Ares/Mars	Divination
wren		Smallness, energy, triumph by cleverness

popular and political symbols usually derive their sense. For instance, our calling a foolish person "cuckoo" descends from an old view that by laying her eggs in other nests, a female cuckoo was somehow being unfaithful to her mate; thus a man foolish enough to be married to such a woman is known as a cuckold. Likewise, the near-ubiquitous association of the eagle with majesty and might can be extended to explain its use as a symbol for Rome, Christ, the United States, and probably even football teams and rock bands.

Further Resources

Forth, G. (2004). *Nage birds: Classification and symbolism among an eastern Indonesian people.* New York: Routledge.

Friedmann, H. (1980). *A bestiary for Saint Jerome: Animal symbolism in European religious art.* Washington DC: Smithsonian Institution Press.

Galaty, J. G. (1998). The Maasai ornithorium: Tropic flights of avian imagination in Africa. *Ethnology, 37,* 227–35.

Lawrence, E. A. (1997). *Hunting the wren: Transformation of bird to symbol.* Knoxville: University of Tennessee Press.

Rowland, B. (1978). *Birds with human souls: A guide to bird symbolism.* Knoxville: University of Tennessee Press.

Russell, N., & McGowan, K. J. (2003). Dance of the cranes: Crane symbolism at Catalhöyük and beyond. *Antiquity, 77,* 445–55.

Willis, R. (1994). *Signifying animals.* New York: Routledge.

David C. Lahti

■ Culture, Religion, and Belief Systems
Blessing of the Animals Rituals

It is unclear when Blessing of the Animals rituals first occurred in the Christian tradition, though most likely they reflect a conflated Christian-pagan practice. Certainly as the roles of animals in human culture shift, so do the purposes of animal blessings. By the early twenty-first century, Blessing of Animals rituals in Western Christianity focused on domestic, companion species (dogs and cats in particular), whereas earlier blessings seem to have incorporated work and agricultural animals, such as mules, oxen, and horses. The earliest evidence is visual, including images of Saint Anthony Abbot (a fourth-century Christian holy man) blessing animals along with poor or inflicted humans. Anthony Abbot, whose feast day is on January 17, is the patron saint of animals. This mid-January blessing ritual, in recognition of his feast day, occurred into the early twentieth-century in cities such as Rome. Reports indicate that humans brought a wide range of animals to the steps in front of Catholic churches throughout the city for the blessing. Written reports, along with images from as early as the fifteenth century, indicate a tradition of animal blessings connected to the saint. It is also possible that Catholic

Rogation Days, which included a blessing of farm fields, also incorporated the blessing of farm animals.

In the twentieth century, Blessing of Animal rituals became increasingly prevalent in Western Christianity, from the United States to Canada to Australia and, to a lesser extent, in Europe. These rituals follow a standard pattern. Often geared to attract families with children, they tend to have a human-focused impetus. They are usually held outside, in front of the church building or in a park close to the religious institution, though occasionally they are held in sanctuaries. Many of the large and influential Christian denominations developed these blessings: Roman Catholic, United Methodist, Presbyterian (USA), Disciples of Christ, and the Episcopal Church. However, they also tend, more than many other religious rituals, to be ecumenical or interfaith in nature—even secular in sponsorship at times.

As the position of "pets" shifts in Western cultures, so does the incorporation of these companion animals into the religious life of the humans who live with them. In other words, as pets become more central to the lives of some humans, these humans seek ways to incorporate their companion animals into all facets of their lives. Thus, Blessings of the Animals/Pets is growing rapidly.

While there is no "standard" ritual, it is helpful to discuss one of the largest and earliest, one that might be the catalyst for the growth of Blessings. A large and influential Episcopal Church in New York City, the Cathedral of Saint John the Divine, holds, arguably, the largest and most impressive Blessing of Animals. It set the stage and provided the model for subsequent rituals. Beginning in the 1970s, the Cathedral held an annual Blessing of Animals on the Sunday closest to the Feast of Saint Francis (October 4). By the beginning of the twenty-first century hundreds, if not thousands, of these blessings occur, mostly in connection with St. Francis's feast day.

The blessing at Saint John the Divine provides a helpful template for understanding the phenomenon. Officially the ritual is entitled "The Holy Eucharist & Procession of Animals." Many years the sanctuary fills to capacity, with over 3,000 humans and as many as 1,500 animals present. Congregants wait in line with their companion animals for hours in order to find a place in the sanctuary. After a formal Eucharistic service, including music and dancing, the central doors of the sanctuary are opened and the procession takes place. It should be noted that these doors are only opened three times each year: Christmas, Easter, and the Blessing of Animals. Myriad animals with differing cultural positions process: camels (exoticized), cattle (usually food in the United States), bees, fish, hedgehogs, and hawks, for examples. Following the Eucharistic liturgy individual humans along with their companion animals move outside and are offered the opportunity for an individual blessing for each animal. The entire event takes several hours.

Other Christian—ecumenical and interfaith—as well as "secular" Blessings take place throughout the year. For example, not-for-profit or municipal entities such as local animal shelters sometimes sponsor Blessings. Often a local clergyperson or group of interfaith leaders presides. Cats, dogs, guinea pigs, ferrets, parrots, turtles, hermit crabs, and snakes are among the ritual participants at these quasi-religious events.

It is difficult to determine the purpose of the Blessings or to conclude with any certainty why they spread so rapidly in the late twentieth and early twenty-first century. This phenomenon probably accompanies the growth of the pet industry, the "ownership" of pets, and other companion animal-related issues of the same time period. However, it should be recognized that the rituals can be problematic, in particular for animals. While ritual, spectacle, and performance are certainly connected, the spectacle and forced performance of companion animals is ethically questionable. It is possible that Blessings of Animals serve no purpose for the animals, but only provide humans with a circus-like atmosphere and a sense of expanding ethical horizons. Or, of even less value for the lives of real animals, these Blessings are simply a way to bring new humans into various religious communities.

Blessings of Animals also fit within the larger "environmental" or "green" movement within some forms of Christianity. So animals become symbolic of a commitment to "God's creation" or to the human "stewardship" component of creation stories. This is indeed a focus at the Cathedral of Saint John the Divine. In the ideal world—the world of the peaceable kingdom that is prophesied variously in Western religious traditions—other-than-human animals are often included.

Cardinal Mahoney blesses a dog in a wagon during a ceremony on Olivera Street in Los Angeles, 2006. ©David Young-Wolff/Alamy.

However, it could also be argued that Blessings of the Animals rituals suggest a shifting attitude toward animals specifically in the early twenty-first century. While they were excluded from "sanctuaries" for generations, now they are being invited to return (at least once a year). They have sacred significance and are worthy of blessing. This is indeed an expansion of the religious sensibility that dominated the Western world in the post-Enlightenment era. Interpreting the cultural impact will take decades. In the meantime, the numbers and variety of Blessings continue to expand. As the roles of animals shift, so do the roles of Blessings from those that acknowledge animals' usefulness to humanity to those that also recognize their role as human companions and, in some cases, to blessings that recognize their own intrinsic value.

See also

Culture, Religion, and Belief Systems—*Francis of Assisi*

Further Resources

Hobgood-Oster, L. (2007). *Holy dogs and asses: Animals in Christianity*. Urbana, IL: University of Illinois Press.

McMurrough, C. (1939). Blessing of animals: Roman rite. *Orate fratres, 14*(2), 83–86.

Laura Hobgood-Oster

■ Culture, Religion, and Belief Systems
The Book of Animals

The highly prolific Iraqi man of letters Abu Uthman Amr ibn Bahr al-Fuqaymi al-Basri al-Jahiz (776–869) is one of the most famous figures in Arabic literature. His incomplete seven-volume *Book of Animals* is only one of some 200 of his written works, but it is perhaps the best known. Typical of Muslim literature, al-Jahiz's use of animals is instrumental: although ostensibly a comprehensive zoological catalog, the *Book of Animals* aims primarily to demonstrate the magnificence of God through a study of his created beings.

In some respects al-Jahiz's ideas were well ahead of their time, anticipating such things as the theory of evolution and the influence of climate on animal psychology. The fact that he describes the kangaroo (or at least some large marsupial) indicates that even in the ninth century, Middle Eastern Muslims possessed information on the fauna of far-off Australia, thanks to their Indian Ocean trade networks.

In other areas, however, al-Jahiz was little more than a transmitter of folklore. He reports, for example, that the giraffe originated as a hybrid between the camel and the hyena. He relies heavily on information gathered from Bedouin nomads, whom he sees as reliable sources since they spend their lives surrounded by wild animals, even though "their memory is sometimes faulty" and their knowledge of animals is "nothing more to them than a source of income." He believes in the theory of spontaneous generation (of such things as flies, worms, fish, frogs, and scorpions) and considers those who disagree to be ignoramuses.

In his scholarly approach al-Jahiz draws heavily on Aristotle's *History of Animals,* although he frequently and mockingly "corrects" the Greek philosopher's information. Al-Jahiz derives his taxonomy largely from that of the Qur'an. He divides animals into three broad categories: "Walking animals," including humans, noncarnivorous quadrupeds (*bahaim*), and carnivorous quadrupeds (*siba*); "Flying things" (*tayr*), including noble, ordinary, "less armed," and "small" carnivorous birds, noncarnivorous birds, and flying insects; and finally, "Crawling things" (*hasharāt*), that is, nonflying insects, snakes, and the like.

In other words, al-Jahiz's classification scheme derives from their habits of eating and locomotion, which allows for a mixing of mammals, birds, reptiles, amphibians, and invertebrates within the same category. He gives little attention to fish, claiming that too little is known about them. Al-Jahiz complains that most information on aquatic animals comes from sailors, who "are people who do not reflect on the implications of what they are talking about and who do not consider the ethics of their acts." His belief—which some would say is not without basis—is that sailors and fishermen are prone to exaggerate their fantastic tales of sea creatures.

The importance given to whether a creature is carnivorous or not suggests an anthropocentric sense of priorities: one of the most important things to know about a given species is whether or not it poses any danger to humans. More important still, from al-Jahiz's point of view, is the fact that all animals—indeed all of Creation—are miraculous signs of God that offer valuable lessons for our own salvation. Al-Jahiz takes ample note of the fact that various animal species are better equipped for a wide range of tasks than humans are; nevertheless, following the standard Qur'anic interpretation of his time, al-Jahiz accepts without question the received cosmic hierarchy that places humans above all other animals because of what he calls their capacity for reason and their "mastery" (*tamkin*).

As an example of the kind of "moral lessons" that animals can offer to humans, al-Jahiz offers the story of the Qadi and the Fly, a tale about a highly respected judge who tries to retain his composure while being harassed by a fly as the courtroom looks on. Eventually he begins to swat in vain at the fly, which repeatedly dodges the judge's hand and just as quickly returns to settle on his nose. Finally the judge blurts out, "God forgive me! I have just understood that whereas I enjoyed such respect and dignity among people, I have been defeated and ridiculed by the lowest of His creatures!" (Souami, p. 311).

Under the heading of "swine" al-Jahiz attempts to deal rationally with the fact that the Scriptures speak of God transforming humans into pigs and monkeys. One question that arises is why the Qur'an prohibits the consumption of pork but not of monkeys. Al-Jahiz believes that this is because the formerly Christian Arab tribes were enthusiastic pork eaters, while monkeys were naturally unappealing enough not to require a prohibition. (He offers the same opinion in regard to dogs.) Al-Jahiz opines that those who love pork are unaware of the fact that pigs consume feces.

Though the information in *Kitab al-hayawan* is of mixed reliability, and despite the fact that al-Jahiz's voluminous work actually contains many digressions that have nothing to do with animals at all, it served as a reference point for virtually all subsequent Muslim writers on animal-related issues and remains widely known and cited today.

Further Resources

Pellat, C. (1971). Al-Djahiz. *Encyclopedia of Islam, 2,* 386. Leiden: Brill.
Souami, L. (1988). *Le cadi et la mouche : Anthologie du Livre des animaux.* Paris: Sindbad.

Richard C. Foltz

■ Culture, Religion, and Belief Systems
Cattle Mutilation

"Cattle mutilation" is a name given to animal deaths and postmortem changes that have particular characteristics. These are summarized by Dr. George Onet as cattle dying suddenly and showing no signs of struggle or tracks surrounding the carcass. The animal is then found with missing body parts that seem to have been precisely cut away with a sharp instrument with little evidence of blood on the skin or ground.

Cattle mutilation was first described in the early 1960s and waves of reports occur sporadically around the world, typically lasting several years and spreading to cover a wide area, such as several American states. Two prominent waves of reports came from the western plains of the United States during the 1970s and Argentina from 2000 until the present day.

Reports of the "mutilations" are often picked up and sensationally reported by the media. The victims are commonly cattle but may include other animals such as horses and dogs. The media are often quick to associate the discoveries with speculation about satanic cults or extraterrestrial visitors.

Humans and Cattle

Humans have a long and complex association with cattle, using them as prey, sacrifices, beasts of burden, objects of sport (rodeos and bull fights), and objects of veneration. As Jack Conrad points out, the bull is a powerful symbol of strength and fertility.

As such we may be particularly sensitive to events involving cattle and predisposed to offer extreme or supernatural explanations.

Only a relatively small proportion of the human population has regular contact with cattle, but the bonds between farmers and cattle remain stronger than those with many other farm animals. Breeding bulls and dairy cows have a relatively long working life of several years or more, and cattle are large animals that are difficult to handle unless you form an amicable working relationship with them.

For several reasons many modern farmers have surprisingly little experience with dying or dead cattle. Farming techniques now have nutritional and veterinary resources to ensure that most cattle experience good physical health. When cattle need to be slaughtered or culled they are often transported away from the farm to specialized slaughter facilities or euthanized and the carcass disposed of immediately. Cattle that die through misadventure or natural causes tend to be discovered quickly in intensive systems and dealt with rapidly (by burial or cremation) for reasons of hygiene. Conversely cattle that die on very extensive ranges are unlikely to be discovered until decomposition is advanced or scavengers have almost totally consumed the remains.

Cattle mutilation deaths have been ascribed to various unusual causes. In Argentina, Mexico, and other places they have been attributed to a mythical creature called the *chupacabra* ("goat sucker"). In America they have been linked to secret government research, satanic cults, and the activities of extraterrestrial aliens. The evidence for each explanation is unreliable and includes helicopter sightings, marks on the ground, and inferences that the carcasses have been drained of blood, dropped from a height, or burned. The diversity of explanations suggests that people are seeking a villain that matches their perception of the mutilations as difficult to accomplish, mysterious, and threatening.

Scavengers and Cattle

Careful investigations have suggested that so-called cattle mutilation is simply the effect of scavengers. An investigation led by Kenneth M. Rommel Jr. in 1979 supported this theory and pointed out several relevant facts. Postmortems on fresh carcasses often identified a natural cause of death. The lack of blood on the scene is entirely normal as blood pools and coagulates within the animal's body after death. The carcass areas mutilated are those accessible to predators such as the eye and ear on the upper surface of the head and soft tissues such as the udder. Also, the marks left by scavengers can create an impression of "surgical precision" that could fool many observers. There were still some apparent inconsistencies with the facts, such as the lack of tracks around the carcass and the precise cuts on some carcasses.

Rommel's conclusions demonstrated that mutilations tend to be seen in two clear ways: either as natural scavenging combined with witness testimony that was distorted by anxiety and rumor or as highly unnatural events that defy simple explanations. These entrenched positions can be seen in one exchange between Rommel and a rancher in which the rancher said, "[If] coyotes did that they did it with knives." Rommel replied "I say that if surgeons did it, they did it with their teeth" (Marshall 1984).

One explanation that fits the vast majority of the phenomena surrounding cattle mutilations is scavenging, not by large canids such as coyotes, but by small rodents such as mice and rats. This explanation was offered by Argentinean officials and fits well with both North American and U.S. mutilations. Rodents are small, light, and have some of the sharpest teeth in the animal kingdom. A very precise and sometimes slightly serrated edge is typical of any object gnawed by rats, and commonly observed on "mutilated" cattle.

Dr. George Onet demonstrated that mutilation cuts are more consistent with a scalpel cut than tearing with pliers (intended to replicate scavenging). However, rodent scavenging would be better simulated using extremely sharp pincers that cut the flesh without tearing it. Rats would not inflict the fatal wound, but they quickly produce the typical mutilation pattern on animals that die of other causes.

Weak Bonds and Strong Reactions

Farmers retain bonds with their cattle as well an awareness of rodent vermin, which remain a key problem for production and storage of agricultural products. However, information about both cattle and rodents that would once have been common knowledge is now open to misinterpretation, suggesting a weakened connection between humans and other animals.

With cattle carcasses, for example, the common mention of the lack of blood around dead animals overlooks that fact that copious blood is rarely found around an animal that dies a natural death because of settling and coagulation. The present of peculiar fluid and absence of brain tissue is easily attributable to the effects of decomposition. The appearance of burns or cauterization is normal with the dehydration of exposed flesh. All of these postmortem changes might seem unusual but are caused by entirely natural processes.

Meanwhile rodents seem to have become an almost forgotten animal; even people who suggest a scavenging explanation for mutilations tend to assume that coyotes or other large animals are the agents. However, rodent species are almost ubiquitous, are too small to leave distinct tracks, and have teeth that are exceedingly sharp. The description of wounds that are both surgically sharp and edged either smoothly or with irregular serrations immediately brings to mind marks typically left by gnawing rodents. Their nocturnal habits and small size also suit them to doing their work unseen by humans.

We now store most products that would be appealing to rodents in impermeable metal or plastic containers and have little opportunities to see their typical gnawing patterns, especially as they appear on hide or flesh. It may be our very success and efficiency in raising cattle that puts us out of touch with the normal appearance of recently scavenged carcasses.

It is interesting to observe that veterinarians—whose expertise provides them with the most familiarity of postmortem changes in animals—often contradict wilder claims about mutilations. Rapid inspection of carcasses by veterinarians specifically qualified in necropsy would probably prevent the growth of more unusual explanations for a spate of animal deaths and mutilation by determining the immediate cause of the deaths and offering the more common explanations for postmortem damage. Cattle mutilation phenomena that still remained inexplicable under these circumstances would certainly warrant further investigation.

Whatever Remains

Why then have explanations that resort to unseen, bizarre agencies such as cults and aliens seemed more plausible to many observers than rodent scavenging? Perhaps through agents such as the media we have become more familiar with the alleged qualities of secret and supernatural beings than with common domesticated and wild animals. Thus it seems more natural to ascribe extreme stealth and precision to a satanist than to a rat.

It is easy to forget that rodents are supremely skilled animals that live successfully in almost every environment on earth, although we rarely see them. Studies in Argentina showed that carcasses left in the open quickly attracted opportunistic scavenging from mice, whereas the alleged human agents of mutilations have never been apprehended and charged despite the many thousands of reported cases.

People who have become out of touch with the behavior of common scavengers have opened themselves to highly speculative explanations of the phenomenon of post-mortem changes to cattle and other animal carcasses. The solution may be to learn more about our relationships with animal like rodents, insects, and other small creatures that we share our environment with, even if we cannot see them.

Further Resources

Conrad, J. R. (1957). *The Horn and the sword: The history of the bull as symbol of power and fertility*. New York: E P Dutton and Company.

Marshall, R. (1984.) *Mysteries of the unexplained*. New York/Montreal: The Reader's Digest Association.

Onet, G. E. (2002). *Animal mutilations: What we know*. Las Vegas, NV: National Institute for Discover Science. Also available at http://www.nidsci.org/pdf/cache_county_mutilation.pdf

Emily Patterson-Kane

■ Culture, Religion, and Belief Systems
Chief Seattle's Speech

In October 1887 the *Seattle Sunday Star* printed the tenth in a series of articles drawing on "scraps from a diary" by Dr. Henry Smith in which he presents a speech allegedly made by the Duwamish leader, Sealth. A small American settlement, established on the shores of Puget Sound in 1852, was named after Sealth, but in the form "Seattle" by which the city and the "Chief" have become famous. Versions of the speech attributed to Chief Seattle have become famous as an environmentalist anthem and as an assertion of a close relationship between Native Americans and nature.

Smith (originally the Ohio Superintendent of Schools and a Duwamish speaker) claims that Chief Seattle, described as noble and senatorial, made this speech in 1854 or 1855 on the arrival of Isaac Stevens, governor of the Washington Territory. Stevens' response is not recorded, except for a comment that there should be another meeting to agree a treaty. Smith's version of Seattle's speech formed the basis for a substantial revision published "to mark World Environment Day" on June 4, 1976. This was written a little earlier by Ted Perry for an environmentalist film for the Southern Baptist Convention. Evidence from other participants in the meeting between Sealth and Stevens makes it unlikely that the speech is genuine. The only document from Sealth in the U.S. National Archives is a brief expression of friendship toward "the Americans" and a desire to see the treaty. It is likely that Smith authored the speech to lament the expected extinction of Native Americans and their harmony with nature. Even before Perry's environmentalist makeover, the speech was part of an American tradition of nature writing that celebrates places and peoples threatened by urbanization and industrialization.

While there is no reference to animals in the National Archive's document from Sealth, and an animate "nature" is represented in Smith's text primarily by sympathetic rocks and compassionate stars that mourn the passing of Sealth and his people, Perry's version is full of animals. To explain why selling the land is unthinkable, Seattle says that insects are "holy" and that "the deer, the horse, the great eagle, these are our brothers." The rivers are "brothers" and the earth is the Mother who should not be bought, sold, or plundered. Seattle mourns the American destruction of buffalos and eagles and chides his hearers for taming wild horses. In one typical phrase Seattle asks, "What is man without the beasts? If all the beasts were gone, man would die from a great loneliness of the spirit. For whatever happens to the beasts, soon happens to man. *All things are connected.*" There can be few more eloquent or lyrical expressions of modern environmentalism than the speech attributed to Chief Seattle. That the real Sealth never saw buffalos or large cities (which both Smith's and Perry's speeches find objectionable) supports other evidence that he never made this speech in honor of human kinship with the world.

Further Resources

Clark, J. (1985). *Thus spoke Chief Seattle: The story of an undocumented speech.* http://www.archives.gov/publications/prologue/1985/spring/chief-seattle.html (accessed March 6, 2007).

Documents Relating to the Negotiation of Ratified and Unratified Treaties with Various Indian Tribes, 1801–69. NARA Microfilm Publication T495, roll 5.

Perry, T. (1976). *Chief Seattle's Thoughts.* http://www.kyphilom.com/www/seattle.html (accessed March 6, 2007).

Smith, H. A. (1887). *Chief Seattle's speech.* http://www.geocities.com/Athens/2344/chiefs4.htm (accessed March 6, 2007).

Graham Harvey

■ Culture, Religion, and Belief Systems
Chinese Youth Attitudes toward Animals

In early 2002, an animal cruelty incident involving a college student and five bears at Beijing Zoo electrified the mainland Chinese media (Wang, Xiong, & Zou, 2002). Using concentrated acid, a Qinghua University student caused severe damage to the eyes, skins, mouths, and internal organs of the bears. An outpouring of sympathy for the victims was coupled with condemnation from all directions at the perpetrator. For months, the Chinese media gave extensive coverage to the incident. One Internet forum reportedly received more comments on the incident than on any other domestic or international event (China Central Television, 2004).

The focal point of the incident was Liu Haiyang, the bear attacker. On the one hand, the public was quick to note Liu's upbringing in a single-child family. Liu's behavior seemed to confirm the public's perceptions of China's 80 million single-child youngsters, who are viewed as self-centered, uncaring, and indifferent to feelings of others (Wang, 2002b). On the other hand, the public expressed shock that the cruel act was perpetrated by someone with all the trappings of a promising professional and political future. Liu was a science student at Beijing's prestigious Qinghua University, the alma mater of many contemporary Chinese leaders, including President Hu Jintao and former Premier Zhu Rongji. (Four out of seven of the Sixteenth Communist Party Politburo Standing Committee members were graduates of Beijing's Qinghua University.) The view that those who are cruel to animals are likely to be indifferent to the welfare of their fellow humans is gaining wider acceptance on the Chinese mainland. Are China's college students and the Chinese youth in general morally compromised in their judgments on issues related to animals?

To gauge youth attitude toward nonhuman animals, a two-part survey was conducted in 2002 and 2003 among college students in China. Since today's college students are China's future political, economic, and opinion leaders, their attitudes toward animals and changes in their attitudes over time will have a direct impact on the nation's animal welfare consciousness. More immediately, ascertaining the college students' attitude toward animals along with a comparative study of other surveys conducted in mainland China will reveal whether mainland Chinese society is philosophically ready for animal-related policy change.

Previous Surveys of Attitudes toward Animals

This study of the college students' attitudes toward animals involved two separate surveys conducted in 2002 and 2003. For comparative purposes, we have included in our analysis the results of a 1998 survey of Beijing and Shanghai residents conducted by the International Fund for Animal Welfare (IFAW). Although the IFAW survey, the first of its kind in mainland China, has its limitations due to the fact that Beijing and Shanghai may not necessarily represent the entire nation, the reference value is helpful because the views of the residents probably do not differ greatly from those of the rest of the nation. One extraordinary accomplishment of the Chinese government over the last half century has been its ability to achieve a high level of uniformity in public opinion across the country through thought reforms, mass political campaigns, state media propaganda, and the educational system. Attitudes toward animals expressed by Shanghai and Beijing residents should mirror the general attitudes in the nation.

Objectives

Our survey of the Chinese college students growing up in a nation richer than ever before aims to determine if their attitude toward nonhuman animals is in any way different from the rest of the society. Is upbringing under better material conditions or in single-child families creating a generation with compromised moral judgments? In their efforts to explain the many alleged "unique" traits of the single-child youngsters, Chinese researchers have focused on the role of the family, schools, and society at large in the socialization of China's youth. According to their studies, families do play an important role in shaping the youth's values, views, and world outlook, but family influence wanes as youths get older (Feng, 2000) They found that the single-child youngsters, many of whom are college students, are not different from the rest of the society in value judgment. One study even found that an overwhelming majority (84 percent) of the surveyed single-child youngsters were more sympathetic to others (research group for studying the character development of China's single-child youth 1998). If society as a whole plays a more prominent role in the socialization, how much do the students resemble society in attitudes toward animals?

Second, shocking incidents involving college students who are science majors have been occurring in recent years. A science major was caught intentionally spreading a computer virus in one case only to be overshadowed by another in which a two-month-old puppy was microwaved alive. People were asking: what was wrong with the science students? Li Yan of Qinghua University attributed these incidents to China's overemphasis on training technical talents while ignoring the need to develop students' moral character (Wang, 2002a). One indicator of this lopsided emphasis on technical education is believed to be the neglect of education in the humanities in China's science and engineering schools (SCUN, 2004). Such neglect in the Chinese education system, some scholars worried, could result in the production of students with technical talents who are deficient in compassion, defiant of state laws, supercilious about morality, and unconscious of their social responsibility (Liu, 2002). Are the science majors different from their nonscience counterparts? Studies conducted by Chinese scholars have found that, despite their different modes of thinking and different approaches to problem-solving, science students have adopted the same values and outlook as the nonscience majors (Qian, 1999; Liu, 2000). Can similar results be expected in attitudes toward animals between the science and nonscience students?

Attitudinal change can be an important impetus to policy change. Our follow-up survey in 2003 was designed to gauge the impact of severe acute respiratory syndrome (SARS) on the attitudes of the college students. In an effort to fight the epidemic, scientists reportedly traced the source of the virus causing SARS to a number of wildlife species, such as civet cats, sold on the live animal markets in South China (Fowler, 2004). Reporting the connections between SARS and China's runaway wildlife exploitation, the media exposed horrendous conditions various wildlife and farm animals were subjected to on the farms, on the markets, and in transport. The follow-up survey was to determine whether SARS had affected the students' attitudes toward animals.

Questionnaires

The questionnaires for the 2002 and 2003 surveys are slightly different. While the 2003 survey adopted all the questions used in the 2002 survey, it contained three additional sets of questions. Both surveys included demographic questions (name [optional], gender, year in college, location of family residence, and field of study [science, engineering, medicine, agriculture, liberal arts, and social sciences]). The main body of both surveys consisted of three sections for the 2002 survey and four sections for the 2003 survey. One section asked whether nonhuman animals were sentient beings to determine if college students were empathetic with nonhuman species. As Tania Singer and her colleagues wrote in a recent edition of *Science,* the ability to empathize with others is a prerequisite for understanding, attachment, bonding, and love (Singer et al., 2004). In this section, students' interest in nonhuman animals explored their level of interest in animal-related literature and broadcast programs.

Students' position on animal exploitation in various settings was the subject of study in the next section of the questionnaires. First, we asked about the students' views on the relations between humans and nonhuman animals. Next, we asked about the respondents' perception of zoos. The last question listed ten acts such as skinning quails alive and using small live animals as shooting targets. We expected students' reactions to the ten acts to indicate the level of their indifference, disapproval, or indignation toward extreme cases of human exploitation of nonhuman animals. The third section of both surveys touched on the students' attitudes toward animal protection activities. We asked if they supported, participated, or intended to participate in animal protection activities. We expected their answers to indicate whether the Chinese youth were supportive of activities for animal protection.

In the 2003 survey, three sets of additional questions were added. We asked the respondents whether eating dog meat was morally the same as consuming meat of other animals. The second set of questions focused on wildlife farming. We asked specifically if the respondents had ever heard of bear farming, a practice for extracting bile from an open wound cut in the bear's stomach. Additionally, we asked those who had heard of bear farming if they considered it acceptable. The final questions were asked to determine if the respondents had experienced attitudinal change as a result of SARS.

Participants

In both the 2002 and 2003 surveys, we resorted with good reason to what some call the "convenient samples of college students." A total of 1,300 students were drawn for both surveys. The 2002 survey was conducted in thirteen universities, whereas the 2003 survey was repeated in ten of them. Both surveys were conducted in three different groups of universities: highly selective national key universities, fairly competitive national key universities, and provincial universities.

Student samples were appropriately stratified so that they represented both sexes, different years in school, different disciplines, and locations of family residence (rural and urban). We expected that the two survey results would mirror the attitudinal and behavioral orientation of the single-child generation since a majority of the college students are from this group. By including students from different academic disciplines, we could compare the attitudes of science students with those of their nonscience counterparts.

Findings and Discussions

For both surveys, we drew 1,300 samples. A total of 1,208 questionnaires were received for the 2002 survey, a return rate of 92.9 percent of the questionnaires issued. However, only 1,081 valid questionnaires, accounting for 83.1 percent of the total questionnaires distributed, were received for the 2003 survey.

Empathy for Animals

In 1998, IFAW funded a survey of residents in Beijing and Shanghai on their attitudes toward animals. The results were surprisingly encouraging in that a majority of the respondents reacted in favor of better treatment of animals. In the words of Merritt Clifton, editor-in-chief of *Animal People*, the results strikingly resembled those conducted in the United States some ten to fifteen years earlier (Clifton, 2000). In Beijing and Shanghai, 93.6 percent of the respondents believed that animals were capable of suffering and pain. Only 4.9 percent believed otherwise, and as many as 93.7 percent agreed that animals had emotions of sadness and happiness (IFAW, 1998).

Our surveys have shown similar attitudes among the college students with insignificant variations between males and females and between students from urban and rural households. In the 2002 survey, 97.8 percent of the respondents said that nonhuman animals had the capacity for pain and suffering, and 96.1 percent of the surveyed agreed that animals were capable of emotional expressions. This perception of animals as sentient beings was confirmed by our follow-up survey conducted in 2003, in which 98.2 and 96.4 percent of the respondents believed that nonhuman animals could feel pain and had emotions, respectively. In terms of empathy for animals, China's college students do not stand as a separate group either. As a matter of fact, our two surveys have shown that the college students scored higher in empathy for animals than did the Beijing and Shanghai residents.

Curiosity about Nonhuman Animals

Interest in animals does not necessarily correlate with high animal welfare consciousness. Yet people who are curious about nonhuman species are likely to be more inquisitive about them. This interest can lead to knowledge about the nonhuman species. The IFAW survey showed that 79 percent of the Beijing and Shanghai respondents were interested in animal-related works and broadcast programs (IFAW, 1998). In comparison, only 70.1 and 70.4 percent of college students, respectively, in our two surveys said that they liked such works or programs. The difference in the level of interest between the college students and Beijing-Shanghai residents could be attributable to the fact that the former had less time for recreational reading and television-watching. However, while 4.4 percent of the IFAW samples said they were not at all interested in reading about animals or watching broadcast programs on animals, only 3.4 percent of our samples in both 2002 and 2003 surveys felt the same.

One question could be raised as to the relation between the level of interest in animal-related works and the increase of such works in recent years. Specifically, to what extent is the respondents' interest a result of the increased number of animal-related publications and broadcast productions? Readers and viewers are likely to read or watch more of such works without actively seeking to do so if such works increase in quantity.

Human-Animal Relations

Anyone who has ever been to South China's live animal markets would be shocked at the conditions assorted animals are subjected to. Annie Mather, media director of Animals Asia Foundation, gave a vivid description of her first impression of a wildlife market in Guangdong:

> My reaction the first time I stepped into a live animal market was one of real horror. I know that animals "go to market" all over the world, in preparation to be sold and slaughtered . . . but what shocked me so much was that so many of the animals in the market in Guangzhou were really suffering and nobody seemed to care. For example, the animals were cooped up in tiny cages with no access to water or food, and worse, often with three legs, dying of gangrene, waiting sometimes for days or weeks in this condition until they were sold. (Mather, personal communication, March 27, 2004)

Mather's observation is echoed by Manab Chakraborty, executive director of Hong Kong's Kadoorie Farm and Botanic Garden, who has for years monitored wildlife trade in China. He wrote that in South China, "wild animals are always poorly treated. Most found in markets are dehydrated, injured and sick. Some traders develop cruel ways of killing them as a 'gimmick' to attract customers seeking the new and exciting" (Chakraborty, 2003). While the live animal markets in South China can be "soulless hellholes" for varieties of animals whose value is dependent on their use value to the humans, how do the college students see animals? To our surprise, our surveys display a very low percentage (2.7 and 2.5 percent) of the respondents who saw animals' existence for human use. In contrast, 92.4 and 93 percent of the respondents in the two surveys believed that animals deserved respect.

One recent development on the Chinese mainland is the sudden appearance of many wildlife parks and private zoos (Li, 2004). Many of these facilities were reportedly maintained in horrendous conditions (Zhang & Hou, 2002; AAF, 2003). Yet aggressive marketing by the zoo owners has portrayed these largely profit-seeking operations as educational and conservational institutions. How did the Chinese youth perceive the zoos? While 53.4 and 52.0 percent of the respondents in the 2002 and 2003 surveys viewed zoos as prisons for animals, a little more than a quarter of the respondents (25.7 in 2002 and 26.1 percent in 2003) saw zoos as the place for human-animal interaction. As Merritt Clifton pointed out, "this is the inverse of most U.S. and European findings" despite the fact that Chinese zoos are approximately fifty years behind the animal welfare standards stipulated by the American Zoo Association

(Clifton, personal communication, March 24, 2004). But why do more than a quarter of the college respondents view zoos uncritically?

We believe that there could be three possible explanations. First, most urban dwellers in mainland China have never even seen squirrels or wild rabbits, animals that one sees in one's backyard or in public parks in the West. Except for their appearance on television, in books, or in the live animal markets, zoos are the only place for urban Chinese to see wild animals. Second, the aggressive marketing of the many newly built wildlife parks and private zoos could have influenced many in the society. Third, public debates on controversial issues have always been discouraged by the government, who sees a forum as potentially destabilizing. Animal welfare activists in China have long fought a hard battle to get their messages across (Li, personal communication, November 20, 2003). Public debate on animal cruelty has been discouraged by the authorities for fear that such a debate would invite more foreign pressure to close profit-making and job-creating businesses charged with animal cruelty (Xiaoxi, 2003).

Eating Wildlife

The outbreak of SARS brought international attention to the rampancy of eating wildlife in China ("Wildlife Trade Faces Tough Curbs," 2003). Though these eating habits were traditionally limited to South China, they have recently spread to the rest of the country like prairie fire (Li, 2003). In one live animal market in Guangzhou, the annual sales value of wildlife amounted to RMB 800 million (about US$100 million). On a normal market day before SARS, one could find almost anything creepy, crawly, feathery, and slippery for sale to the thousands of local restaurants dealing in exotic foods. Many wild animals on sale were either freshly caught in the wild with missing or bloody limbs or artificially bred on the thousands of wildlife breeding farms, often operated in abject conditions. Among the thousands of live, dead, or dying animals were state-protected species such as spotted deer, pangolins, owls, and others (Sui, 2003). Like its manufactured goods, Guangdong's taste for wild animals has spread to the rest of the country. Shanghai residents reportedly consume as many live snakes a year as those in Guangzhou. Northeast China has seen illegal poaching and slaughtering of the state-protected black bears for their paws, a rare delicacy in the local diet (Zheng, 2001). Wildlife eating has continued unabated in many parts of the country despite the SARS epidemic. Very recently, Russian customs authorities seized 800 bear paws bound for China ("Russia Seizes China-Bound Truck with 800 Bear Paws," 2004).

How common is wildlife eating (wild animals including endangered species such as black and brown bears, pangolins, owls, and others such as wild rabbits, snakes, and frogs) among the general public? Contrary to the perceptions of outside observers, both the IFAW and our surveys have found that wildlife eating remains a culinary subculture. In Beijing and Shanghai, in 1998, 37.8 percent of those surveyed said that they had recently eaten wild animals (IFAW, 1998). Comparatively, only 24 percent of the surveyed students in our 2003 survey had eaten wildlife. The lower percentage of college students who had eaten wildlife is perhaps attributable to the fact that they mostly dine at school canteens where wildlife is not commonly served.

Follow-up research is perhaps necessary to find out why a majority of those surveyed had never eaten wildlife. Was wildlife eating not part of their normal diet? Or did they consciously choose not to eat wildlife? A question could even be posed to those who have eaten wildlife about whether they can do without exotic tastes. With or without additional findings, the IFAW and our surveys show that wildlife eating is not part of the mainstream of China's culinary culture.

Animal Cruelty

Animal cruelty happens everywhere. Economic liberalization in post-Mao China has made widespread exploitation of different kinds of animals a byproduct of the national drive for wealth. To attract customers, some restaurants reportedly resort to cruel practices and provide exotic foods to eaters. At many wildlife parks, visitors could purchase live animals such as rabbits, chicken, pigs, and even cows to feed hungry tigers ("Feeding Time at Zoo Terrifies Children," 2004). Lion cubs were chained to poles and peacocks were forced to extend their wings for visitors to take pictures. Most shockingly, visitors can practice live ammunition shooting by firing at small, live animals fastened on the target boards. Some wildlife parks and private zoos are operated in conditions in which animals suffer from severe food and environmental deprivation (Ni, 2003).

In recent years, dog farms have emerged across the Chinese mainland. The dogs are raised in iron cages and are often slaughtered, skinned, and disemboweled in full view of other dogs waiting for their turn. While the mainland Chinese media has increased its criticism of acts of animal cruelty, recent reports revealed that the Beijing Municipal Government was actively considering the introduction of Spanish-style bullfights into China (Fang, 2004).

How do the college students react to acts of animal cruelty? Are they indifferent? In our surveys, the respondents are asked to identify from a list of ten acts those that they consider as cruel to animals (see Table 1). These acts include "raising meat dogs in small cages," "using animals in circus performance," "eating brains from live monkeys," "putting on a monkey show," "skinning quails alive," "force-watering of pigs, fowls, and other livestock before slaughtering," "scaling fish alive," "caging wild birds," "shooting small animals as live targets," and "de-sexing pets."

Table 1. Percentage of Respondents Who Identified Each Act as Cruel

	Choices	Percentage 2002 Survey N = 1,208 MoE + 2.82*	Percentage 2003 Survey N = 1,081 MoE + 2.98*
1	Raising meat dogs in small cages	29.8	32.4
2	Using animals in circus performance	38.2	43.8
3	Eating brains taken from live monkeys	88.5	90.4
4	Putting on a monkey show	56.5	62.9
5	Skinning quails alive	74.5	74.4
6	Force-watering pigs, cattle, fowls and other livestock before slaughtering	60.3	63.3
7	Scaling fish alive	57.3	59.3
8	Caging wild birds	51.6	53.6
9	Shooting small animals as live targets	90.2	89.2
10	De-sexing pets	41.9	44.4

* Based on total college student size of 19 million as of 2003.

We included "de-sexing pets" because it is widely perceived to be a cruel practice in China. People may still associate the procedure of sterilization of cats with the simple and often crude roadside surgery for de-sexing roosters commonly practiced in China's rural areas. Despite the fact that China's cities are increasingly confronted with the growing stray population, few people except animal activists see the need to control the pet population (Lu, personal communication, November 6, 2003).

Our surveys in 2002 and 2003 show that high percentages of the respondents identified shooting small animals as targets, eating brains taken from live monkeys, and skinning quails alive as particularly deplorable. As expected, over 40 percent of the respondents in both surveys saw sterilizing pets as cruel.

It is important to note that, though most respondents concurred that these were all cruel acts against animals, significant percentages of the respondents did not identify acts 1, 2, 4, 6, 7, and 8 as cruel. Yet out of the 1,082 respondents in the 2003 survey, a majority of them selected more than four acts as cruel. Only 2.3 percent of the respondents checked one act, whereas those who checked five to ten acts accounted for 71.4 percent.

Noticeably, we did not include testing on animals in the list. The list we offered in both surveys included the acts that had recently been criticized by the Chinese media. At least the media's position on many of these acts was crystal clear. It was our purpose to determine if the college students would react similarly. Regrettably, animal testing is the least talked about area of animal cruelty. As a result, we did not know the general public's position on animal testing and therefore did not have a reference point for comparison.

Animal Protection Activities

Are China's college students supportive of animal protection activities? Do they participate in such activities? Both the 2002 and 2003 surveys indicate that a high percentage of the respondents (95.0 and 93.7 percent, respectively) philosophically supported animal protection work. In contrast, those who did not express such support accounted for 0.8 and 1.4 percent. By comparison, 97.1, 91.6, and 94.0 percent of the Beijing and Shanghai respondents answered affirmatively to three IFAW survey questions about whether they supported Chinese wildlife protection organizations, international environmental groups, and international animal welfare organizations, respectively (IFAW, 1998).

But philosophical support tends to stop short of direct and personal participation. In the 2002 survey, 48.2 percent of the respondents expressed willingness to participate, as opposed to 51.4 percent who had not seriously considered participating in such activities. A slight improvement was seen in the 2003 survey, in which 51.0 percent said they intended to participate, while 43.9 percent did not intend to. Both the IFAW survey and ours show that expression of support does not equate direct participation.

The gap between expression of support and actual participation seems to be a universal problem. According to Merritt Clifton, few nations offer as many easily accessible opportunities to help animals as the United States. However, "while 31 percent of U.S. residents enjoy watching wildlife, based on U.S. Fish and Wildlife Service surveys, and approximately two-thirds of American households keep dogs or cats, only one U.S. household in four donates to animal protection causes, which receive less than 2 percent of all U.S. charitable donations" (Clifton, 2004).

In mainland China, participation in animal protection activities is discouraged by a combination of economic, ideological, and political factors. In today's China, economic modernization is the government's top priority. Many profit-making enterprises, such as the more than 270 bear farms, are involved in animal cruelty. Protesting against these enterprises conflicts with the local need for growth. (For example, bear farming in

China's Sichuan Province involves the incarceration of more than 2,000 bears suffering from the cruel daily bile extraction from an open wound surgically created in their stomachs. Local forestry officials admitted that bear farming was cruel. Yet, bear farming is directly linked to many pharmaceutical companies in the province and therefore directly or indirectly contributes to the employment of some 10,000 people.) Ideologically, animal lovers have long been smeared as elements who "worship decadent Western bourgeois lifestyle" (Qin, 1994). These elements allegedly love to use animal cruelty incidents to defame their own motherland. Is the Chinese government behind the attack on animal advocates? Not necessarily. Yet, the extremist xenophobic sentiment fostered by years of ideological indoctrination in the prereform era cannot but influence the value judgments of some of the opponents of animal welfare.

Politically, the Chinese government continues to suppress autonomous initiatives from the society. Animal protection activities, particularly protest activities against relevant government policies, are not received well by the authorities. Public questioning of government's animal-related policies is resisted since it may create a precedent, make government liable to potential lawsuits, impact social stability, and ultimately shake the foundation of the one-party rule (Peerenboom, 2003). These factors obstruct autonomous societal participation in activities for animal rights and welfare.

Dog Eating

In the 2003 survey, we added a question on the moral difference between eating dogs and consuming beef, pork, and poultry. We intended to find out if the college students held different views compared with the residents of Beijing and Shanghai. And if they did, what could the survey tell us?

Dog eating has always existed in China. It was traditionally limited to South and Northeast China. In recent years, dog eating has also become more common across the mainland. While more people are keeping dogs as pets, dog eating has also attracted increasing criticism from within and outside China. How do the mainlanders perceive dog eating? Thirty-six percent of the IFAW survey respondents saw no moral difference between dog eating and consumption of other common meats. The IFAW study also showed that the higher the respondents' education (41.6 percent of college graduates, 40.7 percent of senior high graduates, 30.2 percent junior high graduates, 26.3 percent of elementary graduates, and 18.4 percent of illiterates), the less difference they saw between eating dog meat and consuming other animals (IFAW, 1998). How do we explain this finding?

One explanation addresses the political background of the recipients of China's education. The phenomenon of highly educated people seeing no moral difference between eating dogs and beef or pork "could reflect the extent to which educational opportunity in China had long been reserved for loyal Communists, whose views might tend to reflect the opinion of . . . Mao Tsetung that dogs were parasites, better eaten than fed. Educated people born after Mao's 1976 death may develop a different perspective" (Clifton 2000). Our 2003 survey suggested that "a different perspective" on dog eating is yet to arise among the new educated elite. Forty-five percent of the 2003 survey college respondents saw dog eating as morally acceptable.

Merritt Clifton does not see the IFAW and our survey results as totally negative. He wrote in an e-mail to the authors that, "while here in the United States, we are struggling to get people to begin to see eating a pig as being just as cruel as eating a dog, in China the breakthrough may come simultaneously on behalf of all animals: the person who stops eating dogs will stop eating pigs and chickens too because they will all be seen as being moral equals" (Clifton, 2000). The question we pose to Clifton is: Are the Chinese students really suggesting that eating other animal meat is morally wrong? Or, are they simply being defensive of dog eating by suggesting that Western beef eating is just as morally wrong?

We do not know exactly why 45 percent of our respondents believed so. But a Qinghua University professor's open rebuttal of Western criticism of China's dog farms can perhaps explain the feelings of these students (Huang, 2003). Highly educated people in China are knowledgeable about the outside world, including outside criticisms of aspects of Chinese culture. Importantly, they are also more nationalistic and more sensitive than the masses to China's image in the world. Being knowledgeable, they can cite examples of foreign practices to counterattack foreign criticisms of Chinese practices. Being nationalistic, they tend to equate foreign criticisms with a wholesale assault on the Chinese culture. Like educated South Koreans, educated Chinese detest their motherland being portrayed as barbaric because of dog eating or other practices criticized by Westerners.

The high percentage of the educated mainland Chinese seeing no moral difference between dog eating and consumption of other meats remains an open question. Are they trying to defend a culinary subculture in response to perceived hostile foreign criticism? Or are they simply expressing frustration that the moral question of consumption of other animals is not equally criticized?

Awareness of Wildlife Breeding

In the 2003 survey, we added two questions on students' awareness of wildlife breeding in China. The objective was to determine if the students were also aware of the cruelty associated with the wildlife farming activities. Wildlife farming in China involves many species with economic and alleged medicinal values. Among the most intensively farmed species are Asiatic bears, foxes, martens, civet cats, snakes, and frogs (Ran & Zhou, 2000). It is only in the last decade or so that conditions on the wildlife breeding farms have attracted the attention of domestic and international animal welfare groups. We specifically asked our respondents whether they had heard about bear farming, the farming operation that has aroused the most domestic and international criticism for its cruel practice of bear confinement in small iron cages for life and of bile extraction from an open wound cut in their stomach.

According to the 1998 IFAW survey, only 30.2 percent of the Beijing and Shanghai residents had ever heard of it (IFAW, 1998). The China Bear Rescue Campaign launched by the Animals Asia Foundation (AAF) and the personal fight of AAF's founder and CEO Jill Robinson have brought Chinese and international attention to the conditions of China's bear farms. Our 2003 survey showed that 40 percent of the respondents said that they had read about bear farming. More important, more than 87 percent of the IFAW respondents and 90 percent of our 2003 survey subjects saw it as a cruel practice. Students' recognition of the intrinsic cruelty of bear farming contradicts the Chinese official claim that farm bear conditions have been improved and that bear farming is humane.

Science versus Nonscience Majors

Our surveys did not find science students to be different from their nonscience counterparts in their attitudes toward nonhuman animals. While they scored lower on views and attitudes unfavorable to animals, on some questions (raising meat dogs in small cages, support of animal protection activities, and bear farming as an acceptable practice), they scored higher on others (equal treatment of animals and intention to participate in animal protection activities). The interest of science students in readings and broadcast programs on animals was 7 percent higher than the nonscience majors (65 percent). In general, the

Table 2.   Comparison of Science and Nonscience Student Responses (2003 Survey)

Questions	Science majors N = 837	Nonscience majors N = 244
Yes, animals have the capacity for suffering.	98%	98%
Yes, I like readings and TV programs on animals.	72%	65%
Yes, we should consider animal welfare and treat them equally.	94%	93%
Yes, zoos are the place where animals are imprisoned.	51%	51%
Yes, raising meat dogs in cages is cruel.	31%	33%
Yes, I support animal protection activities.	92%	95%
Yes, I intend to participate in animal protection activities.	52%	46%

difference in attitudes between the science and nonscience students was insignificant. (See Table 2 for a snapshot of their responses on a selected number of questions.)

Attitude Change after SARS

Catastrophic events often lead to attitude change. Has the Chinese public changed their attitudes toward nonhuman animals after the SARS crisis? Months after the end of the epidemic, we could see a positive trend in societal opposition to wildlife eating and animal cruelty in general. More people have realized that unbridled exploitation of wildlife and other animals has adverse effects threatening not only China's wildlife species but also the human welfare. The government has for the first time called on the people to adopt different dietary habits (Dong, 2004). More than a thousand chefs across the country responded by publicly vowing not to prepare wildlife food. In Guangdong, where wildlife eating had been most rampant, the government held public hearings for the first time on the fate of wildlife trade in the province ("Proposed Ban on Eating Wild Animals Offers Public Food for Thought," 2003).

In response to our survey question, 13 percent of the students surveyed said that their formerly unfavorable attitudes toward animals have been changed after SARS. Fifteen percent of the respondents said their support for the use and eating of wildlife remained unchanged. A total of 62.6 percent of the respondents opposed wildlife eating. Yet, if we compare the results of 2002 and 2003 surveys, we see noticeable differences in the students' attitudes. Table 3 highlights the changes in views and attitudes.

Conclusions

The two surveys on college students' animal welfare consciousness were the first ever conducted on the Chinese mainland. According to the results, China's educated youth do not stand as a separate group with compromised moral judgments. Since society plays an important socialization role, the attitudes of the Chinese youth toward nonhuman animals are quite uniform regardless of their academic focus, family residence status, years in college, and difference in gender.

Our surveys also shed light on the various obstacles that have hindered the rise of animal welfare consciousness in society. The first of these factors is the nation's priority of eco-

Table 3. Comparison of 2002 and 2003 Survey Results

Questions	2002 Survey Percentage	2003 Survey Percentage
Yes, animals are capable of suffering.	97.8%	98.2%
Yes, I like readings and broadcast programs on animals.	70.1%	70.4%
Yes, humans should treat animals equally.	48.6%	55.6%
Yes, animals exist for human use.	2.7%	2.5%
Yes, it is cruel to raise meat dogs in small cages.	29.8%	32.4%
Yes, it is cruel to force animals to perform in circus.	38.2%	43.9%
Yes, it is cruel to eat brains taken from a live monkey.	88.9%	90.7%
Yes, it is cruel to stage a monkey show.	56.3%	63.1%
Yes, it is cruel to skin a quail alive.	74.5%	74.7%
Yes, it is cruel to force-water pigs, cows and other livestock before slaughtering.	60.3%	63.5%
Yes, it is cruel to scale a fish alive.	57.3%	59.5%
Yes, it is cruel to keep a bird in the cage.	51.6%	53.8%
Yes, I intend to participate in animal protection activities.	48.2%	51.0%

nomic development. On questions that seemed to suggest a choice between supporting economic growth and supporting animal protection, we see divisive opinions or a tendency to lean toward the former. Another factor is the state's preoccupation with social stability and the associated government revulsion against unofficial groupings and activities. The low percentage of students' participation in animal protection activities could be a rational response to that position of the government. Finally, nationalistic sentiment can bring forth defensive postures, causing resistance to external calls for a change of particular Chinese attitudes, behaviors, or policies. New research to study the connection between these factors and students' attitudes is necessary to understand the thought process of the students.

It is important that the surveys confirm that China is philosophically ready for breakthroughs in animal welfare policy making. A majority of the college youth expressed empathy for animals. Most of them believed that animals are sentient and have emotions. They deplored extreme cruelty against animals in the catering business, entertainment industry, wildlife farms, and slaughter houses. They stood for better treatment of nonhuman animals. It is time for a drastic change in China's policy regarding animal protection.

Further Resources

Animals Asia Foundation (AAF). (2003). Emergency help: Starving bears in Yulin zoo, China, January 2003. *The Official Animals Asia Newsletter*, 22.

Chakraborty, M. (2003). Wildlife trade and consumption threaten public health. *South China Morning Post*, June 5.

China Central Television (2004). San wei zhengxie weiyuan tan Qinghua xuesheng shang xiong shijian (Three delegates to the National People's Political Consultative Conference discuss the bear attack incident). Retrieved January 20, 2004, from http://www.gdnet.com.cn/hottopic/2002/qinghua/0305-01.htm

Clifton, M. (2000). Kindness: Where East meets West. *Animal People,* March 2000, 8–9.

Dong, J. (2004). Zhongguo yesheng dongwu baohu xuehui fachu bu lanshi yesheng dongwu de changyi (China Wildlife Conservation Association calls for a stop of excessive wildlife eating). Retrieved January 27, 2004, from http://tech.enorth.com.cn/system/2004/01/17/000718990.shtml

Fang, D. (2004). Is Beijing ready for Spanish bullfights? *South China Morning Post,* March 13.

Feeding Time at Zoo Terrifies Children. (2000). *South China Morning Post,* May 8. Retrieved April 7, 2004, from http://www.wag.co.za/root_news/feeding_time_at_zoo.html

Feng, X. (2000). Dusheng zhinu de shehui hua guocheng jiqi jieguo (The process and result of socialization of China's single-child generation). *Zhongguo shehui kexue (Journal of Chinese social sciences),* 6. Retrieved January 20, 2003, from http://www.usc.cuhk.edu.hk/wk_wzdetails.asp?id=1905

Fowler, J. (2004). WHO traces SARS to civet cats. Retrieved April 18, 2004, from http://story.news.yahoo.com/news?tmpl=story&u=/ap/20030523/ap_on_he_me/sarsvirus&cid=541&ncid=1624

Huang, C. (2003). Gei guoji sheng bona quan zhengjiu zuzhi de fuxin (A response to the international St. Bernard dog rescue group). Retrieved September 20, 2003, from http://www.xys.org/xys/ebooks/others/science/misc/dog.txt

International Fund for Animal Welfare (IFAW). (1998). *Public opinion survey on animal welfare.* Beijing: BMS & Associates.

Kellert, S. R. (1996). *The value of life: Biological diversity and human society.* Washington: Island Press.

Kruse, C. R. (1999). Gender, views of nature, and support for animal rights. *Society and Animals,* 7(3), 179–98.

Li, B. (2002). Zhao Zhongxiang weiyuan: yesheng dongwu yuan buneng zai jian le (Deputy Zhao Zhongxiang: There should be no more wildlife parks). *Xinhua she (New China news agency),* March 7. Retrieved January 28, 2004, from http://www.china.org.cn/chinese/lianghui/115618.htm

Li, P. J. (2003). Animals in China: From 'Four Pests' to two signs of hope. *Animal People,* October, 8.

Liu, H. (2002). Zhi xuexi hao jiu xing ma? You liu hai yang shang xiong shijian shuo qi (Academic accomplishments are not everything: What did Liu Haiyang's bear attack incident tell us?). *Zhongguo jiaoyu bao (China education news),* March 2, 2.

Liu, Y. (2000). Zhiye jishu yuanxiao shishi renwen jiaoyu de celüe (On the tactics of implementing humanities education in professional technical schools). *Zhiye yu jiaoyu (Vocation and education),* 5. Retrieved January 22, 2004, from http://www.tech.net.cn/research/segment/arm/1488.shtml

Ni, C. (2003). Activists claim Chinese zoos abuse animals. *Bradenton Herald,* January 5. Retrieved April 16, 2004, from http://www.bradenton.com/mld/bradenton/news/local/4876659.htm

Peek, C. W., Bell, N. J., & Dunham, C. C. (1996). Gender, gender ideology, and animal rights and advocacy. *Gender and Society,* 10(4), 464–78.

Peerenboom, R. (2003). SARS and civil society. Animal cruelty: A litmus test for mainland political reform? *South China Morning Post,* May 28.

"Proposed Ban on Eating Wild Animals Offers Public Food for Thought." (2003). *South China Morning Post,* July 9. Retrieved April 16, 2004, from http://china.scmp.com/chimain/ZZZFS977THD.html

Qian, L. (1999). Qiang hua zhiran kexue yu renwen kexue de fenli shi yizhong daotui. (It is retrogressive to keep the sciences and humanities disciplines separate in college education). *Zhongguo gaige bao (China reform news),* June 2. Retrieved on January 10, 2004, from http://www.gmw.com.cn/0_wzb/1999/19990613/GB/1653^WZ3-1314.htm

Qin, G. (1994). You quan duo wei huan shuo kai qu (On the peril of too many dogs). *People's Daily* (overseas edition), August 3, 3.

Ran, M., & Zhou H. (2000). *Zhongguo yaoyong dongwu yangzhi yu kaifa (The farming of wildlife animals for use in Chinese traditional medicine)*. Guiyang: Guizhou Science and Technology Press.

Russia Seizes China-Bound Truck with 800 Bear Paws. (2004). *Reuters,* March 23. Retrieved April 5, 2004, from http://www.enn.com/news/2004-03-23/s_14242.asp

Singer, T., Seymour, B., O'Doherty, J., Kaube, H., Dolan, R. J., & Frith, C. D. (2004). Empathy for pain involves the affective but not sensory components of pain. *Science, 303,* 1157–62.

South Central University for Nationalities (SCUN). (2004). Daxuesheng suzhi jiaoyu de xianzhuang yu chengyin (On the current state and origin of quality education for the contemporary college students). Retrieved June 12, 2004, from http://www.scuec.edu.cn/hkb/info/ztjz4.htm

Sui, C. (2003). China raids wildlife markets, seizes thousands of animals in SARS fight. *Agence France-Presse,* May 28.

Wang, J., Xiong, Y., & Zou, G. (2002). Qinghua xuesheng xiadushuo liusuan shaoshang wu gouxiong (A Qinghua University student viciously attacked five bears using sulphuric acid). *Jinghua shibao (Beijing times),* February 25, 32.

Wang, L. (2002a). Hushi renwen jiaoyu de eguo (The consequence of the neglect of education in humanities). *Zhongguo xinwen wang (China online news network).* Retrieved February 20, 2004, from http://news.sina.com.cn/c/2002-02-26/1726486711.html

Wang R. (2002b). Zai canren xingwei de beihou (Behind the cruel act). *Jingzhou wanbao jiaoyu zhoukan (Jingzhou evening news education weekly),* April 7, 1.

Wildlife Trade Faces Tough Curbs. (2003). *China Daily,* May 28. Retrieved April 27, 2004, from http://www.china.org.cn/english/environment/65596.htm

Zhang, W., & Hou, Y. (2002). Dongwu shijie zhuchiren Zhao Zhongxiang: renlei shi yesheng dongwu de zhongjiezhe (Humans are the terminators of wild animals: an interview with Zhao Zhongxiang, host of CCTV's Animal World program). *Liaosheng wanbao (Liaosheng evening news),* May 28. Retrieved September 19, 2003, from http://www.china.org.cn/chinese/TR-c/151177.htm

Zheng, Y. (2001). *China's Ecological Winter*. Hong Kong: The Mirror Books.

Zhongguo dusheng zinü renge fazhan ketizu (Research group for studying the character development of China's single-child youth) (1998). Zhongguo chengshi dusheng zinü renge fazhan xianzhuang ji jiaoyu yanjiu baogao (Report on the status and character development of China's urban single-child youth). *Zhongguo jiaoyu yanjiu* [Journal of education studies], October. Retrieved January 23, 2004, from http://www.sunyunxiao.net.cn/sunnyunxiao/e_edu/e_a_report/e_a_report05

Peter Li

■ Culture, Religion, and Belief Systems
Christian Art and Animals

Donkeys and oxen, sheep and lions, and dogs and doves all abound in the visual arts of the Christian tradition. Sometimes representing real animals, and other times symbolic in nature, these animals serve pedagogical and aesthetic purposes. They also suggest the possibility that animals were included in Christian communities throughout its history. Animals have always held an amazingly prominent position in Christian art, a position not mirrored in Christian theological works. An overview of animal representations in three areas provides a general foundation for an understanding of the place of animals in Christian art: animals from Biblical stories, bestiaries, and animals associated with particular saints and legends.

Christian artwork is often a way of telling and interpreting biblical stories, particularly for those people who cannot read the text themselves, which has been the case throughout most of Christian history. Because animals are present as characters or actors in a number of these stories, they are portrayed in the visual imagery as well. Some of the most common stories depicted, a few of which are outlined later in this entry, include the Creation (Gen. 1–2); Noah's Ark (Gen. 6–9); the Sacrifice of Isaac (Gen. 22); the Peaceable Kingdom (Isa. 11 and 65); Jonah and the Whale (Jon. 1–2); Daniel in the Lion's Den (Dan. 6); the Nativity of Jesus (Matt. 1–2, Luke 1–2); Jesus' Entry into Jerusalem (Matt. 21, Mark 11, Luke 19, John 12); Jesus' Last Supper (Matt. 26, Mark 14, Luke 22, John 13).

In Raphael's "Creation of the Animals" (1518–1519; Vatican, Rome), a smiling lion walks at the right hand of a bearded male God, and unicorns dance in the background. Of course, carnivores, omnivores, and herbivores stroll peacefully together in the Garden of Eden, reflecting the idea that all animals (including humans) were vegetarians before the Fall. Master Bertram's "Creation of the Animals" (1383; Grabow Altarpiece, Hamburg) reflects the idea that Jesus was present in the Garden, so he is central to the image. Mammals line up at his right, water creatures and birds at his left. Again, most of the animals have smiles on their faces as Jesus gazes lovingly down at them.

The Peaceable Kingdom imagery, made famous in the American setting by Edward Hicks, reflects a messianic vision from the book of Isaiah of the end of violence, not only between people but also between wild and domestic animals. So, the "wolf shall live with the lamb, the leopard shall lie down with the kid, the calf and the lion and the fatling together, and a little child shall lead them" (Isa. 11:6). Hicks, a Quaker for whom the theme of peace was central, may have painted this same vision up to one hundred times.

Abraham's near sacrifice of Isaac portrays a different, and much less peaceable, ending for animals than either the creation theme or the peaceable kingdom theme. God tells Abraham that he must sacrifice his son, Isaac, in order to prove his faith in God. Abraham cuts the wood for the burnt offering, binds his son, places him on the altar, and is about to slaughter him when an angel appears. Luckily for Isaac, the angel tells Abraham to stop and provides a ram, stuck in a thicket, instead. Abraham takes the ram and offers it up instead. Caravaggio's painting (1601; Uffizi, Florence) is a typical portrayal showing the angel stopping Abraham's lifted, knife-wielding hand while the ram's head emerges from the bushes.

Stories of the Nativity, or birth of Jesus, are found in two of the canonical Gospels: Matthew and Luke. Although animals are mentioned only indirectly in these accounts, certain portions suggest that animals must have been present. In Luke, Jesus is born and laid in a manger (a feed bin for domestic animals), and shepherds, who are tending their flocks, come to see him. In Matthew, magi or wise men come to visit the newborn child, and one can assume that they were riding on some form of animal. But it is actually the artwork dating as early as the fourth century CE and extending throughout Christian history that places animals in the Nativity scene. This is based in part on a prophesy found in Isaiah 1:3 and in part from the reports from Luke of Jesus' placement in a manger. Eventually, an ox and an ass gazing at the baby Jesus began appearing in almost every piece of art, and they are present in sculpted nativity scenes as well. Sheep join the shepherds, and camels transport the wise men. But the ox and the ass are oftentimes the most interesting of these animals who witness the birth of Jesus. A fourth-century Roman sarcophagus (Vatican, Rome) includes the birth of Christ among the many carvings decorating its exterior. The infant Jesus reclines in the manger, and it appears that the ass is actually licking him. From this point forward, these two figures were central to most depictions of the Nativity. A fresco in the Basilica of Saint Francis (thirteenth century, Assisi) places the ox and ass closer to Jesus than any other figures, and the two animals are the only ones looking directly at him. Botticelli's *The Mystical Nativity* (1500; National Gallery, London) shows Jesus reaching up to the ass, who is bent down directly over him central to the picture. A donkey and her colt show up again in the stories and images of Jesus' life. All four canonical gospels relate that Jesus rode the colt into Jerusalem on the day traditionally celebrated as Palm Sunday in the liturgical calendar. This colt appears in artwork throughout Christian cultures as well.

Interestingly, dogs and cats show up in another scene, even though there is no mention or even suggestion of them in the biblical text. The scene is the last supper that Jesus shares with his disciples before his crucifixion. Famous portrayals of this scene, from Tintoretto's Last Supper images in Venice (1592–94, 1578–81) to Rosselli's image in the Sistine Chapel (1481–82, Vatican, Rome), include dogs and cats. The Last Supper portrayal in the Basilica of Saint Francis places a dog and a cat in the room next to Jesus and the disciples; the animals are happily licking the extra food off of the plates.

An entirely other purpose is served by bestiaries. Based on the fourth-century Latin "Physiologus," a genre describing real and fantastical animals with allegorical interpretations, bestiaries are illustrated collections of animal fables. From as early as the eighth century, though emerging prominently by the eleventh, through the late fifteenth century, this particular pedagogical genre was popular throughout Europe. They were particularly popular in England, and several extant pieces, such as the Aberdeen Bestiary, provide excellent resources for research into this aspect of Christian culture. Bestiaries sought to teach moral lessons through tales about and images of animals. The animals exemplified virtue or vice, served as allegorical models, or simply provided lessons about the ways of creation.

Some consistent bestiary images and their symbolic associations include the lion or the pelican as Christ (a connection used in literature as well by such authors as C. S. Lewis in *The Chronicles of Narnia*), the serpent as the Devil, the dog as faithful, the phoenix as the resurrection, the unicorn as the Virgin. Some associations are more complex and probably relate a natural history as part of their story. So, for example, oxen not only are the strongest of the beasts, but also can predict the weather; they refuse to leave their stalls if rain is coming. A dog's ability to heal wounds translates into the bestiary as a representation of the wounds of sin being healed by confession. The overall view of animals in this form of Christian art places them in symbolic categories primarily. They certainly do not make choices or develop

Statue of Saint Roch with his dog, 18th CE. Saint Roch cared for plague victims during his wanderings through Italy. It is said that when he himself was infected, a dog brought him food in his lonely retreat in the woods. ©Erich Lessing/ Art Resource, NY.

Creation of the Animals by Raphael. ©Bildarchiv Preussischer Kulturbesitz/Art Resource, NY.

characteristics themselves; rather, they embody traits as designated by God at the time of Creation, thus becoming a type of second book in Christian theology—the book of nature.

In addition to bestiaries and stories of animals from the biblical texts, images of animals often accompany the hagiographical and iconographical images of saints. For example, Saint Roch can always be identified by the dog at his side. According to his story, Roch was stricken with the plague while on a pilgrimage. A dog befriended him, bringing him bread to eat so that he would survive. Thus, his representation usually includes a dog holding a loaf of bread in his mouth. Another saint for whom a specific animal figures prominently is Jerome. He spent thirty years revising the Latin text of the Bible and thus is credited with writing the Vulgate, a version of the Bible still in use. But central to his story and the imagery associated with him is a lion. This lion came to the monastery where Jerome lived, and although all the other brothers fled in fear, Jerome allowed the lion to approach him. The lion had a thorn in his paw, which Jerome proceeded to remove. From that point forward, the lion served as Jerome's constant companion until he died. Scenes such as Albrecht Dürer's *Saint Jerome in His Study* (1514) show the lion curled up at the saint's feet as he works on his translation. In a number of images with gatherings of saints, Jerome can be identified by the lion at his side.

Of all saints pictured with animals, Saint Francis of Assisi (c. 1181/2–1226) is the most prominent. He is one of the most popular saints in Christian history, and his feast day is associated with blessings of pets or other animals. His iconography shows birds, the wolf of Gubbio, rabbits, crickets, and fish. Whether he is preaching to the birds or making peace between the wolf and the residents of Gubbio, the stories of his life as portrayed in various pieces of art almost always include animals.

Animals began to disappear from Christian art during the Renaissance and, following the Protestant Reformation with its focus on the "word," became increasingly rare in Christian sanctuaries. But the centrality of their depiction for over a thousand years secured the place of animals in the visual world of Christianity.

See also

Art—Animals in Art

The Peaceable Kingdom *by Edward Hicks, ca. 1844–45. ©Art Resource, NY.*

Further Resources

Ameisenowa, Z. (1949). Animal-headed gods, evangelists, saints and righteous men. *Journal of the Warburg and Courtauld Institutes, 12,* 21–45.

Baker, S. (2001). *Picturing the beast: Animals, identity, and representation.* Urbana: University of Illinois Press.

Bowran, E., et al. (2006). *Best in show: The dog in art from the Renaissance to today.* New Haven: Yale University Press.

Cohen, S. (1988). Animals in the paintings of Titian, a key to hidden meanings. *Gazette des Beaux-Arts, 132*(1558), 193–212.

Klingender, F. (1971). *Animals in art and thought to the end of the Middle Ages.* London: Routledge & Kegan Paul.

Laura Hobgood-Oster

■ Culture, Religion, and Belief Systems
Christian Saints and Animals

In the history of the Christian tradition, saints play a central role. They are healers, miracle workers, intermediaries between the human and the divine, martyrs, and, most significantly, examples of the way to live a holy life. Saints, whether officially canonized by the Church or not, often serve as the focus of local ritual and tradition. Parts of their bodies fill the reliquaries of European Catholic basilicas because their remains are believed to be efficacious for those close to them. The feast days of individual saints mark the liturgical and popular calendars of believers.

The stories of saints, their hagiographies (the edifying lives of the saints), are central to Christianity. These accounts, preserved in visual, oral, and written form and covering over nineteen centuries, serve a primary pedagogical purpose. Interestingly, though, human saints are not alone in their journeys. Oftentimes, other-than-human animals accompany them, filling various roles and functions. The stories of these animals edify the masses along with the stories of the human saints; indeed, the animals are subjects or actors in their own right, setting examples and proving themselves holy. The animals are central in so many of the stories that, without them, the lives and images of myriad saints would be unrecognizable. A glimpse at a few of these hagiographies, images, and festivals opens the window into the relationships between animals and saints in the history of the Christian tradition. Some of the accounts survive in written texts, others in visual representations that fill sacred spaces throughout Europe, and still others in the enacted rituals that make up feast days. A glimpse into several of these saints' lives reveals the significance of other-than-human animals in their stories.

St. Anthony Abbot (d. 356), later known as Anthony the Great and the "Father" of Western monasticism, is hailed as one of the primary patron saints of animals. Anthony lived in the third century CE and is connected to traditions of blessings of animals in Christianity. Some images of Anthony show him blessing the sick, the poor, and the animals. Early in his solitary life, he learned of another ascetic, Paul the Hermit, who had chosen the way of the desert before Anthony did. On a journey to meet Paul, Anthony wandered lost in the wilderness until, finally, a wolf led him to the hermit's cave. But this was only one animal in the life of Paul. A crow had fed Paul for over sixty years, bearing bread each day, reminiscent of the story of the ravens feeding the prophet Elijah in the Hebrew Bible. During Anthony's visit, the crow brought enough sustenance for two. Anthony returned to visit Paul again, but found him dead. Two lions approached and dug his grave. As a matter of fact, this particular theme is repeated in stories of St. Mary of Egypt, another fourth-century desert ascetic, who was also buried by two friendly lions. Anthony's iconography often includes a pig because, according to some accounts, Anthony healed a pig early in his life. The healed swine then accompanied him for years; thus, Anthony is also the patron saint of swineherds.

Saints Anthony Abbot and Paul the Hermit *by Diego Rodriguez Velazquez, 1642. ©Erich Lessing/Art Resource, NY.*

Central to stories of Christianity and animals is St. Francis of Assisi (d. 1226), patron saint of ecology and of animals. Francis's hagiographies and traditions are among the most popular in the history of Christianity. He was born into a wealthy family, his father a cloth merchant, in the late twelfth century in Assisi, a small hill town in the middle of Italy. A variety of events led him to choose a life of humility, poverty, and preaching. Eventually, a large religious order grew out of his countercultural movement. Among his emphases was the idea that all animals praised God. Hagiographical accounts and visual images of Francis include stories of him preaching to birds, who then take flight in the shape of a cross, sing loudly, and continue praising God. He called the birds his "little sisters." As a matter of fact, Anthony of Padua, a second-generation Franciscan, is credited with preaching to fish as well. But birds were not the only pious creatures that the saint praised. After a late-night snowfall, Francis woke up early to begin his daily prayers, but the rest of his brothers continued to sleep, avoiding the cold. Francis, however, was not alone. A cricket followed him and joined him in the morning liturgy, leaving a cricket path in the snow as he walked. When the other brothers finally woke and came to the chapel, Francis pointed out the cricket prints and proclaimed this little creature more devoted than the human monks.

In addition to examining their acts of piety, Francis advocated a peaceful relationship between humans and other animals. The most famous example is the incident with the wolf of Gubbio. In this small town close to Assisi, a wolf was attacking some of the people, and they pleaded to Francis for help. Rather than kill the wolf, Francis spoke to the canine and tamed him so that he would no longer kill and so that the people would not kill him. The saint then asked the townspeople to feed their resident wolf from that point forward so that he would not need to hunt anymore. They lived in peace from that point forward.

Many more stories are recounted in works such as *The Little Flowers of St. Francis*. By the late twentieth century, blessings of animals were celebrated at numerous churches in North America and Europe. These blessings are usually held in connection with the Feast of St. Francis, October 4. More than any other saint, Francis of Assisi makes the direct connection between humans and other animals and recognizes the purpose and value of other-than-human species.

In addition to serving humans, as in the case of Paul the Hermit's crow, and to living a life of praise and piety, as did Francis's cricket, animals are portrayed as important companions to saints. St. Jerome, a doctor of the church, is known primarily for translating the Bible from Hebrew into Latin, a version known as the Vulgate. But Jerome had a constant companion while undertaking this monumental task—a lion. As the story goes, a lion wandered into the courtyard of the monastery close to Bethlehem where Jerome was living and working. All the other monks fled in fear when they saw this massive animal approach. But the saint noticed that the lion was limping and offered both hospitality and healing. The lion remained with Jerome for the rest of his life, eventually dying from grief after his human companion died.

Another group of images includes animals as mouthpieces for the divine or representatives of God. St. Brigit, an Irish saint and early leader of the Celtic church, is connected to a variety of animals. But some of the most intriguing stories include her compassion toward stray dogs. While she was cooking a meat stew, with just enough meat for the men who had joined her father to eat, a dog came to her. She fed the dog a portion of the meat, so there was not enough left for the human guests. But miraculously, the meat in the stew multiplied, and there was plenty. A reading of the story suggests that the dog might be representative of an angel or of the divine coming in the form of the most impoverished, thus testing Brigit's compassion.

In another story of animals and divine revelation, St. Eustace, a Roman soldier who was pursuing Christians when the religion was still illegal, was out hunting one day when a huge stag appeared. After a chase, the stag turned to face him, and Eustace saw a crucifix between his antlers. Then the stag began to speak, pleading with Eustace to convert and to stop killing Christians (the story might also imply that he should stop hunting other animals). Eustace converted to Christianity and later died a martyr. Interestingly, he became the patron saint of hunters.

As mentioned previously, animals appear on the feast days of many saints as well, such as St. Zopito in Italy. On his feast day, the Monday after Pentecost, a procession recalls events surrounding the transfer of his relics from Rome to Loreto Aprutino. On that day, a farmer continued to plow his fields rather than bow before the relics as priests carried them past. But his oxen took note of the martyr's relics, stopped plowing, and knelt in pious observance. So each year, a large white ox wearing a colorful headdress and carrying a young girl on his back moves through the streets of the city. Numerous religious symbols process in front of him. But when the relics of the saints pass by, the ox kneels again. At that point, the entire assembly follows the ox and the remains of the saint back to the church. So although symbolic or storied animals often appear in the hagiographical accounts, real animals are still connected directly to saints in some of their rituals.

Finally, animals give visual clues to the identity of a number of saints. St. Clare of Assisi is frequently pictured with a cat (a common companion for female solitaries), St. Roch with a dog (both are seen as healers in this relationship), St. Giles with his companion hind whom he saved from a hunter, and St. Martin de Porres with dogs and cats (for whom he established a shelter in Lima, Peru). The list is lengthy.

Clearly, animals are closely connected to many saints in the history of the Christian tradition. These animals help identify the saints, provide stories of humility and piety from which others can learn, and exemplify what is often the ideal in Christian life.

Further Resources

Armstrong, E. A. (1973). *Saint Francis: Nature mystic, the derivation and significance of the nature stories in the Franciscan legend.* Berkeley: University of California Press.
Duchet-Suchaux, G., & Pastoureau, M. (1994). *The Bible and the saints: Flammarion iconographic guides.* Paris: Flammarion.
Heffernan, T. J. (1988). *Sacred biography: Saints and their biographers in the Middle Ages.* New York: Oxford University Press.
Hobgood-Oster, L. (2007). *Holy dogs and asses: Animals in Christianity.* Urbana: University of Illinois Press.
Kieckhefer, R., & Bond, G. D. (1988). *Sainthood: Its manifestations in world religions.* Berkeley: University of California Press.
McKinley, P. (1969). *Saint-watching.* New York: Viking Press.
Voragine, J. (1993). *The golden legend: Readings on the saints* (Vols. 1–2, W. G. Ryan, Trans.). Princeton: Princeton University Press.

Laura Hobgood-Oster

■ Culture, Religion, and Belief Systems
Christian Theology, Ethics, and Animals

Christianity is a diverse religious tradition, with many historical and regional, as well as denominational, variations in practice and belief. This entry does not pretend to exhaust Christian thinking about nonhuman animals but rather offers a survey of different attitudes and ideas, highlighting some of the most important themes that have emerged in 2,000 years of Christian thinking about the created world.

As an emerging system of practice and belief, early Christianity drew on Jewish and Greek roots. From Judaism, early Christians inherited biblical narratives of human origins as the source of claims about both humans and other animals. In the first biblical account of the creation, in Genesis 1:26–28, humans alone are created in God's likeness and given dominion over the rest of creation:

> 26. And God said, "Let us make man in our image, after our likeness, and let them have dominion over the fish of the sea, and over the birds of the air, and over the cattle, and over all the earth, and over every creeping thing that creeps upon the earth."

> 27. So God created man in His own image, in the image of God he created him; male and female he created them.

> 28. And God blessed them, and God said to them, "Be fruitful, and multiply, and fill the earth and subdue it; and have dominion over the fish of the sea and over the birds of the air and over every living thing that moves upon the earth."

Early Christians brought the notion of human superiority over other animals suggested in Genesis 1 into dialogue with Greek philosophy, and especially with its tendency to devalue the physical body, along with nonhuman animals. Among the best-known versions of this approach in the first few centuries CE were Gnosticism and Manicheanism, both of which viewed nonhuman creation, including the Earth and other creatures, as evil. Mainstream Christians came to reject this view and to affirm the overall goodness of creation, as an act of a benevolent and all-powerful God.

This view is expressed most powerfully in the writings of Augustine (354–430 CE), who struggled to balance his conviction that humans' ultimate good lay in heaven, on the one hand, with his belief that God's creation was essentially good, on the other. Augustine was a Manichean prior to his conversion to Christianity, but he rejected the notion that the earth was essentially fallen and argued that sin stems from failures of will and not from the mere fact of physical embodiment. All creation, including the human body, is a revelation of God's goodness, Augustine wrote in his *Confessions,* because God created "the earth which I walk on" as well as the human body, the "earth which I carry." However, Augustine still insisted that this earth is ultimately insignificant in relation to the eternal good offered in Heaven. Ultimately, Augustine reinforces the idea that what really gives value to humans is their link to the divine, which is not shared by other animals.

This tension stands at the center of Protestant ideas about human nature, human relations to God, and the value of the rest of creation, including nonhuman animals. For Martin Luther, this tension meant that humans have, as he wrote in "The Freedom of a Christian," "a twofold nature, a spiritual and a bodily one." The first, spiritual nature is owed to God, and the second, bodily nature is a result of humans' embodiment in a flawed material world. The two natures never harmonize completely, and when they come into conflict, duties to God must come first. Ultimately, humans' spiritual and material natures remain separate and unequal. Although Luther and other early Protestants did not write explicitly about nonhuman animals, their theologies reflect a hierarchical view of creation in which humans, while deeply flawed, rank much higher than other animals. This reflects, of course, the generally utilitarian view toward nonhuman animals in medieval and Reformation-era Europe.

More recent Protestant thinkers have written more directly about animals and the relations between human and nonhuman natures. In his influential book *The Nature and*

Destiny of Man, Reinhold Niebuhr asserted that human life is distinguished from animal life by the former's "qualified participation in creation. Within limits it breaks the forms of nature and creates new configurations of vitality." Humans, for Niebuhr, are not entirely subject to their "creatureliness" as are other animals. People are unique in that despite their limitations, they share something of God's creativity, unlike any other creature.

Niebuhr provides a good example of major Protestant theological and ethical assumptions applied to nonhuman animals. On the one hand, the creation, including nonhuman creatures, must be good because the creator God is good. On the other hand, all of creation is deeply flawed and fallible. Within this creation, humans have a unique position because of their possession of an eternal soul, which connects them to the divine in a way not available to other animals.

The Roman Catholic tradition has generally viewed the created world much more positively and as more closely connected to the divine. This Catholic position was systematized in the work of Thomas Aquinas (1225–74). Thomas wrote in the context of a revived interest, in the Middle Ages, in the notion of a "Great Chain of Being," which joined all creatures in a harmonious hierarchy. Thomas Aquinas summarized the harmonious relationship between God and creation and among both human and nonhuman elements of that creation in his notion of natural law. "The whole community of the universe," Thomas proclaimed in his *Summa Theologica,* "is governed by the divine reason." For Thomas, "everything that in any way is, is from God." And because "all things partake in some way in the eternal law," all aspects of creation are linked together. Natural law is the eternal law imprinted on creatures, especially in rational creatures (humans and angels), but also in "dumb creation," which sits lower in the hierarchy than humans but still reflects the overall goodness and harmony of all of God's creation. Thus, natural law mixes harmony and hierarchy. As Thomas wrote, "In natural things, species seem to be arranged in a hierarchy: as the mixed things are more perfect than the elements, and plants than minerals, and animals than plants, and men more than other animals." Lower creatures can approach divine goodness only through their relationship to higher ones, and humans are superior to and dominant over other animals. According to Thomas, "the subjection of other animals to man is natural." Despite Thomas's conviction of the fundamental goodness of creation, he reinforces a hierarchical and human-centered ethic.

An alternative view comes from the life and writings of Saint Francis of Assisi (1182–1226 CE). Popular stories about Francis emphasize his love for nonhuman animals, as in the story of the wolf of Gubbio, whom Francis educates through kindness. In written works such as the "Little Flowers," Francis speaks of animals as brothers and sisters, and emphasizes that God loves and cares for all of creation. Francis thus expresses a perspective in which nonhuman animals are not valued only instrumentally—because they are necessary and good for human ends—but also intrinsically. Animals receive care and blessings from God, who values them in and of themselves. In turn, animals are grateful to God, suggesting that they have a greater degree of agency than is usually allowed.

Francis, however, is an exception in Roman Catholic thinking about nonhuman animals. Generally, Catholic theologians, like Protestants, affirm the superiority of humans over other species, while also insisting on the goodness of creation and human embodiment. The Catholic bishops who gathered for the Second Vatican Council (1962–65), for example, asserted that "man is the only creature on earth that God willed for itself"—that is, the only creature that is an end in itself rather than a means for others (*Gaudium et Spes*). Further, God made man "master of all earthly creatures that he might subdue them and use them to Christ's glory." In 1980, Pope John Paul II echoed these themes in his important encyclical *Redemptor Hominis,* a systematic treatment of the Roman Catholic vision of human nature.

John Paul celebrates the "unrepeatable reality" of each person, chosen by God for "grace and glory" and also given the earth and nonhuman creatures to subdue and dominate.

Roman Catholicism continues to be deeply humanistic, meaning that it affirms human uniqueness and the right of humans to dominate the rest of creation. However, this affirmation is made in the context of a cosmic hierarchy that assigns a purpose and value to all of creation, including nonhuman animals. This comes from the assumption, inherent in natural law thinking, that there is continuity among different parts of creation and between creation and the Creator. This contrasts sharply with the dominant Protestant idea that there is a radical break between creation and the creator and between body and spirit. For some Christian eco-theologians and animal advocates, then, natural law provides a strong foundation for asserting the value of nonhuman animals within God's creation.

Despite humanistic tendencies, one example demonstrates recent attempts to move beyond anthropocentrism. In 1999 the Roman Catholic Bishops of North America's Pacific Northwest convened to write the "Columbia River Watershed: Caring for Creation and the Common Good." What stands out as unique in this project is the attempt to extend the circle of concern beyond the common good of humans and to that of nonhuman nature, or creation. In many ways, this document builds on the Roman Catholic natural law tradition, stressing the relationship of the entire created community to that of the governance of the creator. Although much of the document deals with issues beyond animals, it does pay attention to the necessary relations of animals to humans in the health of the watershed or ecosystem, most notably in the case of salmon. More importantly, however, this document demonstrates gradual steps toward inclusion of nonhuman animals in the sphere of ethical concern within mainstream Christianity.

Most latent in Catholic writings on environmental and animal ethics is the notion of theocentrism, which makes God the ultimate center of reference when relating to the world. From a theocentric perspective, animals may be equally imbued with divinity as the rest of creation. Therefore, one must relate to animal as one would want to relate to God, given that God is the ultimate measure to whom all of creation is tethered. Through right relations to animals and creation, the Christian is in turn relating to God.

In recent years, a number of contemporary Christian theologians and ethicists have undertaken explicit and detailed examinations of their tradition's perspectives on nonhuman nature, including nonhuman animals. Environmental approaches within Christian thinking, often short-handed as "eco-theology," encompass a variety of models, including those that emphasize human stewardship of nature, on the one hand, and others that take a more egalitarian, biocentric stance, on the other. Although Christian writing on nonhuman animals is often grouped under the heading of eco-theology, not infrequently, tensions arise between ecological concerns and animal rights and welfare issues. Disagreements usually center on the tensions between the ecological focus on the well-being of wholes and the "animal rights" focus on concern for individual creatures. Despite the tensions, much of the thinking about nonhuman animals within Christianity is strongly shaped by eco-theological approaches. Another important influence has been secular philosophical reflection about animals, including Peter Singer's utilitarian model and Tom Regan's right-based approach, among others.

Drawing on these diverse sources, Andrew Linzey has articulated the best-known Christian theological analysis of nonhuman animals. In his book *Animal Theology*, Linzey argues that Christian theology can provide a way out of the "philosophical straitjacket" of contemporary animal rights philosophy, particularly the tensions between the approaches of Regan and Singer. Drawing on the Trinitarian doctrine of incarnation, he argues that human dominion over animals needs "to take as its model the Christ-given paradigm of lordship manifest in service" (1994, viii). Linzey proposes a "Christo-centric" animal rights theology, which gives moral priority to the weak. According to Linzey, "The weak and the oppressed constitute a special moral category based on our special relationship of power over them. I propose that animals constitute a special category of moral obligation, a category to which the best, perhaps only, analogy is that of parental obligations to children" (36). By stressing animals as "the weak," requiring moral obligation, Linzey proposes a position that works against the human uniqueness that has characterized most of Christian theology.

Although he strives to combat perceptions of human uniqueness that lead to domination over nature, Linzey does admit a degree of human elevation over the rest of creation. However, this position obligates one to care for creation as steward rather than tyrant. This privileged position, he maintains, obligates humans to be sensitive to the suffering of others, both human and nonhuman. Only once one enters into sensitivity to the suffering of creation, Linzey argues, can one claim to have entered into a full understanding of the priestly nature of Christ and the stewardly obligations of humanity.

Another contemporary Christian theologian who has focused on nonhuman animals is Stephen Webb, who both builds on and critiques Linzey's work. In his 1998 book *On God and Dogs,* Webb argues that ultimately, Linzey focuses too much on rights and autonomy; instead, Christian attitudes toward animals should focus on relationships and connection. In more recent work, Webb has elaborated practical consequences of proper Christian attitudes toward nonhuman animals and nature more generally, especially in relation to food production and consumption.

Other Christians have spearheaded similar practical projects, including vegetarian advocacy and animal rescue and rehabilitation groups, among others. Some of these projects find common ground with Christian eco-theological approaches more broadly, whereas others tread a more singular path. All share a common assertion, however, that Christian values include an appreciation for and proper treatment of nonhuman animals.

See also

Culture, Religion, and Belief Systems—Blessing of the Animals Rituals
Culture, Religion, and Belief Systems—Christian Art and Animals
Culture, Religion, and Belief Systems—Christian Saints and Animals
Culture, Religion, and Belief Systems—Francis of Assisi
Culture, Religion, and Belief Systems—Noah's Ark
Culture, Religion, and Belief Systems—Religion and Animals
Culture, Religion, and Belief Systems—Religion and Human-Animal Bonds

Further Resources

Bishops of the Pacific Northwest. (1999). The Columbia River watershed: Caring for creation and the common good. [Pastoral Letter].
Francis of Assisi. (1998). *The little flowers of Saint Francis.* New York: Random House.
Linzey, A. (1984). *Christianity and the rights of animals.* New York: Crossroad.
———. (1994). *Animal theology.* Champaign: University of Illinois Press.
———. (1999). *Animal gospel.* Louisville: John Knox Press.
Northcott, M. (1996). *The environment and Christian ethics.* Cambridge: Cambridge University Press.
Santmire, P. (1985). *The travail of nature.* Minneapolis: Ausburg Fortress Press.
Webb, S. (1998). *On God and dogs: A theology of compassion for animals.* New York: Oxford.
———. (2001). *Good eating: The Christian practice of everyday life.* Grand Rapids: Brazos Press.

Anna Peterson and Samuel Snyder

Global Warming and Animals

As temperatures rise annually, cranes return to the northern United States earlier and earlier each spring.

Chart by Terry Root. Crane image Courtesy of Shutterstock.

Global Warming and Animals

Some species in danger of extinction caused by global warming are those that cannot move to a different location as the temperature increases. For example, the quite sedentary Mallee emu-wren (Stipiturus malle) has a small range, threatened by fires. It cannot move until its habitat moves, unless humans intervene, which would require a difficult, costly, and complex process. The bird will probably become extinct within the next 25–50 years.

Courtesy of Shutterstock.

▦ Conservation and Environment

The Passenger Pigeon

As many as 3–5 million passenger pigeons populated eastern North America before the arrival of Europeans in the 1600s. By 1914, hunting and deforestation had driven them to extinction.
©Mary Evans Picture Library/Alamy.

Whale Watching

Public fascination with marine life prompted by the rise of the environmental movement has spawned a lucrative whale-watching industry. Although the experience may increase the environmental awareness of whale watchers, frequent encounters with tourists can be highly disturbing or potentially harmful to whales.
Courtesy of Fabian Ritter.

Wolf and Human Conflicts: A Long, Bad History

Wolves prey on large ungulates, such as deer and cattle, and this has created competition with humans, which has sometimes led to wolf extermination campaigns.
Courtesy of the National Park Service

Animal Symbolism in Native American Cultures

The Bear Dance, *from George Catlin,* North American Indian Portfolio, *1844. Some tribes consider bears the animals most similar to humans, as both can walk upright and possess an inherent element of ferocity in their nature.*

©*The Art Archive.*

Animism

A brightly painted shaman's mask modeled after a peacock. Modern theories contend that animist cultures do not worship animals; rather, they acknowledge nature as an integral part of their community. Shamans act as mediators between humans and the nonhuman beings with which they seek to communicate.

Courtesy of Shutterstock.

■ Culture, Religion, and Belief Systems

Dolphin Mythology

A bottlenose dolphin and her calf off the coast of Bermuda. Because of their cooperative, communal life-style and occasional friendliness to humans, some New Age groups have come to view dolphins as a model of perfect love and moral superiority.

©Blaircwh/Dreamstime.com.

India and the Elephant

At many temples in southern India tuskers are donated by wealthy devotees and during festivals these elephants are painted and dressed up with ornaments to lead processions. The size and majesty of an elephant, as well as its companionable nature, inspires both awe and devotion.

Courtesy of Shuttestock.

■ **Culture, Religion, and Belief Systems**

India's Holy Cows and India's Food: Issues and Possibilities

The role of cattle in India has become the subject of intense debate, as wealthy farmers expand the size of their herds to increase milk production. While the cow's religious role as the "mother provider" has earned it sacred status, the growing herds have hurt the poorer population by diverting grain and farmland to feed livestock. And though protected by the tenet of ahisma, or nonviolence, which prohibits the harming of animals, in arid regions, where water and fodder are scarce, poorly nourished cattle are weak and emaciated.

Courtesy of Michael W. Fox.

Islam's Animal Fables: *Kalila and Dimna*

A page from the fables Kalila and Dimna in which the jackal tries to persuade the lion to stop devouring the beasts and devote himself to pious acts.

©Werner Forman/Art Resource, NY.

Monkeys and Humans in India

An Indian woman prays near the statue of Hindu monkey god Hanuman at the Basistha temple in Guwahati, 2004. Monkeys lurking at the ancient Kamakhya temple, in the northeast Indian state of Assam, have attacked up to 300 children over three weeks.

©AP Photo/Anupam Nath.

Noah's Ark

Mosaic of birds entering Noah's ark. The story of saving animals from a giant flood is common to several Mesopotamian civilizations.

©Scala/Art Resource, NY.

Religion and Animals

Hindu scripture states that all living beings, including animals, have souls. As fellow divine, though inferior, beings, animals are worthy of ethical treatment. Here, even monkeys are allowed to sit unmolested in an Indian temple.

©Zeddez1975/Dreamstime.com.

Culture, Religion, and Belief Systems

Shamanism

Bolivian shaman Carmen Acho performs a traditional purification ritual holding a quirquin-cho over Juana Ita Gomez, 2006. According to the shaman, the ritual cleans the participant from evil spirits and brings good luck for the new year.

©AP Photo/Karel Navarro.

Snakes as Cultural and Religious Symbols

Buddhist mythical figure of Naga guarding the entrance to a Temple in Chaing Mai, Thailand. Traditionally, Nagas are a race of supernatural beings sharing both snake and human attributes.

©Petdcat/Dreamstime.com.

Totemism

Two Tinglit totem poles overlooking Vancouver, Canada. An eagle sits atop one. Totem animals are linked to social groups, who look to the animal spirit for aid and identify themselves with certain qualities of that animal.

Courtesy of Shutterstock.

Enrichment for Animals

Animal Boredom

An orangutan sits behind a cage in Bangkok, Thailand, 2006. Boredom is a serious issue for animals in captivity and affects their well-being significantly.

©Pornchaikittiwongsakul/AFP/Getty Images.

■ Culture, Religion, and Belief Systems
Dolphin Mythology

"Diviner than the Dolphin is nothing yet created." Oppian of Silicia, Greek Poet 200 CE

No other animal has attained the mythological status of the dolphin. Dolphins and other cetaceans (whales and porpoises) populate many of the earliest records of ancient civilization and presently enjoy an iconic status as the ultimate "New Age" animal.

The human affinity for dolphins probably results from a combination of characteristics of the animal. Much of the fascination comes from our recognition that these are highly intelligent and curious beings living in a strange hidden environment. The dolphin "smile," which is not a smile but rather just the physical configuration of its jaw, has intrigued many and has likely led to the notion that dolphins are friendly benign creatures.

There are three prominent themes in dolphin mythology from ancient times to the present. All of these themes are bound together by the common belief that dolphins occupy a special elevated status in nature. These three themes, or propositions, are as follows:

1. Dolphins and humans have a special kinship.
2. Dolphins are higher minded than and morally superior to humans.
3. Dolphins have special powers that can be used for healing.

Theme 1: Dolphins and Humans Have a Special Kinship

One of the most robust themes in dolphin mythology is the idea that humans and dolphins share an intimate and unique bond. This concept takes many forms, including the belief that humans share a common ancestor with dolphins, which is based on the motif of human-dolphin and dolphin-human transformations. For instance, many stories of common ancestry can be found in ancient mythology from Australia, where the Noonuccal tribe of Minjerribah (Stradbroke Island, Australia), traditionally fished for mullet with the help of dolphins, who "rounded up" fish for the men to catch in exchange for some of their own. The members of this tribe believed that the dolphins were so helpful because humans and dolphins shared a common ancestor in the past, namely the cultural hero Gowonda, who was transformed into a dolphin and thereafter helped the members of his human tribe to catch fish. All Australian tribes hold similar beliefs that they are direct descendants of dolphins and practice a deep meditation called Dolphin Dreamtime, a spiritual communion with their dolphin ancestor that provides answers to their tribal problems.

In Papua New Guinea, there is the tale of the Dudugera. This story is not one of transformation but rather human-dolphin consort—another common variant of the human-dolphin bond theme. The story concerns a child who was conceived by a woman who was swimming playfully in the sea one day when a god, disguised as a dolphin, appeared and swam through her legs, brushing against her skin and magically impregnating her. When the child was born, he was given the name Dudugera, or "leg child," to denote the unusual way in which he had been conceived, and for this he was mocked as he grew older. In his anger, Dudugera soared into the sky and began hurling fire toward the earth—thus becoming the sun. His mother tried to soften his anger by creating the first clouds to obscure the sun.

In Cambodia and Laos, where the dolphin is historically revered, the Me Kong river dolphin (also known as the Irrawaddy) is thought to be either a young pregnant girl who drowned in a river and was reborn as a dolphin or, in Laos, the reborn male companion of a young woman drowned in a river. In North America, native tribes such as the Tlingit, Nootka, and Haida view the killer whale (orca) as an ancestral spirit, and the Inuits of the arctic tell a tale of a young woman named Sedna, Inuit goddess of the sea, who was thrown overboard a ship and tried to hold on with her fingers. The men on the ship cut her fingers off and each digit fell into the water and became the first dolphins and whales.

In South America, thousands of tales are told about the native Indian legend of the pink river dolphin, or Boto. The Boto was thought to take human form in order to tempt men and woman into their underwater world from which there was no escape. Interestingly, across the world, similar stories exist about the Baiji, the Chinese river dolphin.

Perhaps the best-known examples of ancient dolphin mythology come from Greco-Roman civilization. Aristotle was the first to recognize that dolphins are mammals. Indeed, the root of the word dolphin, delphis, means womb. In ancient Greece, Rome, and Mesopotamia, dolphins adorned frescoes, artwork, jewelry, and coins, and the killing of a dolphin was punishable by death. The important Minoan palace of Knossos from 7000–3000 BCE features one of the earliest and best-known ornamentations in a dolphin fresco on the wall of the queen's bathroom. Dolphins were closely linked with the gods in ancient Greece. The myth of Dionysus from the sixth to seventh century BCE provides one of the earliest accounts of the special transformational relationship that humans share with dolphins. Dionysus, the god of wine and ecstasy, was captured by pirates and forced aboard ship. When he used his powers to throw the pirates overboard, Dionysus showed compassion and instead of drowning the pirates, changed them into dolphins so that they could live out their lives in harmony. The Greek sun god Apollo often exchanged his god-like status to assume dolphin form. He founded the Oracle at Delphi on the slopes of Mt. Parnassus, where a respected goddess blessed travelers from far and wide.

As part of their sacred status, dolphins were also seen as messengers or mediators between two worlds—that of the god and that of the mortals—or even between life and death. In ancient Syria, the dolphin was associated with the goddess Atargatis, the nourisher of life and receiver of the dead who would be born again. In later myths, particularly from Rome, the dolphin carries virtuous souls to the "Islands of the Blest," and around the Black Sea, images of dolphins have been found in the hands of the dead, perhaps to ensure their safe passage to the afterlife. The ancient South Pacific Maori viewed dolphins as messengers of the gods, and in many ancient Greek myths, the role of the dolphin as messenger is central. According to the well-known tale, when Poseidon (the god of the sea) decided on Amphitrite as a wife, he sent a dolphin messenger to persuade her to marry him. Amphitrite (who was originally reluctant) was so taken with the dolphin's charms that she agreed. Poseidon showed his gratitude by placing the dolphin among the stars as the constellation Delphinus.

Clearly the ubiquity of the human-dolphin and dolphin-human transformation theme across the world shows the importance of this concept to humankind as a way to explain our beginnings, as well as a way to find redemption in the form of rebirth as a revered life form.

Theme 2: Dolphins Are Higher Minded Than and Morally Superior to Humans

Another important mythological trait attributed to dolphins is that they are intellectually and morally superior to humans. Dolphins are considered completely altruistic, peace-loving, and wise. The essence of these tales is that dolphins are all that humans generally *are not,* and they provide a model for us to aspire to. The notion that dolphins are strongly motivated to help people is reflected in the worldwide recurrence of tales about dolphins cooperating with humans in various occupational activities and saving them from drowning or shipwreck. Tales of dolphins cooperating with fisherman have been, not surprisingly, ubiquitous among seafaring cultures for thousands of years. As mentioned previously, Australian aboriginal tribal myths abound of dolphins helping fishermen to catch their quarry. These ancient tales are given credibility by numerous documented examples of modern cooperative fishing between dolphins and humans around the world. For instance, in the Irrawaddy River of Myanmar in Southeast Asia, the Irrawaddy dolphins have been cooperating with fishermen for generations by herding mullets into their nets and obtaining stray fish in return. A very similar mutualistic form of fishing, this time with bottlenose dolphins, is practiced on the Mauritanian coast of Northern Africa. These kinds of activities are very similar to the kind of cooperative feeding behaviors observed within dolphin populations themselves and also across dolphin species.

The earliest account of a dolphin saving the life of a human comes from the ancient Greek tale of the poet and musician Arion from 600 BCE, who was rescued from drowning by being carried on the back of a dolphin to safety. The motif of the dolphin carrying a human on its back is found throughout the world. In ancient Greece, Poseidon's son, Taras, was said to be rescued from a shipwreck by a dolphin. And again in ancient Greece, there is the well-known tale of a young boy who befriended a wild dolphin named Simo, who carried the boy on his back to school every day. When the boy fell ill and died, the dolphin beached himself and died also—presumably from a broken heart. In Italy there is a very similar tale. These stories have been depicted time and time again in various forms of ancient artwork, including the famous "Boy on a Dolphin" coin from third-century BCE southern Italy.

In modern times, documented stories of dolphins protecting humans from sharks or guiding ships across treacherous waters are numerous. Cases of "wild solitary sociable dolphins"—that is, wild dolphins that choose to stay in close contact with humans—also abound and support the notion that dolphins can be motivated to befriend humans. One famous example is JoJo, a wild Atlantic bottlenose dolphin from the Turks and Caicos Islands who has sustained a close relationship over the past fifteen years with island inhabitant Dean Bernal, who is now JoJo's official caretaker. Bernal has claimed that JoJo protected him once when a shark threatened. All told, there are about seventy documented wild solitary sociable dolphins around the world.

The ancient notion of dolphins as peaceful beings living in a morally superior society to humans persists to this day in the New Age movement. Dolphins have become the ultimate New Age symbol. To New Agers, dolphins are a model for unconditional love and, just as in ancient times, represent a path toward transformation and entrance into a utopian society. Dolphins are omnipresent in New Age music, Web sites, magazines, books, and stores. In the most extreme New Age assertions, dolphins are not just helpers, but saviors of humankind, and the devotion to dolphins becomes one of spiritual worship. Jeff Weir, Director of the Dolphin Research Institute in Australia, aptly refers to these extreme spiritual beliefs as "Dolphinism."

The person perhaps most responsible for these New Age notions of dolphins is Dr. John Lilly, who, in the 1960s, conducted research with captive dolphins. Although Lilly's early scientific work on dolphin brains and behavior was groundbreaking and important, he and his followers eventually began to mix science with spirituality and claimed that dolphins were intellectually and ethically superior to humans. Lilly also attempted to set up a formal program of interspecies communication and cooperation between humans and dolphins called the Cetacean Nation.

The mythological status of the completely benign dolphin has been challenged in recent years by documented reports of dolphins and whales killing fellow marine mammals for no apparent survival purposes. Although it is no surprise to scientists that cetaceans are complex and sometimes aggressive creatures, New Agers have largely ignored these observations.

Theme 3: Dolphins Have Special Powers That Can Be Used for Healing

The myth of dolphins as lifesavers of humans is connected with notions that dolphins have magical powers that can be used for healing. The ancient Celts attributed special healing powers to dolphins, as did the Norse. In aboriginal Australian tribes, humans could commune with ancient dolphin ancestors telepathically. Throughout time, people as far apart as Brazil and Fiji have traded in dolphin and whale body parts for medicinal and totemic purposes. Clearly, the notion that dolphins have special powers that could be applied to heal the sick is ancient. In modern times, this notion has been invigorated by the New Age movement and alternative medicine circles. An entire worldwide commercial industry in "Dolphin-Assisted Therapy" (DAT) has been founded on the idea that dolphins have the power to heal by touch, sonar, or some other channel. In DAT, sick, injured, or mentally ill people swim with captive dolphins under the guise that there will be a therapeutic effect (treatment or cure) from this close interaction with a dolphin. Needless to say, there is no scientific support for these outlandish claims, but many hopeful people continue to pay for these expensive DAT programs and similar "swim-with-the-dolphin" programs in the hope that the experience will, in some way, be curative or transformative. DAT and other swim-with-the-dolphin programs encourage the notion that these activities are research-based when, in fact, they are little more than public entertainment.

What Are the Consequences of Modern Dolphin Mythology for Human-Dolphin Interaction and Welfare?

Unfortunately, the human affinity for dolphins and other cetaceans has not immunized them against exploitation by many cultures. Some cultures, such as those in North America and Alaska, the Japanese, and the Norwegians, have relied on dolphins and whales for sustenance in the past. The relationship between these cultures and cetaceans, therefore, has been a highly complex one of venerating an animal that is a source of life sustainment much in the same way that Native Americans have revered the buffalo. But in the present, there are no cultures that depend on the killing of dolphins or whales for survival, and yet thousands of these animals are slaughtered all over the world for profit. Also, local populations of many species are at risk from human activities such as fishing, hunting, chemical pollution, habitat degradation, and sound pollution. For instance, the Irrawaddy, the revered Mekong dolphin, is currently one of the most severely endangered cetacean species in the world because of river traffic, bycatch by fisherman, and other human activities.

Ironically, New Age beliefs and the desire for close human-dolphin interaction have led to practices that ultimately harm dolphins. Public entertainment, DAT, and "swim-with-the-dolphin" programs promote dolphin captivity and the deprivations that are associated with that industry—not to mention the real danger to people who swim with these large animals. In order for people to indulge in such New Age activities, dolphins are required to be continually present and responsive, that is, conveniently incarcerated.

The irony of viewing dolphins as highly intelligent and socially complex beings while subjecting them to the "convenience" of captivity is lost on many people who engage in these activities. Although the more benign ecotourism industry has provided a way to view dolphins and other cetaceans in their own habitat, it is still important to assess and monitor the effects of this practice.

> The human fascination with dolphins has led us through a long history of imputing dolphins with supernatural qualities, such as deification, telepathy, and healing powers. However, modern science has confirmed that dolphins, and other cetaceans—while not magical—are indeed highly intelligent, communicative, socially complex, self-aware animals. These real qualities are, in the end, immensely more intriguing than any traits that humans might imagine and project onto these creatures.

Further Resources

Alpers, A. (1961). *Dolphins: The myth and the mammal.* Boston: Houghton-Mifflin.
Malins, D. (1986). The dolphin is not like a human with flippers. *Cetus, 6*(1), 5–8.
Marino, L., & Lilienfeld, S. (1998). Dolphin-assisted therapy: Flawed data, flawed conclusions. *Anthrozöos, 11*(4), 194–199.

Lori Marino

■ Culture, Religion, and Belief Systems
Fish Fighting in Southeast Asia

For centuries, humans have engaged in the sport of fighting animals, though many do not consider it a sport (acknowledging that not all readers will agree on the use of the word "sport" in this context). The fighting of beetles, crickets, roosters (cocks), dogs, and bulls are all popular in Asia, particularly in Southeast Asia, where many of these sports originated. One practice that holds particular importance for the people of Southeast Asia is fish fighting. The Siamese fighting fish (*Betta splendens*) is the best-known species of fighting fish, though the fighting of other species (*Betta smaragdina*, emerald betta; *Trichopsis vittatus*, croaking gourami; *Aploceilus panchax*, blue panchax; *Dermogenys pusillus*, pygmy halfbeak) occurs on a smaller scale. The fighting of *Betta splendens* is particularly popular in Thailand (formerly Siam), and thus, we focus our discussion on this species in this country.

It is believed that betta have been raised by humans for the purposes of sport fighting for more than 650 years, beginning with King Lithai Tammaraja I in the Sukhothai period. It is unclear which species—*Betta splendens, Betta smaragdina,* or *Betta imbellis*—was first domesticated for fighting because many people use the term "plakat" or "biting fish" to refer to all three of these species. Few written records regarding fighting fish exist before the records of the nineteenth century. In 1840, King Rama III of Siam (brother of Rama IV, immortalized in *Anna and the King* and *The King and I*) gave away some of his fighting fish, which found their way into the hands of a Danish physician named Theodore Edward Cantor. Cantor later assigned the fish the incorrect scientific name of *Macropodus pugnax,* but it was renamed by the British ichthyologist Charles Tate Regan in 1910 as *Betta splendens,* or the "beautiful warrior."

Male Betta splendens *engaged in combat. Courtesy of Cheung Kwok.*

Betta are small fish (maximum total length of 6.5 cm) that prefer warm, shallow, and still or slow-moving waters. They are native to Thailand and perhaps parts of Vietnam, Laos, and Cambodia. Thanks to movement by humans and their ability to adapt to drainage ditches, canals, and rice paddies, their range has expanded to include Malaysia and Singapore. Feral populations of *Betta splendens* are known to exist in Brazil, Colombia, and the Dominican Republic. The former population was established as part of mosquito-control efforts (betta are predators on aquatic insect larvae), whereas the latter population was likely established by fish that escaped from a tropical fish farm in 1979 during Hurricane David. Occasionally, betta populations become established in the United States (primarily in Florida) by fish that have escaped from tropical fish farms or have been released by hobbyists.

Male betta are highly territorial toward other males; in the wild, they defend an area of approximately 1.5 square meters against all intruders. In this territory, they construct floating nests of bubbles made from salivary secretions, into which they place eggs released by the female during spawning. Males are solely responsible for guarding the eggs, which helps explain their pugnacious behavior. Upon seeing an intruder, males erect their dorsal fins and their gill covers (opercula); the latter display increases the apparent size of the head. Competing males approach each other and their chromatophores (pigment cells) expand, causing their coloration to intensify. Fighting usually escalates with a series of lateral (side-by-side) displays that include tail beating and attempted bites at their opponent's fins and tail. The two fish also batter each other with repeated strikes to the flank or lock jaws for up to thirty minutes. The loser is bitten and chased repeatedly and, if there is no refuge or intervention, eventually killed.

Fighting fish that are purchased from a breeder are referred to as "plakat luk morh" or "the biting fish that belongs to the clay pot," and those caught from the wild are called "plakat luk pah" or "the biting fish that belongs to the jungle." Regardless of its origin, a fish must be conditioned before beginning its fighting career. In "luk morh" fish, this process usually begins at seven to nine months of age. Fish are isolated and maintained on a restricted diet of live food. Leaves from plants such as the tropical almond (*Terminalia catappa*), whose extract hardens the skin, are added to the fish's tank. After a week or two in isolation, fish are subjected to a series of conditioning exercises. These exercises include chasing females (to reduce body fat and increase alertness), water agitation (to increase stamina and provoke aggression), repeated scooping with a net (to habituate the fish to being moved from tank to tank), and antagonizing with a black-tipped stick (to elicit direct attacks). This period of conditioning can last one or more weeks before the owner decides his or her fish is ready to fight.

Breeders capitalize on the intrinsic aggressiveness of betta and, through the process of artificial selection, breed fish based on characteristics related to fighting. Fighting styles differ throughout Southeast Asia, and thus, the desired fish characteristics may also vary by region. Generally speaking, however, breeders look for well-proportioned fish with shiny, dark skin and smooth scales (indicators of general health) in addition to their aggressive temperament. Much as with other forms of animal domestication, sophisticated breeders and purchasers of betta pay attention to a fish's lineage as an indicator of its future fighting success. Long-finned strains of betta, widely cultivated for their beautiful appearance, are rarely used in competitive fighting. Instead, long-finned betta are often displayed in competitive fish shows. The popularity of this pastime is spreading rapidly worldwide, promoted by groups that actively oppose the practice of fish fighting (e.g., the International Betta Congress).

Fighting fish is illegal in many Southeast Asian cities, such as Bangkok and Singapore, and may be punishable by substantial fines. Therefore, it is primarily a rural, or at least a covert, pastime. In rural areas, people gather together to sell, purchase, fight, and wager on betta. These groups may convene near a breeder's ponds or in a building dedicated to the sport, in which case the structure is often shared with cock fighters. In rural areas, these arenas are usually operated with the approval of provincial governors. Attendance at these events is dominated by men fifteen to seventy-five years old, though women and children are not explicitly banned from attending. Men twenty-five to fifty years old make up the majority of participants; many are professional breeders, gamblers, or small-business owners.

Before a fish fight, the combatants are placed in tall, flat-sided glass jars alongside one another. Flat-sided jars are used to reduce visual distortion that might misrepresent the size of the fish. The fish immediately react to each other, using the previously described displays. Based on the initial reactions of each fish, wagers are then made by the observers. Although gambling on sporting events (with the exception of horse racing) has been illegal in Thailand since the Gambling Act of 1935, the practice remains widespread, particularly in rural areas. The fish are then moved to a shared jar and allowed to fight. Fights are usually two to four hours long, and they end when the judge proclaims a winner. Winners are determined in one of several ways: by the extent of physical injuries suffered by each fish, by the unwillingness of one combatant to continue fighting, or by the withdrawal of one fish by its owner, usually under pressure from escalating wagers against his fish. In some cases, the owner of the "loser" may petition the judge to put his fish alongside a third, "referee" fish. If the loser continues to display aggressively, the fight may be continued. If this happens twice, the fight ends in a draw.

Both winners and losers are treated with almond leaf to heal their wounds. Winners may be fought several more times and are used also as breeding stock to produce another generation of fighters. In keeping with the Thai proverb "the fallen fruit goes not beyond its root," losers are rarely bred or fought again and instead are released into the wild or kept as pets. The effects of the large-scale introduction of domesticated betta into the wild are unknown. Some biologists claim that truly wild betta no longer exist near urban areas because of hybridization with released fighters. The recently discovered *Betta sp. Mahachai*, a potentially new species, may be driven genetically extinct by hybridization with domestic *Betta splendens* before it is ever described by scientists. Several other species of betta that are threatened with extinction may also be negatively affected by competition from domesticated *Betta splendens*.

Today, it is rare for people to fight wild-caught fish or "plakat luk pah." Instead, intensive artificial selection by breeders supplies the demand for increasingly aggressive, dominant fish. Small-scale breeders typically sell their fish domestically or use them for their own enjoyment. One common practice among families is to use the bathroom cistern (from which water is taken to flush the toilet) as a breeding tank. Large-scale breeders have expansive operations with hundreds of cubic meters of concrete or clay tanks and are capable of supplying international markets. For example, some breeders have the capacity to raise tens of thousands of fish at a time and produce thousands of fish per week for export. Today, some breeders sell directly to their customers worldwide via the Internet, and others ship their fish to Singapore, the hub of the international aquarium fish industry. For a typical male betta, the retail price is roughly ten times what a breeder in Thailand sold it for; for highly prized fighters, the final retail price may be 100 times higher. Despite this disparity, the breeding and selling of betta represents a significant source of income for many people in rural areas. A minor economic benefit of fish fighting comes through Thailand's multibillion-dollar tourism industry—some tour guides capitalize on the curiosity of foreigners who want to see sports such as "muey thai" (Thai kickboxing) and fish fighting that are unique to this region.

Many people in Thailand, and in Southeast Asia in general, have mixed views of fish fighting. Some consider "blood sports," such as animal fighting, contradictory to the non-violent doctrines of Buddhism, the predominant religion. Many regard the practice of animal fighting with some ambivalence; they view the sanctity of animal life entirely differently from that of human life. Regardless of their religious views, most Thai recognize betta or "plakat" as an important part of their cultural heritage. Legal restrictions on gambling have made fighting fish arenas less common than they were thirty or forty years ago. However, the prevalence and cultural significance of fish fighting in Southeast Asia, and among immigrants from these countries living abroad, is likely to persist far into the future.

See also

Sports

Further Resources

Jaroensutasinee, M., & Jaroensutasinee, K. (2001). Sexual size dimorphism and male contest in wild Siamese fighting fish. *Journal of Fish Biology, 59,* 1614–21.

Lucas, G. A. (1997). *Siamese fighting fish.* Neptune, NJ: T.F.H.

Regan, C. T. (1909). The Asiatic fishes of the family Anabantidae. *Proceedings of the Zoological Society of London, 44,* 767–87.

Simpson, M. J. A. (1968). The display of the Siamese fighting fish, *Betta splendens. Animal Behaviour Monographs, 1,* 1–73.

Smith, H. M. (1945). *The freshwater fishes of Siam, or Thailand.* Washington, DC: United States Government Printing Office.

Vierke, J. (1988). Bettas, gouramis, and other anabantoids. Neptune, NJ: T.F.H..

Ethan D. Clotfelter and Maureen G. Manning

■ Culture, Religion, and Belief Systems
Francis of Assisi (1181/82–1226)

Born Giovanni Bernardone in Assisi, Italy, Francis of Assisi is one of the most beloved of Catholic saints and one of the most familiar among non-Catholics. He is most frequently noted for his relationship with nature and animals. It is important to recognize, however, that his relationship and bond with animals and nature was an expression of his religious faith and the ascetic lifestyle he chose for himself and his followers. It is unfortunate that he is too often reduced to the role of holding a birdbath in our gardens. His life and message have much to contribute to our current world and how we relate to one another and to the world around us.

Francis's father, Pietro, was a wealthy cloth merchant in Assisi. As a young man, Francis was playful and merry, enjoying the many benefits that his father's wealth provided. He was a leader among the young nobles of Assisi, and his wit and good nature typically put him at the center of revelries and festivals. He was proficient in the use of arms and considered a military career. A couple of bouts with illness thwarted this career and marked a change in his personality. He became more contemplative and took part in celebrations with his friends increasingly less often. He now dressed in simple attire and frequently gave whatever possessions he might have with him at any time to beggars he met on the road. A true turning point came in his life while he was praying in the decaying chapel of St. Damian's just outside Assisi. Francis heard a voice, "Go, Francis, and repair my house, which as you see is falling into ruin." At first, Francis took this as a literal command and endeavored to restore the chapel. Eventually, he rebuilt the walls of stone with his own hands. In time, Francis would fulfill this charge by revitalizing a medieval church awash in political controversy and bereft of spiritual authority.

Francis began preaching in the surrounding areas and helping the poor, attracting followers impressed with his holiness. He and his followers formed the Friars Minor. The Friars took vows of poverty, chastity, and obedience. They dressed in rough robes tied at the waist with knotted ropes, lived in simple huts, and spent substantial time in prayer. However, they did not cloister themselves and avoid contact with the world. Francis believed that the way to holiness was through prayer, both through the kind of prayer normally recognized and through a "prayerful" life expressed in work and service to others. Francis took it as his special calling to nurture the spiritual life of the poor and outcast members of society. He preached in the vernacular instead of Latin or Greek and filled his sermons with similes based on the day-to-day experiences of the common people. Francis stood in stark contrast to the nobility projected by many clergy of the day. People flocked to hear him preach in his simple robe, in a way that touched them in a personal fashion.

One of Francis's revelations was that the work of God could be discerned and celebrated in all aspects of creation and was not restricted to a narrow range of Church activities. In this way, Francis became known for his relationship with animals and the natural world. Several specific examples of his care and concern have become a part of Franciscan lore. In one example, Francis and several other friars were traveling near the town of Bevagna when they came upon a large, diverse flock of birds. Inspired, Francis approached the birds and asked them if they would stay and hear him preach the Word of God. When the birds stayed and gathered about Francis, he began by addressing them as brother and sister birds. When he concluded, he blessed the birds, and they flew off in a cloud of song. It is said that injured and frightened animals would seek his comforting touch. He would hold the animal gently and speak to it softly before releasing it with an admonition to take greater care in avoiding the dangers of the world.

One of the most oft repeated tales of Francis concerns the wolf of Gubbio. The people of Gubbio were terrorized by a great wolf that preyed upon their flocks and fellow citizens. When Francis arrived to preach in the town, the townspeople prevailed upon him to help. Francis and his companions bravely went out from the city to meet the wolf. When the wolf came charging out at them with fangs bared, Francis made the sign of the cross, and the wolf halted. "Brother Wolf, I want to make peace between you and the people of Gubbio." The wolf showed his assent by giving his paw to Francis. Francis returned to the city with the wolf at his side. The people saw this as a great miracle and listened intently as Francis preached a sermon of repentance. When he finished, Francis asked the townspeople to agree to peace with the wolf. They would provide the wolf with food and not try to hurt him, and he in turn would be gentle and harm neither the animals nor the people of Gubbio.

One of the most enduring contributions of Francis to Christian culture was his creation of the Christmas crèche. In 1223 at Greccio, he reproduced the scene of Christ's birth in the stable at Bethlehem. Members of the community played the roles of Mary, Joseph, and the shepherds who were the first witnesses of the birth. His purpose was to demonstrate to the people that Christ's birth was for all people, most especially the humble among them. True to his sense that God's creation was indeed unified, Francis did not forget to include the animals that populated the stable. Francis's thoughtful inclusion of the animals at the crèche inspired a range of traditions that include consideration and prayer for animals during the Christmas season.

Saint Francis of Assisi preaching to the birds, *by Di Bondone Giotto. ©The Art Archive/Musée du Louvre Paris/Dagli Orti (A).*

Francis's inclusion of animals in his preaching and worship stood in stark contrast to the teachings of Augustine and Thomas Aquinas. They held that animals were created by God for the benefit of humans and that because animals were not rational beings, humans had no moral obligations specifically regarding their treatment. Any moral obligation related to the treatment of animals arose out of any relevant association they may have in relation to humans. For example, it was wrong to kill a man's ox, not because it represented mistreatment of the ox, but rather because it was theft of another man's property. Similarly, beating an animal in anger was wrong because humans as rational beings should control their emotional outbursts. Francis recognized animals and all nature as part of God's overall plan of creation and believed they should be recognized and honored for whatever role, however small, they might play in this plan. Francis clearly rejected the assumption that animals were irrational, or he would not have preached to the birds or expected the wolf to be an equal partner in an agreement with the people of Gubbio.

Francis's masterpiece was *The Canticle of Brother Sun*. In it he praises God and thanks him for all of creation. Written near the end of his life, it was widely and rapidly accepted, and it brought him substantial comfort while a variety of ailments were causing him great suffering. When his death came, Francis was surrounded by those he had inspired to join him in his life of prayer and sacrifice. As his final statement of humility, he asked to be stripped of his garments and laid naked on the ground in the embrace of Mother Earth. He was canonized as a saint in 1228, just two years after his death. In the years since, he was been widely admired by people of many nations as creeds. He is revered as the patron saint of animals and the environment. His feast on October 3 is frequently celebrated with a blessing of animals.

See also

Culture, Religion, and Belief Systems—Blessing of the Animals Rituals.

Further Resources

House, A. (2000). *Francis of Assisi: A revolutionary life*. Mahwah, NJ: Hidden Spring.
Robson, M. (1997). *St. Francis of Assisi*. London: SCM Press.

Stephen Zawistowski

Culture, Religion, and Belief Systems
Gandhi, Mohandas Karamchand (1869–1948)

Like all devout Hindus, the Gandhis were strict vegetarians. In an 1891 interview with *The Vegetarian*, however, Mohandas Karamchand (later "Mahatma") Gandhi revealed that in his youth he had been "betrayed into taking meat about six or seven times" at a period when he allowed his friends (particularly a vicious youth called Mehtab) to do his thinking for him. In the end, he gave up meat-eating because he could not go on lying to his mother, and before leaving Gujarat for London in 1888, Gandhi vowed never to touch meat again.

Once in London, determined to keep his promise, Gandhi ignored the advice of new acquaintances who told him that he could not survive its cold climate without eating meat and went searching for a vegetarian restaurant in London, finally finding one on Farringdon Street (the "Central") a few months later. There he bought some vegetarian literature, including H. S. Salt's *A Plea for Vegetarianism,* and finally gave up the idea that eating "flesh food" might be better for one's diet, converting to vegetarianism as a matter of principle. It was also there that he discovered the Vegetarian Society of Manchester (which published the *Vegetarian Messenger*) and the London Vegetarian Society (LVS), which Gandhi joined in 1890 and in whose weekly journal (*The Vegetarian*) he would soon come to publish his first pieces of work in 1891.

Gandhi's first writings focused on Indian vegetarians and their festivals, distinguishing between those whose vegetarianism was voluntary and strict ("pure vegetarians"), those who were willing to take meat but were too poor to buy it, and those (in Britain) who followed a V. E. M. (vegetables, eggs, milk) diet. Pure vegetarians, he pointed out, "argue that to eat an egg is equivalent to killing life; since an egg, if left undisturbed would, *prima facie,* become fowl," adding that he is sorry to say that he himself had recently been eating eggs. Unlike "extreme vegetarians" (i.e., vegans), he added, vegetarians do not abstain from eating dairy products because they do not believe that cows (who are an object of worship among Hindus) suffer when milked.

Gandhi also argued that the cause of the physical weakness of Indian vegetarians could not be attributed to their diet, using the example of certain Indian shepherds to argue that, if anything, a vegetarian diet was "conducive to bodily strength." As for mental strength, Gandhi pointed out that Buddha, Pythagoras, Plato, Porphyry, John Ray, Daniel (from the Old Testament), John Wesley, John Howard, Percy Bysshe Shelley, Sir Isaac Pitman, Thomas Edison, and Sir W. B. Richardson were all vegetarian (as was George Bernard Shaw, whom Gandhi would soon befriend). Later that year, he started the Bayswater branch of the LVS, and he continued to support the society when he briefly returned to India and upon his subsequent move to South Africa a couple of years later. In 1931 he gave a talk at the LVS on the "Moral Basis of Vegetarianism," which later formed the inspiration for a posthumously published collection of related writings.

Gandhi's vegetarianism nicely complements—and might even be thought to have motivated—his general advocacy of nonviolence that marked India's struggle for independence from British colonial rule. His philosophy of compassion clearly extended beyond people to all creatures, and it is noteworthy that apart from vegetarianism, Gandhi also championed humane farming and in the 1920s protested against vivisection, believing it to be his duty to speak out on behalf of animals.

Be that as it may, there is considerable tension between Gandhi's writings on the nonkilling of animals from the 1920s and those from the late 1940s. In the former, he takes a more extreme pacifist view (though he does allow for mercy killing; see *Young India* newspaper, November 18, 1926, p. 396). For example, in his 1925 *Autobiography* (translated in 1927), he wrote, "To my mind the life of a lamb is no less precious than that of a human being. I should be unwilling to take the life of a lamb for the sake of the human body" (p. 172; cf. article in *Young India,* May 18, 1921, p. 156).

In a 1927 newspaper article, he wrote,

> I do not want to live at the cost of the life even of a snake. I should let him bite me to death rather than kill him . . . If in not seeking to defend myself against such noxious animals I die, I should rise again a better and fuller man. With that faith in me, how should I seek to kill a fellow-being in a snake? (*Young India,* April 14, 1927, p. 121)

As noted, the Gandhi of the 1920s was also a stern critic of vivisection:

I abhor vivisection with my whole soul. I detest the unpardonable slaughter of innocent life in the name of science and humanity so-called, and all the scientific discoveries stained with innocent blood I count as of no consequence. If the circulation of blood theory could not have been discovered without vivisection, the human kind could well have done without it. (*Young India,* December 17, 1925, p. 40)

In the late 1940s, by contrast, Gandhi appears to have abandoned the view that it is always wrong to kill an animal no matter what. Thus, in the second volume of his book *Non-Violence in Peace and War* (published posthumously in 1949) he wrote,

I am not able to accept in its entirety the doctrine of non-killing of animals. I have no feeling in me to save the life of these animals who devour or cause hurt to man. I consider it wrong to help in the increase of their progeny. Therefore, I will not feed ants, monkeys or dogs. I will never sacrifice a man's life in order to save theirs. Thinking along these lines, I have come to the conclusion that to do away with monkeys where they have become a menace to the well-being of man is pardonable. Such killing becomes a duty. The question may arise as to why this rule should not also apply to human beings. It cannot because however bad, they are as we are. Unlike the animal, God has given man the faculty of reason. (Vol. II, p. 67)

And in two newspaper articles in 1946, Gandhi wrote the following:

The emphasis laid on the sacredness of sub-human life in Jainism is understandable. But that can never mean that one is to be kind to this life in preference to human life. While writing about the sacredness of such life, I take it that the sacredness of human life has been taken for granted. The former has been over-emphasized. And while putting it into practice, the idea has undergone distortion. For instance, there are many who derive complete satisfaction in feeding ants. It would appear that the theory has become a wooden, lifeless dogma. (*Harijan,* June 9, 1946, p. 172)

True Ahimsa [nonviolence] demands that, if we must save the society as well as ourselves from the mischief of monkeys and the like, we have to kill them. (*Harijan,* July 7, 1946, p. 213)

Despite such changes of opinion, one underlying view remains clear and constant in Gandhi's thought: that "the greatness of a nation and its moral progress can be judged by the way its animals are treated" (*Moral Basis of Vegetarianism*).

Further Resources

Gandhi, M. K. (1940). *Autobiography: The story of my experiments with truth* (M. Desai, Trans.). Washington, DC: Public Affairs Press. (Original work published 1925, translated 1927/1929).

———. (1959). *The moral basis of vegetarianism.* Ahmedabad: Navajivan.

———. (1959–1994). *Collected works, Vols. 1–100* [Includes reprints of all newspaper articles]. New Delhi: Publications Division, Ministry of Information and Broadcasting, Government of India. Available online at http://www.gandhiserve.org/cwmg/cwmg.html and on CD-ROM.

———. (1960–1962). *Non-violence in peace and war.* Ahmedabad: Navajivan. (Original work published 1942–1949).

Mahatma Gandhi Foundation–India. *The Official Mahatma Gandhi eArchive & Reference Library,* http://www.mahatma.org.in/

Richards, G. (1982). *The philosophy of Gandhi*. London: Curzon Press.

Wolpert, S. (2001). *Gandhi's passion: The life and legacy of Mahatma Gandhi*. Oxford: Oxford University Press.

Constantine Sandis

■ Culture, Religion, and Belief Systems
Human-Animal Relationships in Japan

Human involvement with animals is seen with a variety of animals both on land and in the sea, but this entry deals specifically with an overview of the historical relationship between the Japanese and land animals.

Japan is an island nation comprising a chain of large and small islands running roughly north to south, and animal life from the alpine regions to the plains is diverse, including bears, wild boars, deer, macaques, cranes, and swans.

As surmised from information such as the oral traditions of the Ainu in Hokkaido, the lives of primitive Japanese were governed by animism, a belief that all things have spirits. Japan's largest island of Honshu and those south are dominated by Shinto, which is based on the myths of national founding, and Buddhism, which originated in India and came to Japan from the continent. Even now, the religious norms of the Japanese in general are a compromise belief in both Shinto and Buddhism. In accordance with the Buddhist norm that "mountains, rivers, grass, and trees all have Buddha-nature," meaning that all things in existence are in the image of the Buddha as everlasting life, the Japanese view of animals has basically been that the lives of animals should be respected like those of humans because they are a part of nature, and since the time of Emperor Tenmu (the 670s), edicts against eating meat have been issued many times.

Such thinking reached its apex during the Edo period under the rule of Tokugawa Tsunayoshi, the fifth Shogun, who issued the "Order to Have Pity on Living Things," which, for example, banned the abuse of horses, bovines, and dogs; the sale of fish for consumption; and the raising of birds and the strangulation of foul. However, some see this as a complex situation that included the political aim of controlling the use of rifles that was then beginning to spread among the populace. As such, one cannot jump to overall conclusions about the relationship between the Japanese and animals based on Tsunayoshi's act.

Records of various kinds suggest that there was no excessive killing of animals in Japan as in the West, but at the same time, contemporary accounts reveal that meat was eaten under a variety of pretexts, such as medicinal or health purposes. However, throughout Japan, one can find Buddhist prayer gatherings for insect memorial services, monuments erected to the souls of animals, and other expressions of compassion for living things that people killed.

It was during the Meiji period, which started in 1867 at the end of the Edo period, that big changes occurred in the custom of compassion for living things based on Buddhist teachings. The Meiji government abandoned its isolation policy, which the Shogunate had continued for the 300 years up to that time, and switched to an "enrich the state, strengthen the military" policy that aimed to catch up to the West and make Japan more powerful. This entailed actively absorbing the institutions, science, and technology of Western societies.

Along with emulation of Western societies came the imperialistic view of nature by which humans act violently against nature. Because the Japanese at that time did not understand what effect that new view of nature would have on Japan's natural environment and fauna, without a second thought they discarded their view that all things had spirits, which had until then been the usual teaching of Japanese society. This brought about a drastic change in the relationship between Japan's animals and humans.

A typical example is the extinction of the Hokkaido wolf (*Canis lupus hattai*). In the mid-1800s Japan was expanding its horse farms in Hokkaido to produce more horses for the military, but because wolves attack horses, the Meiji government invited Horace Capron (see Russell, 2007) from the United States and sought his guidance in an effort to protect its horses. Capron had been a general in the Civil War and commissioner of agriculture under presidents Johnson and Grant. In Japan, Capron implemented the same wolf extermination operation as in the United States and in no time brought about the extinction of the wolf from all of Hokkaido. Wolves disappeared from the other Japanese islands in the 1900s, but this was a result of rabies and other contagious diseases, not hunting pressure.

But on the main island of Honshu and other places, wildlife such as macaques, the Japanese serow (*Nemorhaedus crispus*), cranes, the Japanese crested ibis (*Nipponia nippon*), and storks began disappearing because of the rifle. Rifles had come to Japan in 1534, long before the Edo period, with the crew of a Portuguese ship that was wrecked and drifted ashore at Tanegashima Island south of Kyushu. In time, rifles became widely known throughout Japan by the name "Tanegashima," and during the Warring States period, the daimyo increasingly used them as a new battlefield weapon. Rifles further made their way into the hands of the populace, where they were used for hunting and other purposes. For this reason, the Meiji government, in its first year of 1868, issued a Cabinet proclamation that prohibited the general public from using rifles in cities and from freely shooting birds in the wild.

In 1873, the Rifle Hunting Control Regulation was upgraded to the Hunting Law, thereby strictly controlling the use of rifles by the general public and, on a number of occasions in 1876 and thereafter, banning the hunting of the Yezo deer (*Cervus nippon yesoensis*), which was endangered at that time. However, this ban was meant to revive the venison industry and had nothing to do with respecting the lives of deer. After that, Japan was busy warring with Russia and China, so animal-killing and environmental damage continued until Japan's 1945 defeat in World War II, but it was in 1934, during the early years of this war and turbulence, that the Buddhist monk Godo Nakanishi founded the Wild Bird Society of Japan.

Nakanishi opposed the contemporary social trend for keeping small birds in cages and called for letting wild birds live in the wild. His efforts included the abolition of mist nets, which indiscriminately capture wild birds; establishing wild bird preserves; amending the Hunting Law; and passing the Wildlife Protection and Hunting Law. His activities steadily grew with the support of well-known people in Japanese society, and he played a major role in launching subsequent environmental protection activities.

In post–World War II Japan, environmental damage increasingly worsened as rapid economic growth proceeded and industrial pollution spread throughout the country, but in the 1960s and 1970s, strong environmental protection movements arose from among the citizenry in tandem with the government's pollution-control measures, and people formed many organizations to protect Japan's disappearing flora and fauna.

Activities also began to protect animals' right of existence, such as opposition to animal experiments and criticism of zoos and aquariums. Inspired by U.S. lawsuits on the rights of nature, Japanese activists have filed many lawsuits on behalf of wild Amami rabbits (*Pentalagus furnessi*), foxes, dragonflies, and other wildlife against environmental damage. This is a new effort arising out of the lively flow of information from abroad in conjunction with globalization. Catalyzed by the passage of primate-protection laws in New Zealand, the United Kingdom, and the United States, there is now a wide variety of new involvement with animals in Japan, such as efforts by university researchers to protect primates' right of existence.

There is also legislation on animal protection and welfare, and in 2001 a number of citizen organizations held Japan's first "animal summit," which adopted a declaration that called for recognizing the basic right of existence for all other living things, just as for humans.

Although it is desirable to popularize the practice of refraining from meat-eating and of practicing vegetarianism to determine what action humans should take to assure animals' right of existence, efforts of this nature in Japan are moving very slowly. The practice of meat-eating has grown with the influx of Western culture into Japan, and there is no sign of it abating. Meanwhile, joint efforts by the government and citizens to increase the numbers of rare animals and return them to the wild are slowly realizing successes, and in September 2005 the government made the first attempt to return the near-extinct stork to the wild.

See also

> Culture, Religion, and Belief Systems—Japan's Dolphins and the Human-Animal Relationship

Further Resources

Russell, H. S. (2007) *Time to become barbarian: The extraordinary life of General Horace Capron.* Lanham, MD: University Press of America.

Eiji Fujiwara

■ Culture, Religion, and Belief Systems
India and the Elephant

Elephants and human beings have coexisted in Asia for millennia. The relationship between the two species has historic, social, political, artistic, and religious significance. From a human perspective, the elephant in Asia (*Elephas maximus*) has always represented power, dignity, and grace. These symbolic characteristics place the elephant in a unique position within the Asian cultural imagination. As a species, they are worshipped and revered in countries such as India, Myanmar, Cambodia, and Thailand.

On the other hand, the relationship between man and elephant has not always been sacred or benevolent; it has often been marked by cruelty and exploitation. In India and other parts of Asia, the forests on which the elephant's existence depends are severely threatened. As the human population has grown and pressures on the land have increased, the number of elephants living in the wild has dropped dramatically. Conflicts between elephants and human beings have been exacerbated by the loss of forests and the spread of agriculture. Wild elephants raid fields on the perimeter of national parks and sometimes kill human beings that come in their path. More often than not, they themselves fall victim to the guns of farmers protecting their crops. Even in the largest sanctuaries and national parks, the elephant's habitat is often inadequate, for a herd in its natural habitat can roam over hundreds of square kilometers, and an adult can eat more than 300 kilos of green fodder every day.

Though it is difficult to determine exact numbers, there are approximately 50,000 Asian elephants in the wild and roughly 15,000 in captivity. *Elephas maximus* is distributed throughout south and southeast Asia, with by far the largest concentration, about 30,000, living in India. Conservation efforts are underway in many countries, supported by international organizations such as the World Wildlife Fund. In 1992 the Indian government initiated Project Elephant, establishing a nationwide network of sanctuaries where these animals remain protected and dedicating funds for research, for the monitoring of national parks, and for enforcing wildlife laws.

More than any other animal, except perhaps tigers, elephants are a symbol of India's rich forest resources, which are rapidly being destroyed. Not only is the elephants' natural habitat depleted, but ivory poaching also continues to take its toll. Though the trade in ivory is strictly prohibited through international conventions and national laws, it continues illegally and threatens both the Asian and the African elephant (*Loxodanta africana*). Although the male and female of the African species both bear tusks, only male elephants in Asia produce ivory. In many parts of India, poaching has caused a disproportionate ratio of male and female elephants. It has also led to the predominance of makhnas (male elephants without tusks), a phenomenon that scientists continue to study, suggesting that human pressures of poaching may ultimately lead to the evolution of elephants without tusks.

In India, elephants have been captured and tamed from at least the second millennium BCE onward. Clay icons from the ancient Indus valley civilizations depict images of captive elephants. Ironically, man has often trained the elephant to destroy its own habitat, using the elephant's strength and dexterity to assist in logging operations. Working elephants are still employed in India by the forest department. Occasionally, they are even used for shunting wagons on the railways. Many wildlife parks in India employ tame elephants and their mahouts (handlers) to carry tourists into the jungle.

Some of the best places to see India's elephants in the wild are Corbett National Park in the northern state of Uttaranchal, Kaziranga National Park in the northeastern state of Assam, and Nagarhole National Park in southern Karnataka state. Scattered herds can be found in other parts of India, but one of the problems of conservation is that the population of elephants is fragmented and isolated in remote pockets of the country.

Captive elephants in India that live outside the forest are generally owned by wealthy landowners who keep them as status symbols, tour operators who use them to give joy rides at monuments, and Hindu mendicants and temple authorities who exploit their religious significance. Every year, in India's northern state of Bihar, dozens of elephants are bought and sold at the Sonepur Mela, one of the largest cattle fairs in the world. The treatment of these animals is not always humane, and they often suffer abuse and neglect at the hands of their owners. Veterinarians from organizations such as the Wildlife Trust of India operate special clinics during the fair to treat diseases and injuries. The relationship between man and elephant is full of paradox and pathos. The mahout shares a special bond with his elephant that often lasts a lifetime. When riding an elephant, the mahout sits behind the elephant's ears and directs it with pressure from his knees and toes, as well as a set of verbal commands that are used throughout India for training elephants. One of the most famous mahouts, and the only woman who currently trains and handles elephants in India, is Parbati Barua. She lives in Assam and is sometimes called "Elephant Queen." Though there is a ban on capturing wild elephants in India, Barua uses traditional methods to subdue and relocate elephants that are raiding crops and threatening farmers.

Although many captive elephants live under restrictive and inhumane circumstances, they often share in the celebrations and tragedies of their masters. They are considered an auspicious presence at weddings and religious festivals. At the same time, when an elephant falls sick or dies, its fate is lamented as if it were another human being. For example, Guruvayoor Kesavan, one of the most famous temple elephants in India's southern state of Kerala, had thousands of mourners attend his funeral in 1976. Kesavan's life-sized statue stands within the temple grounds, and a movie was made about his life.

Elephants are an integral part of Hindu mythology, assuming the anthropomorphic form of Ganesha or Ganapati. A benevolent deity with a human body and an elephant's head, Ganesha removes all obstacles. In the city of Mumbai, there is an annual celebration of Ganapati Chathurthi, during which images of Ganesha—some of them thirty feet tall—are paraded through the streets and eventually immersed in the Arabian Sea. At many temples in southern India, tuskers are donated by wealthy devotees, and during festivals, these elephants are painted and dressed up with ornaments to lead processions. The size and majesty of an elephant, as well as its companionable nature, inspire both awe and devotion.

Historically, the elephant has always been a symbol of power and authority. The armies of ancient India marched into battle behind their war elephants. When Alexander the Great and his troops finally reached the subcontinent in 326 BCE, they were confronted by a phalanx of these armored beasts. Mughal emperors in the sixteenth and seventeenth centuries commanded their troops from elaborate howdahs mounted on the backs of tuskers. They also used elephants for hunting tigers and occasionally even as executioners, crushing prisoners to death beneath their heavy feet. In much of Indian art, however, the elephant presents a gentler and less threatening visage. Buddhist frescoes from the fifth century CE that decorate the walls of caves at Ajanta depict the elephant as a playful, docile creature. Carved images in forts and temples portray the solid, enduring qualities of the elephant, a symbol of indomitable strength. Arguably, the most beautiful drawings and paintings of elephants were done by Rajasthani miniature painters in the eighteenth and early nineteenth century. These tiny artworks capture both the grandeur and the grace of elephants within the space of a few square inches.

Europeans have always been fascinated by the Asian elephant for it represents the exotic east. Under British rule, hundreds were shipped to Europe and America to live in zoos and perform in circuses. They provoked curiosity and wonder in countries all over the world, capturing the imagination of both children and adults with their affable antics and solemn demeanor. But even as these elephants served as living trophies of the empire, they also became a metaphor of oppression. In his famous essay "Shooting an Elephant," George Orwell condemned the colonial adventure by describing his own tragic encounter with an elephant that he was forced to kill.

Too often, the relationship between human beings and elephants is marked by violence. Elephant-human conflict can lead to vindictive retribution when crop raiders and rogue elephants are shot, electrocuted, and poisoned because they are seen as a threat. The complexity and contradictions of human-elephant relationships are illustrated by the story of Moorthy, a makhna, or tuskless male elephant, in southern India, who declared a rogue and captured by the forest department in 1998. He became the focus of an international controversy in which American animal rights activists fought with the state authorities over the proper treatment and care of this elephant. Known to his Western advocates as Loki, the former rogue now lives in Theppakkadu elephant camp and is docile enough to be fed by hand.

Despite many questions raised about the harsh methods used to subdue and train an elephant like Moorthy, Theppakkadu remains one of the best places in India for captive elephants. Located within the margins of a national park, along the Moyar River, the camp provides as close to a natural habitat as any captive elephants enjoy, with plenty of fresh water and an abundance of fodder. The Theppakkadu elephant camp is home to about thirty animals. Some of these elephants are "pensioners" who worked for the forest department until the age of fifty-five, when they are officially retired from service. (The average lifespan of *Elephas maximus* is about the same as *Homo sapiens*.) Theppakkadu also serves as an orphanage for young elephants who have been separated from their mothers and herds. Rescued by the forest department, they are raised in the camp and trained by mahouts from the Irula tribal community, who are elephant handlers by tradition.

In order to understand the relationship between elephants and human beings in India, it is important to balance a symbolic, often romanticized vision of *Elephas maximus* with scientific descriptions of its biology and behavior, as well as the harsh realities of an endangered species. These differing perspectives are not always contradictory and often complement each other. For instance, the prevalent myth about an elephant's prodigious memory raises more complex questions about how we remember an animal in the wild. As a tusker stands at the edge of the forest, a busload of tourists aim their cameras in his direction. What are the images we record, and what do they mean? How will we recall those images in the future? And perhaps most important, as we try to preserve elephants in the wild, what memories will we choose to guide us?

In trying to separate our sentimental stereotypes of the elephant from the true nature of the beast, it may be necessary to debunk the myth of an elephant's memory, but it is

People prepare a giant statue of Ganesh in preparation for the annual celebration of Ganapati Chathurthi in Mumbai, India. Courtesy of Shutterstock.

also essential for us to explain what makes them unforgettable. In this way, we can define our responsibilities to the elephant and look to the future of both our species. Ultimately, we are a part of the elephant's story, just as it is a part of ours.

Further Resources

Alter, S. (2004). *Elephas maximus: A portrait of the Indian elephant.* New York: Harcourt.
Gee, E. P. (1964). *The wildlife of India.* London: Fontana.
Menon, V. (2002). *Tusker: The story of the Asian elephant.* Delhi: Penguin.
Sukumar, R. (1994). *Elephant days and nights: Ten years with the Indian elephant.* Delhi: Oxford.
Williams, J. H. (1955). *Elephant bill.* London: Rupert Hart-Davis.

Stephen Alter

■ Culture, Religion, and Belief Systems
Indian and Nepali Mahouts and their Relationships with Elephants

The Mahouts' Unique Relationship with Elephants

Elephant keeping is a practice of great antiquity stretching back millennia, playing a key role, both symbolically and logistically, in the life of the state and its people. In fact, elephants were so important that a specialist Sanskrit literature even emerged, which encoded practical knowledge concerning the keeping of elephants. For Hindu and Buddhist cultures such as in India and Nepal, the elephant is a sacred creature, which influences elephant-keeping practices.

No other human-animal connection has the duration and intensity of a man and his elephant. The elephant mystique may stem in part from its possession of the largest brain of any terrestrial mammal, with considerable sophistication in social connections, geographic knowledge, and mobility. Elephants' abilities to forge and maintain relationships with a large number of individual elephants may carry over to their capacity to form relationships with people as individuals. In the characteristic pattern, a boy guided by his father would grow up beside his elephant, sharing a life together, including the royal processions, religious veneration, and the status derived from managing such a powerful animal. This intense relationship is changing rapidly with the declining use of elephants.

Throughout South and Southeast Asia, elephant handlers are known by different terms. *Mahout* is the most widely recognized designation for an elephant driver, as in (Southern) India, but it is only one term within a more elaborated system of elephant-handling ranks and roles. In contemporary Nepal, for example, each elephant is ideally serviced by a three-man team. Mahout is the lowest rank given to a neophyte, and handlers are referred to generically as *hattisare,* meaning those who work in an elephant stable or *hattisar.*

Sources of Elephants

Elephants are long-lived, having a long infancy and adolescence resembling humans' before they reach adulthood and are able to perform work. Throughout elephant-keeping history, humans have captured wild adult elephants when needed. The Mughul Emperor Akbar was an exception, innovating captive breeding in his imperial stables in the sixteenth century. Generally, elephants filled roles of economic utility, meaning that there was little incentive to breed elephants and care for young in captivity.

The kheddah is one of several methods of elephant capture, as recorded in ancient Sanskrit texts on elephant science. In a sort of roundup, elephants are driven into a confined area where they are corralled. Large-scale kheddahs were used frequently for about one hundred years in Southern India, until 1971. These elephants were domesticated to fulfill working roles in railroad operations, road building, and logging. Similarly, the last wild capture in Nepal, using the lasso method, was in 1970. With increasing habitat loss and dwindling numbers of wild elephants, this situation is now changing. Except for a handful of births or orphaned wild elephants, the domesticated elephants remaining in Southern India date from the kheddahs. However, in contemporary Nepal, where captive elephants are used in park patrolling, conservation research, and tourism, a breeding center based at the Chitwan National Park now plays an increasingly important role in maintaining the captive population used in all Nepal's lowland protected areas.

Typical markets for trade and purchase of elephants have been in Assam and Bihar of Northern India. Traders are renowned for devious tactics such as lying about an elephant's history and temperament and using drugs to make elephants seem particularly healthy. Handlers have expert knowledge with which to identify desirable elephants. Because of Convention on International Trade in Endangered Species (CITES) regulations, it is now technically illegal to trade in elephants across international borders. This has prompted Nepal to establish an elephant-breeding center. Reciprocal gift exchanges between states, such as rhinos for elephants, has been a method for bypassing these regulations.

The Traditional Relationship

The hours spent together every day by a mahout and elephant make for a unique human-animal relationship unlike that experienced with a family member or a companion animal. Such relationships continue in some villages in Karnataka in Southern India. The mahouts live with their families near the elephant stables. The elephants provide a focal point of village activity and family involvement. Female elephants freely forage overnight in the forest and are retrieved in the mornings, often by young boys who ride them back to the stables. Bull elephants are more closely managed, remaining in the stable area except when accompanied by a mahout or assistant and only handled by adult men. Mahouts are vigilant when other people are in the stable area, actively supervising the young boys for safety around the elephants. The boys slowly assume responsibilities with the female elephants; many aspire to become mahouts. Today's mahouts typically grew up around elephants with mahouts—their fathers, grandfathers, and uncles. They recount a rich history of particularly noble elephants, the current elephants, and incidents of injuries with other mahouts they have known.

In both India and Nepal, elephant stables were previously maintained by maharajahs as a status symbol, primarily to facilitate lavish hunting expeditions. By the 1970s, however, captive elephant resources were being deployed in service of the newly established national parks. Once the democratic Indian state of Karnataka was established in 1973, ownership and care of elephants that resided at the palace or within the royal hunting grounds reverted to the government. Karnataka National Park was established, and the Forest Department assumed responsibility for the domesticated elephants and the hiring of their mahouts. Similarly, in Nepal, King Mahendra laid the ground for tourism and conservation when he declared Chitwan a wildlife sanctuary for rhino conservation in 1962, and in 1963 he gave a concession to the professional hunter John Coapman to establish Tiger Tops, the first tourist resort to provide elephant-back safaris. Tiger Tops helped draw up the plans for the Royal Chitwan National Park, which was also formally announced in 1973. As in Karnataka, the royal stables were now administered through government departments, where they came to play a pivotal role in park management.

Again, in both India and Nepal, the mahouts and assistants that manage the elephants have been hired as park employees. In Karnataka, mahouts were formerly Hindu and Muslim, but since India became a democracy, these jobs have increasingly been assumed by tribals who have traditionally lived in the parks. The Hindu, Muslim, and tribal mahouts today all use the same practices in dealing with the elephants, including the specific verbal commands delivered to the elephants. Traditional elephant-keeping practices also exist in Nepal, where handlers live and work in the elephant stable. Only occasionally do they visit their families, who typically live far away. In what was previously a family tradition, elephant handlers are recruited from areas that were once forested and no longer have their own stables. Although some sons still follow their fathers into the profession, most handlers hope their sons will not become mahouts, a profession that affords little prestige and only a meager income. However, those without land may have no other viable option except for serving as an agricultural laborer. Today, most people in or around Nepal's lowland national parks remain indirectly dependent on elephants as a commodity crucial to management of the national park and its attendant tourism.

In Nepal, rather than a Hindu caste, the indigenous Tharu were involved with elephant handling in the lowland strip of Nepal known as the Tarai from at least 1783, and probably much earlier. A complex system of ranks and roles developed in the eighteenth, nineteenth, and early twentieth centuries at numerous government stables throughout the Tarai, staffed by Tharu people. When not needed for elephant capture or regal hunting, the elephants could be used by local people for transport, agriculture, and logging. Nowadays, although the low-status Tharu still constitute the majority of elephant handlers, other groups have also begun to take up the profession and must adapt to the Tharu-style life of the government stable. The Nepali elephant stable is a Tharu institution, with its own ritual practices, as well as veneration of the elephant as the living embodiment of the Hindu deity Ganesh.

An Emerging Relationship

Elephants today in India and Nepal have little conventional work to do because machinery is available for logging, agriculture, and road building. Four-wheel-drive vehicles ford shallow streams during monsoons virtually as well as elephants. Although ancient practices continue, traditional keeping practices are being shaped by the modern objectives of natural resource management and tourism, leading to some new roles for elephants. Myanmar, as an exception to this pattern, still employs more than 2,000 elephants in the logging industry. In neighboring Thailand, some elephants and their handlers previously employed in the timber industry now beg for a living on the streets of Bangkok, a lifestyle and an environment particularly ill-suited to the welfare of elephants. These contrasts reveal that significant challenges face the endeavor of captive elephant management, ones that impact both human livelihood and elephant welfare.

In Southern India where elephants reside at villages within the park, extended families of men have continued as mahouts for generations. The men have grown up with elephants, having a lifetime of experience with elephants, some remaining with the same elephant for many years. The high status of working the large tuskers, who formerly were the best for doing work and were selected for leading the processional celebrations, is diminished today. Working the females now offers advantages of easier management, participation in the birth of young elephants, and providing some rides to tourists, albeit less extensively than the Nepali tourist lodges.

In Nepal, a different strategy is evolving than in India for addressing the changing elephant and mahout situation. In addition to elephants owned by the government and residing in longstanding stables, elephants are privately owned by the lodges adjacent to the Royal Chitwan National Park. Typically, they are acquired from India, often with their handlers, and are managed according to a common Indian system of a first and second mahout. Some hotels even lease elephants from India for the tourist season. With this system, the elephants' personal biographies are less well-known because they are purchased from a trader or market, without any long-term knowledge of their background or past behavior. These lodges can only take their tourists for elephant safaris in the forests adjacent to the national park and have no opportunities to take their elephants to graze. Seven other safari lodges have concessionary licenses to operate within the national park, with full grazing rights. These generally maintain their own stables, managed according to the Nepali system, which include a phanet (driver), patchuwa (grass-cutter), and mahout (junior assistant) for each elephant. One of these lodges, the Temple Tiger resort, has pioneered a new system for flexibility in human resource management—all handlers work with each of the elephants. The elephants of the Sauraha government stable are also available for hire for tourist safaris. The elephants of the seven lodges spend most of their time restrained at the stables, except for periods when they graze in the forest while accompanied by a mahout, or when they are carrying tourists in a howdah through the forest. Residing a considerable distance from their families, the phanets go home on leave only occasionally and are with their elephants virtually full-time. Yet another category are other private handlers who have no interactions with the handlers of these major stables and about whom little is known. Inexperienced men may apply for elephant-handling jobs at lodges with only one or two elephants and receive an impoverished apprenticeship without the benefits of participating among mahouts, from whom they would learn a rich body of elephant lore and technique.

At the lodges, the elephants are around a variety of tourists every day. Although the integrated strategy of tourism and conservation in Nepal has created new jobs for elephants and their phanets, the overall context lacks the long-term stability and familiarity in the relationship that was characteristic in the Karnataka tradition, where the elephant over a lifetime is surrounded by a group of familiar people and managed by the same mahout and assistants. On the other hand, the stables managed by government and the concession-holding lodges are subject to greater external regulation governing handling practice and elephant welfare. Privately owned elephants in Nepal are often kept alone; their handlers do not get the opportunity to participate with other mahouts. In smaller lodges, the elephant may be faced with a succession of phanets giving directions, as well as a new set of unfamiliar tourists each day. Some handlers work with the same elephant for as long as twenty years or so, whereas others work with as many as six elephants in a career of similar duration. This is not considered either novel or undesirable according to traditional Tharu elephant-keeping wisdom, but may be adverse for the elephant. Males especially should have continuity in their care staff, though females of good temperament can happily adapt to new handlers.

The management preferences of certain regions may have been shaped by the specificities of their particular uses for elephants. For example, in India, Kerala's elephants are kept by temples for use in religious festivals and processions (both Hindu and Christian), for which only male tuskers are desired. In an elephant stable composed entirely of male tuskers, a lifelong bond and commitment between a handler and his elephant is especially important because a handler ideally needs long-standing and intimate experience to

A Nepali phanet is likely to have a longterm relationship with his bull elephant. Courtesy of Lynette Hart.

deal with his elephant when the elephant is in musth.

Elephants in Nepal residing at lodges with park concessions are allowed to spend some time foraging in the forest, while they are accompanied by their phanets. The females are not given the opportunity to live as free semi-captives during the night as are the ones in Karnataka, and as was formerly done in Bengal and Assam. Thus, elephants' social lives with other elephants are more controlled in Nepal. They have little tactile contact with other elephants. Although chained at night, Nepali elephants with full rights to enter Chitwan can socially interact while grazing. Handlers often dismount and let their elephants wander freely. As learned in a new handlers' apprenticeship, elephants who do not like each other and who have previously fought are kept at a distance. Those handlers whose elephants have cordial relationships with each other go off together, to allow the elephants to socialize. Elephant interaction is, then, valued and encouraged, albeit in a tactically determined way. Whereas the female elephants in Karnataka are allowed freely to mate with wild bulls and to bear and care for young for a while, such an opportunity in Nepal is more orchestrated for female elephants. Wild males are valued in Nepal for their role in impregnating females at all government stables because there is a shortage of available elephants. Once approaching full term, a pregnant female is then sent to the specialist facility at Khorsor, where she will live with her new born young until it is three years old, at which time it will be separated from her to be given training.

Conclusion

Although human factors and the working roles of elephants in Southern India and Nepal somewhat differ, there is considerable overlap and similarity in the set of commands and general management style of the two places and across the mahout cultures. Traditional systems in both countries foster an extended community and culture of mahouts. The Nepali tourist lodges that own elephants follow an entrepreneurial model in which being a mahout is more of a job than an ongoing profession. Female elephants in Karnataka have a more permissive lifestyle than those in Nepal, but elephants in both locations are given special daily baths and food treats that are a part of fostering the unique relationship of human and elephant.

Further Resources

Edgerton, F. (1931). *The elephant lore of the Hindus: The elephants-sport (Matanga-Lila) of Nilakantha.* Delhi: Motilal Banarsidass.

Hart, B. L., & Hart, L. A. (2006). *Evolution of the elephant brain: A paradox between brain size and cognitive behavior.* In J. Kaas (Ed.), *Evolution of nervous systems.* Oxford: Elsevier.

Hart, L. A. (1994). The Asian elephants-driver partnership: The drivers' perspective. *Applied Animal Behaviour Science, 40,* 297–312.

———. (2005). The elephant-mahout relationship in India and Nepal: A tourist attraction. In J. Knight (Ed.), *Animals in person: Cultural perspectives on human-animal intimacy* (pp. 163–189). Oxford: Berg.

Hart, L. A., & Sundar. (2000). *Family traditions for mahouts of Asian elephants. Anthrozoos, 13,* 34–42.

Namboodiri, N. (1997). *Practical elephant management: A handbook for mahouts.* Elephant Welfare Association of Kerala and Zoo Outreach Organization. Available online at http://www.elephantcare.org/mancover.htm

Phuangkum, P., Lair, R., & Angkawanith, T. (2005). *Elephant care manual for mahouts and camp managers.* Bangkok: Forest Industry Organisation, Thai Ministry of Natural Resources and Environment & Food and Agriculture Organisation of The United Nations, Regional Office for Asia and The Pacific (RAP).

Sukumar, R. (1994). *Elephant days and nights: Ten years with the Indian elephant.* Oxford: Oxford University Press.

———. (2003). *The living elephant: Evolutionary ecology, behaviour and conservation.* Oxford: Oxford University Press.

WWF-Nepal. (2003). *Hattisars: Managing domesticated elephants in Nepal.* Kathmandu: WWF.

Lynette Hart and Piers Locke

This orphaned, wild baby elephant is being raised by mahouts' families within their village in Karnataka. Courtesy of Lynette Hart.

■ Culture, Religion, and Belief Systems
India's Holy Cows and India's Food: Issues and Possibilities

India has the largest concentration of livestock in the world (250–300 million cattle, 60 million water buffalo, 120 million goats, and 40 million sheep), with one-third of the world's cattle on approximately 3 percent of the world's land area. The economic and social values of cattle are so great that cattle have long been seen as religious symbols and are regarded as sacred.

India's "white revolution" began in 1970, with a nationwide dairy cooperative scheme called "Operation Flood" that was initiated to increase milk production and to help the poor with low-interest loans for purchasing milk cows. The World Bank and the World Food Program provided most of the funds, but this scheme has caused many problems (Crolty, 1980). Less grain and lands are available to feed people because more are diverted to feed dairy cattle owned by the rich. Also, fodder prices have increased, creating difficulties for poor cattle owners and landless cattle owners.

But now, according to Professor Ram Kumar of the India Veterinary Council, there is only sufficient feed for 60 percent of India's cattle population. This means that some 100–120 million of these animals, especially in arid regions, and elsewhere during the dry season and droughts when fodder is scarce, are either starving or chronically malnourished.

Although Muslim, Christian, Sikh, and other Indians eat meat (buffalo, sheep, and goats, whose slaughter is permitted), the majority of Indians are Hindus, and for most of these people, the killing of cattle and eating of beef is unthinkable because this species is regarded as the most sacred of all creatures.

Cow Worship

Cow and bull worship was a common practice in many parts of the world, beginning in Mesopotamia around 6000 BCE and spreading to northwestern India with the invasion of the Indus Valley in the second millennium BCE by Aryan nomadic pastoralists who established the Vedic religion. What is remarkable is that such worship has persisted uniquely in India to the present day. Lodrick (1981) concludes that revulsion against sacrifice, the economic usefulness of cattle, and religious symbolism (especially of the cow as the Mother-provider) were factors contributing to the formulation of the sacred cow doctrine, but it was ahimsa (the principle of nonviolence or nonharming) that provided the moral and ethical compulsion for the doctrine's widespread acceptance in later Indian religious thought and social behavior.

India can be seen as two nations in one: a majority of Hindus, for whom vegetarianism is linked to caste and ritual purity; and the meat-eating Muslims, who are seen as unclean and their touch polluting (Simoons, 1961). From ecological and economic viewpoints, Muslims and Hindus are highly complementary when it comes to cattle: one eats the male calves, and the other takes the calves' milk.

Cow protection has become a highly politicized core of the Hindu religion. What was once a compassionate, symbiotic human-animal bond linked with virtuous behavior (personal purity) that brought with it such principles as ahimsa and vegetarianism for Hindus and ritual codes of animal sacrifice for Muslims that helped affirm community and family ties has now come to serve political ends.

Religious beliefs that ultimately contradict nature's reality and that see the nature of other creatures as unclean or immoral, become life-negating rather than life-affirming and cause great harm (Fox, 1996).

Cattle Welfare Concerns

There a widespread belief that there is no real cattle surplus and that India would do better with even more cattle because their organic manure is so valuable to agriculture. The environmental damage in some regions from overgrazing is especially caused by "scrub" cattle that are kept simply as manure-makers before they are driven to slaughter or they die. Their sad existence as semi-starved and often also chronically sick will continue without mass public education and government action.

The overall cattle population must be reduced; health and productivity must be enhanced through genetic improvement and by better nutrition, through the establishment of emergency fodder banks and sources of water to see them through the dry seasons; and alternative sources of income must be provided for farmers who are reliant on cattle manure as a major product, such as by raising milk-goats and producing more fodder.

According to *India Today* (January 11, 1996), "As long [ago] as 1955, an expert committee on cattle said in its report: 'The scientific development of cattle means the culling of useless animals . . . by banning slaughter . . . the worthless animals will multiply and deprive the more productive animals of any chance of development.'"

Shepard (1996) criticizes one anthropologist who wrote a long article defending the sacred cow on "ecological" grounds as a consumer of weeds and plant materials that otherwise went to waste because this view of the sacred cow is a flagrant but familiar abuse of the concept of ecology as maximum use instead of a complex, stable, biocentric community.

Seeing the increasing desertification of pasture lands caused by overgrazing, the loss of wildlife and habitat, and the cattle having less and less grazing land as good land is put under cultivation, environmentalist Valmik Thapar foresees that if the cattle problem is not soon corrected, "Finally there will be a clash because the land mass of the country can't sustain the growing human and animal population. Then the question will arise as to who is going to eat. Man or cow?" (*India Today*, January 11, 1996).

Cattle Shelters

The first animal shelters in India began with the advent of Buddhism, to whom King Ashoka (269–232 BCE) converted. Ashoka ruled over much of the Indian subcontinent, converting millions to accept Buddhism, and was the first to set up shelters and animal hospitals, although some historians believe that Buddha himself was the first to do so. Ashoka put compassion into action, by caring for animals in need, and into the law also, setting up wildlife preserves and punishments for those who abused and killed animals.

Gowshalas and pinjrapoles are located throughout India and are supported by taxes and charitable donations from the business community. Gowshalas are refuges for cattle, often linked with the Hindu Krishna movement, whereas pinjrapoles serve as a refuge for a more diverse animal population, including birds, other wild animals, and even insects and microorganisms in collected piles of household dust. A 1955 government census found that there were 3,000 animal shelters maintaining some 600,000 cattle and thousands of other animals, from deer to dogs and camels to cats.

Even though Indians know that the buffalo is a better-quality milk producer than most varieties of cows, buffaloes are rarely found in gowshalas because they are considered unclean and not worthy of the same respect as cows.

The prevailing view that such a fate as starvation is better than having cattle defiled by the butcher's knife does little to encourage local public support. Levying a tax on milk, hides, manure, bone, and meat meal fertilizer and taking a percent of the profits from wholesalers of these cattle products to help defray the costs of running a gowshala that serves the community is the kind of initiative that is needed, but one that politics in many regions would preclude.

According to Lodrick's study (1981), all gowshalas that keep dry cows and cattle that cannot be rehabilitated for draught work operate at a deficit. Attempts to make them more productive are not likely to significantly reduce this deficit, and so without adequate community and government funding, as is the case throughout much of India, cattle suffer a fate surely worse than the butcher's knife.

Cattle Death Drives

Millions of old, spent cows, exhausted bullocks, and young male calves are driven on foot up to 300 miles or are crammed into trucks for transit into Kerala, or in railroad cars to West Bengal, the two states where cattle slaughter is legal. Their often bleeding, worn-down hooves make hardly any sound as they pass by. Veterinarian Dr. Ghanshyam Sharma from Sikkim, in the Northeast of India where cow slaughter is also legal, sees cattle coming in from Jamma, Kashmir, Bihar, and Nepal. He observes, "Often entire hooves of these animals are snuffed out and gunny bags are tied around the wounded stumps and this way they walk. Many sustain injuries being loaded and off-loaded during part of the journey or die in transit. Some collapse on the way, are beaten, and even have salt and hot chilies rubbed into their eyes and have their tails hammered, twisted, and broken to make them get up and keep walking. Some of those being transported get trampled and suffocate or have an eye gouged out by another's horn. Water and fodder are rarely provided during their long journeys, even at rest stops. An estimated one million cattle are taken every year into Kerala from other southern states to be slaughtered" (*India Today*, January 11, 1996).

Journalist Subhashini Raghavan, in his exposé of these cattle death marches, found a complex network of middlemen traders, "who are calloused by constant exposure to cruelty" and who develop the attitude that "if an animal is slotted for slaughter, it ceases to be a living being with pain, hunger and terror." Raghavan found that vast numbers of cattle are made to walk hundreds of miles through pedestrian side roads to escape checkpoints, en route to regional markets from local markets and then on to transfer points where they may then be put into trucks. He concludes his article stating that "throughout the length and breadth of this birthplace of Ahimsa, the tragic march of the condemned continues unabated—a poignant symbol of our callousness, in even denying the last comforts and dignity of those who lived their lives serving us." (*The Hindu,* April 16, 1995).

Cattle Slaughter

Belief in *ahimsa* (not harming) and in *aghnya* (not killing) possibly arose as a reaction against the Vedic religion and social order that sanctified animal slaughter, the Brahmins being the highest priestly order in the Hindu caste system that supervised the killing, according to Harris (1991).

Between the eighth and sixth centuries BCE, a new wave of philosophical treatises emerged that included references to ahimsa as well as reincarnation and karma, treatises that were not included in the Vedas. These treatises, along with the emergence of the religious traditions Buddhism and Jainism, which espoused ahimsa, were a challenge to orthodox Hinduism and may have led to the Brahmins prohibiting cow slaughter and promoting ahimsa. Yet still today, thousands of animals—buffalo, sheep, and goats especially—are slaughtered in Hindu temples.

Except in West Bengal and Kerala, where cattle slaughter is permitted, the Cow Slaughter Act prohibits the killing of cattle under sixteen years of age. The penalty for illegal slaughter of cattle is rigorous imprisonment for two years and a fine. The Constitution of India, Part IV, Directive Principles of State Policy, Article 48—Organization of Agriculture and Animal Husbandry, says, "The State shall endeavor to organize agriculture and animal husbandry on modern and scientific lines and shall, in particular, take steps for preserving and improving the breeds and prohibiting the slaughter, of cows and calves and other milch and draught cattle."

According to one government study, 50 percent of small-animal slaughtering and 70 percent of large-animal slaughtering is illegal, taking place in clandestine facilities where there is no supervision of hygiene, animal welfare, or meat-safety inspection (Report of the Expert Committee, 1987).

Of the 3,600 licensed abattoirs in India, only two are mechanized and hygienic, and these are facing strong public opposition (*India Today*, January 11, 1996).

Article 51-A (g) of the Constitution of India states, "It shall be the fundamental duty of every citizen of India to protect and improve the natural environment . . . and to have compassion for all living creatures." This is not in keeping with the predominantly religious sentiment that interprets compassion for living creatures as "rescuing" cows and other abandoned cattle from slaughter and putting them into death camps where they starve to death or die slowly from infections and parasites.

Euthanasia of suffering animals, according to the Prevention of Cruelty to Animals Act, is allowed if "it would be cruel to keep the animal alive," but only if the court or other suitable persons or police officers above the rank of the constable concur. Because of the religious opposition to euthanasia, even of dying animals in severe pain, there is no legal requirement that the owner of such an animal should have it killed. Many orthodox Hindus and Jains oppose the killing of animals for any reason because they feel it is wrong to interfere in any way with another's karma or destiny. It would seem that the doctrine of ahimsa as it relates to the treatment of cattle has been corrupted to serve the interests of social status, caste distinctions, and politics because lower Hindu castes, tribal peoples, and others do kill and consume cattle and other animals, be they healthy or in a condition that calls for immediate euthanasia. Some Indians have reasoned that killing a sick cow is like killing your own mother, and that is unthinkable (see also Simoons, 1961).

The Animal Welfare Board of India, the chronically understaffed and underfunded government agency without any power to enforce animal protection laws, does help subsidize local Blue Cross and SPCA animal shelters and hospitals. But without more support from the central government and from foreign animal protection organizations, the plight of India's animals will worsen as the human population increases and resources become ever more scarce and costly.

Vegetarianism, Religion, and Politics

Vegetarianism in India, like ahimsa, has as much, if not more, to do with concerns about reincarnation, one's personal degree of spiritual purity, and place in society (caste) as with immediate concern for animals. But it is not total vegetarianism because most Hindus and Jains consume dairy products. Few are pure vegan (eating no animal products). Some Jains have acknowledged that to be consistent with their religious beliefs and with the ecological and economic dictates of the current situation, veganism is an ethical imperative. Abstaining from all dairy products would be more consistent with the principle of ahimsa that they hold so dear than "saving" spent dairy cows, calves, and bullocks from slaughter and condemning them to slow death by starvation in gowshalas or pinjrapoles.

Although it was in Jainism that the principle of ahimsa was first espoused, contemporary Jains who are landowners get around the problem of violating ahimsa by having others do the farming, clear the land, kill wild creatures, plough the soil and kill worms, and use all manner of pesticides.

Jainism reached its peak between the fifth and thirteenth centuries CE, spreading across much of India, and then was superseded by Hinduism, and then again by Islam following the invasion of the subcontinent by the Moguls in the eleventh century. Muslims killed and ate cattle, which was anathema to the non-tribal, upper castes of Hindu society. Cow protection and worship then gained political importance and popularity in opposition to Muslim rule and influence. Hindus and Jains will confide today that it is better to put a calf in a gowshala than have a Muslim eat it.

Cow protection became a political icon for Hindus in their conflicts with Muslims and also when under British rule. Muslims settled in India around the thirteenth century and can trace their roots to Mogul pastoralists and Arab-Islamic values. Their ritual slaughter of buffalo, sheep, and goats is looked down on by Hindus, some castes of which, nonetheless, eat meat. According to Srinivas (1968), the whole Brahminic caste is vegetarian. Of the nonvegetarian castes, fish-eaters look down on those who eat goats and sheep, who in turn look down on eaters of poultry and pigs, who look down on beef-eaters.

Muslims, under British rule, fought successfully to have their religious freedom of ritual slaughter upheld. The British wanted preslaughter stunning for humane reasons, but this was not part of sacrificial ritual slaughter under Islamic law. Preslaughter stunning eliminates the need to cast the animal onto the ground prior to having its throat cut, thus eliminating much fear associated with being cast.

For Mohandas Gandhi, cow protection was an important aspect of Indian independence from British colonial rule, figuring in the return to traditional values. He wrote,

> The central fact of Hinduism is cow protection. Cow protection to me is one of the most wonderful phenomenon [sic] in human evolution. It takes the human being beyond his species. The cow to me means the entire subhuman world. . . . Man through the cow is enjoined to realize his identity with all that lives. . . . Cow protection is the gift of Hinduism to the world. And Hinduism will live as long as there are Hindus to protect the cow. (Gandhi, 1954)

Lodrick (1981), in reviewing this history of animal care and shelters in India, concludes, "Buddhism, although the major vehicle for the spread of the ahimsa concept throughout India and indeed throughout much of Asia, never carried the doctrine to the extremes of Jainism. In Buddhist thinking, ahimsa became a positive adjunct to moral conduct stemming from the cardinal virtue of compassion, rather than the all-encompassing negative principle of non-activity of the Jains." Srinivas (1968) contends that humanitarianism is an alien concept in his native country of India because it is the antithesis of those mores that embrace caste, religion, social position, and other human-centered values.

Agricultural Modernization and Humane Progress

Humane concerns over animal slaughter and attempts to modernize slaughtering facilities to make them more humane, sanitary, less wasteful, and less polluting have been opposed by both Muslims and Hindus for religious and political reasons. Muslims see it as threatening their religious freedom (by the adoption of pre-slaughter stunning), and many Hindus see slaughter modernization as a threat to traditional values, totems, taboos, and even national identity and security.

Such opposition is reminiscent of the Hindu cow protection movement that arose in opposition to British rule and the proposed slaughter of cattle as part and parcel of economic development and modernization. Now under the pressures of trade liberalization and an emerging global market economy that is being pushed by the World Trade Organization, efforts to modernize livestock slaughter are being renewed; and opposition intensifies.

Some believe that it is unwise economically and ecologically, as well as socially unjust, to raise any species of farm animal in India (or in any other country for that matter) primarily for meat, eggs, or dairy products and that more animal fat and protein for the rich means less bread or grains for the poor (Fox, 1997). A major goal, some think, should be to reduce the overall livestock population to facilitate ecological restoration. Increasing the productivity and health of milk cows and goats through selective breeding and husbandry improvements also needs more concerted and effective attention and financing. In this way, meat from male offspring and nonproductive females could be a by-product rather than a primary product and could either be consumed locally or be marketed to the meat-consuming sectors.

As India shifts to a more capital-intensive industrial agriculture, countless native cows become surplus and urban scavengers for their impoverished owners, and rare breeds become extinct. Many native peoples have been made landless by agricultural "modernization," and they migrate in increasing numbers to the cities along with their few animals and possessions. The high cattle population in the nation's capital, Delhi, is evidence enough. In 1995 some fifty cattle per day were killed or severely injured by traffic (Kare Newsletter, 1995).

The Prevention of Cow Slaughter Act of 1955, which prohibits the slaughter of cattle that are diseased, disabled, or less than fifteen years old, allegedly resulted in young, nonproductive cows having their legs hacked and broken so that they could be legally slaughtered.

The Bharatiya Janata Party (BJP) banned all slaughter of the bovine species when it gained control of Delhi in 1994, purportedly to tighten various laxities in the prohibition of cow slaughter. The BJP invoked Mohandas Gandhi, who told all India in 1921, "Hindus will be judged . . . by their ability to protect the cow."

During the tumultuous 1996 elections, the Vishnu Hindu Parishad (VHP) party, "ignoring the facts and problems" of cattle overpopulation, starvation, disease, and suffering, according to *India Today* (January 11, 1996), launched an anti–cattle slaughter campaign.

India is now at a crossroads where the choice is between rural sustainability and industrial growth and productivity. It is clear which road India is now taking. India exports much animal produce—millions of tons of dairy products, hides, bones, horns, hooves, meat, poultry, and eggs—even to developed countries such as the United States, Australia, and the United Kingdom. The toxic chemicals that most of India's tanneries continue to discharge into rivers and watersheds cause serious ecological and human health problems. Although some 200 million people are malnourished in India, the country exported US$625 million worth of wheat and flour and US$1.3 billion of rice in 1995 (Lappe et al., 1998).

Some who have studied the problems of the symbolic, material, economic, emotional, social, and spiritual components of the human-cow/cattle relationship believe that a unified sensibility is called for, which becomes a mutually enhancing symbiosis. They believe that the human side of the relationship is more balanced and equitable when the rights, interests, and welfare of animals are given equal and fair consideration. The ethical inconsistencies in the religious and secular communities' attitudes toward and treatment of cattle and other animals is more evident in India than in other countries precisely because India is the birthplace of the highest spiritual principles pertaining to animal welfare, and yet those principles are not always put into practice, creating an essentially disjointed situation between the ideal and the real.

Caring for animals and caring for people, for the poor and the hungry, go hand in hand as part of the humane agenda of any democratic society. It is not unreasonable to consider that the plight of India's cattle mirrors the plight of the poor. There are no miracle remedies for hunger and poverty from advances in technology, science, or medicine. Some who have studied these issues believe that the miracle will come not via genetic engineering of animals and plants but through the transformation of humanity into a compassionate, empathic, and responsible life form. A mutually enhancing symbiosis with the Earth community of plants and animals, both wild and domesticated, suggests our most viable future. Our hope lies in our capacity to reconnect empathically with all living beings and to use sound science and policies as instruments and compassion as the compass.

Further Resources

Crolty, R. (1980). *Cattle, economics and development*. New York: Oxford University Press.

Fox, M. W. (1996). *The boundless circle: Caring for creatures and creation*. Wheaton, IL: Quest Books.

———. (1997). *Eating with conscience: The bioethics of food*. Troutdale, OR: New Sage Press.

———. (2001). *Bringing life to ethics: Global bioethics for a humane society*. Albany: State University of New York Press.

Gandhi, M. K. (1954). *How to serve the cow*. Ahmedabad: Navajivan, pp. 3–4.

Harris, M. (1991). *Cannibals and kings: The origins of cultures*. New York: Vintage Books.

Kindness to Animals and Respect for the Environment (KARE). (1995, July). *Expose Newsletter*, 4(1).

Lappe, F. M, Collins, J., & Rosset, P. (1998). *World hunger: Twelve myths* (2nd ed.). New York: Grove Press.

Lodrick, D. (1981). *Sacred cows, sacred places. Origins and survivals of animal homes in India*. Berkley: University of California Press.

Report of the Expert Committee on Development of the Meat Industry. (1987). New Delhi: Ministry of Agriculture and Co-operation

Shephard, P. (1996). *The others: How animals make us human*. New York: Island Press.

Simoons, F. J. (1961). *Eat not this flesh*. Madison: University of Wisconsin Press.

Srinivas, M. (1968). *Social changes in modern India*. Los Angeles: University of California Press.

Michael W. Fox

Animals in India: A Personal Narrative

Pradeep Kumar Nath

I have spent much time trying to save and protect animals in my native India. During that time, I have discovered that the animal's level of tolerance and faith is very special. They easily adapt to many different and difficult situations and develop deep loyalty to the humans who help them. They are amazingly forgiving.

My entry into animal protection and my pledge to strive to fight on their behalf goes back thirty-five years, when, sitting in front of my house, my friend suddenly threw a stone at a sparrow hardly twenty feet away. His aim was perfect and the sparrow was down instantly, blood oozing from his beak. After a few minutes of shivering and trembling, the sparrow collapsed. Those few minutes touched my heart and changed my life immediately.

This change was very natural for me. From my schooldays onward, when I took food from my own plate food to feed the hungry dogs on the streets, I built an everlasting relationship with street animals. I loved to spend my time with them wherever and whenever I was able to. I continue to work with street dogs because they are innocent and abused.

During the years that I've worked to protect and save animals, local authorities have tried to prevent me from doing so. Even as a child, I witnessed numerous atrocities such as poisoning, bludgeoning and clubbing, strangling, and starving dogs and even injecting strychnine into dogs at local pounds. I even have had to fight dog catchers who would rather kill dogs than try to save them. I tried every means to free them, ranging from scaling the walls of the dog pound, to going through the drainage pipes to reach the pound, to bribing the watchman, to removing the latches and wire-tied gates of the trucks, to paying money to release the dogs. Many times, I chased vehicles carrying impounded dogs and visited the pound in the evenings after school when I found the dogs missing on my feeding rounds. The pain in the eyes of the dogs facing death will forever be fresh in my heart.

After working on my own for twenty-five years, I came to realize that there was a way to fight legally for the animals, and I formed the Visakha Society For Prevention of Cruelty to Animals (VSPCA). I was working, and still work, in a bank, but much of my time is devoted to saving animals.

The three-legged cow Lakshmi and the disabled dog Tiger stand as testimonials for the significance of life despite their severe handicaps.

Lakshmi is now eleven years old. I saved her ten years ago and had to amputate her front leg so that she would not die. I did this myself because the government veterinarians would not do it without being paid. As a calf, Lakshmi had no value to anyone, even the butchers, and left to her fate in a huge drain with the broken leg, she has since been under my care and lives a full life.

Tiger was hit in the back by a speeding car as a puppy, and he has since been paralyzed. Tiger is now fourteen years old, but he looks the same as when I found him thirteen years ago. We named him Tiger after the famed Britannia Tiger biscuits he liked so much. The innocence and purity of these and other rescued animals appeal to me. The truth is in their eyes.

VSPCA has had an immense impact these past ten years. Through various means and methods, VSPCA continues to protect all kind of animals.

Further Resources

Visakha Society for Prevention of Cruelty to Animals (VSPCA). http://www.VisakhaSPCA.org/

■ Culture, Religion, and Belief Systems
India's Sacred Animals

The Hindu religion generally considers all life forms as sacred, and various species, ranging from the tiny insect to the big elephant, are regarded as equally sacred. The long list of sacred animals covers major taxonomic groups, including invertebrates (bee, butterfly, mollusk, spider), reptiles (crocodile, lizard, snake, squirrel, turtle), birds (eagle, falcon, crow, crane, goose, hawk, owl, peacock, swan, dove), and mammals (bat, bear, wild boar, buffalo, bull, cat, cow, dog, deer, elephant, fox, goat, horse, leopard, lion, monkey, rat, tiger, rabbit). The Hindu religion has developed sanctity by association, such as the swan, eagle, and bull that serve as the vehicles of the major Hindu deities Brahma, Vishnu, and Shiva, respectively. Some of the animals are sacred themselves, such as Hanuman, the monkey god, Naga, the snake god, and Ganesha, the elephant god. All major Hindu temples across India have maintained captive elephants for centuries because they play a role in the religious rituals in temples.

The Hindu religion recognizes the animals' right to coexist with humans and, therefore, people are taught to love, nurture, and worship animals. The religion promotes the belief that various Hindu gods and goddesses incarnate in different animal forms. In the past, kings and emperors used different kinds of animals in their emblems to show their respect for and the divine powers of animals. Many festivals in India are still celebrated to honor various types of animals.

The concept of the reincarnation of the god Vishnu represents, in a way, the theory of organic evolution involving animals. In order to indicate the aquatic origin of life forms, Vishnu incarnates in the form of Mathsya, a fish, followed by an amphibious Kurma, a turtle. The next incarnation is Varaha, the wild boar, a terrestrial animal, depicting how life transferred from the aquatic to terrestrial habitat. Subsequently, Narasimha represents a beast's attempt to attain a human form, followed by Vamana, a pigmy human. In the incarnation of Ramachandra, perfect human qualities are identified, while the last one, Kalki, represents the human destruction of the planet, giving scant attention to our fauna and environment.

The prehistoric crocodile is also given importance in the religion. It is believed that the river Ganges depends on a crocodile and her frequent visits to the Bay of Bengal from the Himalayas. The mythical elephant story is well known—when the elephant Gajendra visited a river, a crocodile attacked one of its hind legs. The elephant screamed for God's help, and the hungry reptile was destroyed by God to free the elephant.

Several mythical stories are associated with the sacred animals of India. For example, Garuda, the eagle, is worshiped, and pigeons are the favorite animals of Yamaraja, the god of death. Yamaraja rides a bull water buffalo whenever he intends to visit the earth. Karthikeya, the son of Shiva, employs a peacock for his transportation, and the goddess Saraswati, who possesses the powers of speech, wisdom, and learning, rides a swan. The crow is very well versed with what happens in heaven and, therefore, people who wish to go to paradise try to please it. The deer is associated with many mythical stories, as well. Lord Shiva gets wrapped up in deer skin. Vayu, the air god, has his chariot pulled by a pair of deer. The god Indra employs Ucchaishrava, a snow-white, seven-headed flying horse, for his transportation. Later he changed to Airavata, an elephant with multiple trunks. The sun god's chariot is pulled by seven red horses.

Although there are about 238 species of snakes found in India, only the cobra, with its two-eyed hood, is worshipped widely. Snake worship in India is intriguing. Adisesha, the king of snakes, is the couch of Narayana, a form of Vishnu as he lies on the ocean. A snake can be commonly seen around the neck of Shiva, the Hindu supreme God, in most temples. Also, snake stones are found in almost all Hindu temples. In the western state of Maharashtra, a celebration called Naag-Panchami is specifically devoted to the worship of cobras. Similarly, snake worship is called Jhampan in the eastern state of West Bengal.

The most sacred of all the animals in India is the cow, which is regarded as a substitute mother, Kamadhenu, the giver of all desires. The sanctity of cows may have been based on economic reasons. The Vedic Aryans were a pastoral tribe, and cattle were a major source of wealth, similar to the Masai tribe in East Africa today. Cows provide milk, which helps sustain the life of adults and children. The by-products of milk—yogurt, buttermilk, butter, and ghee—are an integral part of people's daily diet in India. Cow dung is used year-round for fuel. The dung, also used to clean houses, has recently been proven scientifically to have antiseptic value. Cows are also an excellent beast of burden, pulling carts and plowing fields to plant crops. Even after death, cow hides are useful to humans.

Not all India's sacred animals live in peace nowadays. India's livestock population, the world's largest, is estimated at 500 million. More than half are cows, buffalos, and bulls. Once they become unproductive, many of them are sold by their owners, usually

A captive elephant in a Hindu temple in India. Courtesy of G. Agoramoorthy.

subsistence farmers, and sent to slaughter houses. Cow slaughter is permitted only in two Indian states, West Bengal and Kerala. The cows go through horrendous cruelty when they are transported for slaughtering.

Human population pressure is increasing, and forest habitats are shrinking alarmingly, creating unavoidable conflict between man and sacred animals in India. Wild elephants do not have enough forest to roam free in search of food, and they encounter angry farmers who want to protect their farmlands from encroaching elephants. Their captive counterparts in temples are not always happy either; some temples house elephants in very poor conditions. Monkeys that live in urban areas have no access to natural food sources and thus are forced to beg and steal from city dwellers. Poisonous snakes that roam around the forest are now moving into populated areas due to expansion of agriculture and habitat loss. Consequently, deaths due to snake bites number in the thousands each year. Tigers are hunted to near-extinction due to the extensive demand for their body parts in traditional Chinese medicine.

Some animals, however, manage to adapt to the fast-changing environment. Crows have learned to build their nests using steel wires and plastic and are able to survive in the urban landscape. The future survival of India's sacred animals lies with the people of India.

Further Resources

Majupuria, T. C. (2000). *Sacred animals of Nepal and India: With reference to gods and goddesses of Hinduism and Buddhism.* New Delhi: Adriot Publishers.

Govindasamy Agoramoorthy

Animal Protection in India's Bishnoi Community

Govindasamy Agoramoorthy

Vegetarianism, conserving sacred forest groves, and protecting wild animals have been part of the culture and tradition among India's religious communities for centuries, and these practices are signs of ethical use of natural resources. While living in Jodhpur, Rajasthan State of India, during the early 1980s, I was fascinated by the Bishnoi religious community and the way that they respect plants and animals that dwell among the harsh arid environment.

About 6 million Bishnoi have spread over the north Indian states of Rajasthan, Gujarat, Haryana, Uttar Pradesh, and Madhya Pradesh, and they consider wild antelopes such as the Indian gazelle, nilgai, and blackbuck as reincarnations of their ancestors—they allow these amazing creatures to feed and roam free and brave on their farms. They protect *Orans,* which are islands of vegetation that harbor various desert animals and plants set aside by the community for worship in their villages. Nobody is allowed to cut a tree or kill any wild animal, following old traditions to promote nature conservation for over half a millennium.

The Bishnoi community is one of the most revered communities in India, whose informal laws for preserving nature and wildlife have now become a guideline for conservation in many areas. Their teachings lay the roots of the chipko (meaning to "cling") movement.

The word Bishnoi comes from the number of teachings; bis (twenty) and noi (nine), which refers to an eclectic collection of commandments compiled by a saint—Jambhoji, who may be considered as the foremost Indian ecologist. Born in 1451, Saint Jambhoji created the principles to preserve biodiversity and encourage animal husbandry, healthy social behavior, personal hygiene, good health, and guidelines for worship. He banned killing of wild animals, to the extreme extent of brushing the firewood they collect to see

Blackbucks and other antelopes freely roam in the Bishnoi agricultural land. Courtesy of G. Agoramoorthy.

that it is devoid of insects. Wearing blue clothes is prohibited because the dye for coloring is obtained by cutting a large a quantity of shrubs. They rear only cows and buffalo because goats and sheep devour desert vegetation. They bury their dead, often in sitting position, to save precious trees and wood that otherwise would be used for a cremation. The deaths of animals are mourned as if they are the people's relatives.

The Bishnois became prominent in 1730 when the King of Jodhpur's men started to cut down trees, especially the *khejadi*, for fuel. A woman and her three children who saw the trees being destroyed by the king's men attempted to stop them by hugging the trees, but the four were executed. The stunned community was electrified into action, as one by one hugged the tree, and all were massacred; a total of 363 lives were lost on the spot. After hearing the news of the carnage, the king apologized and had a copper plate engraved, which stated that trees will not be cut, and wild animals in all Bishnoi villages will not be hunted. A monument lingers as testimony to the sacrifice in the village of Kejarii, near Jodhpur, where the incident occurred.

Further Resources

Ann, G. G. (2002). Children and trees in North India. *Worldviews: Environment, Culture, Religion, 6,* 276–299.
Malhotra, S. P. (1986). Bishnoi—Their role of conservation of desert ecosystem. In K. A. Shankarnarayan & V. Shankar (Eds.), *Desert environment and management* (pp. 23–24). Jodhpur: Central Arid Zone Research Institute Publication.
Weber, T. (1988). *Hugging the trees: The story of the Chipko Movement.* New Delhi: Viking.

■ Culture, Religion, and Belief Systems
Indigenous Cultures and Animals

Animal-Dependent Cultures

The picture of indigenous cultures is immensely varied, perhaps leading to the overall conclusion that animals are less objectified and still have more subject status than in modern Western cultures. The idea of human supremacy is also less pronounced. But despite their often-considerable knowledge of animal ways, non-Western cultures tend to be anthropocentric insofar as they include animals in their own—human—socio-cultural domain. Animals may be recognized as humanlike persons while being denied their own worldview and ultimately their "animalness."

Anthropologists have written extensively on the role that animals play in human ceremonial and religious life. In hunting and gathering societies people tend to have an organic world view, which leads them to place themselves *within* rather than *above* the natural world.

In various Australian Aboriginal belief systems animals are thought of as kin. Certain humans and certain animals are taken to be part of the same totemic group. A mythical Dreamtime ancestor, for example a kangaroo, has made "dream tracks" upon the earth, thereby creating the features of the landscape (Newsome, 1980). For an instructed and initiated Aboriginal person, Australia is covered with a mosaic of cultural compositions in each of which a portion of earth (a place), a portion of life (a species), and a portion of society (a human group) are joined together. A clan and territory are connected with certain species whose prototypical powers were locally active in the Dreaming in creation (Maddock, 1982). A kangaroo totemic group includes not only kangaroos and human men and women but also other species and perhaps the rain and the sun. It is much more than a classificatory device: implicated is a moral and ritual relationship. The human members of the group refrain from killing, plucking, or eating their totem, and if extreme hunger constrains them to do so they commonly express their regret and perform some ritual act. The people actually care for the animal species to whose totem they belong. Other persons with different animal totems may and do, of course, kill kangaroos (Elkin, 1967).

Hunting Animals

Numerous anthropologists have looked upon hunting as a mere technical device for making a living. Hunting, however, implies chasing, frightening, wounding, and killing other living beings. Hunting peoples have to inspire fear, wound, and inflict pain and death upon animals in order to survive, but this does not stop them from strongly acknowledging their interrelatedness with the animal. There exists abundant evidence that people committed to hunting, torturing, and killing animals often do so with a certain reluctance and feeling of guilt. Throughout the procedures of hunting an attitude of ambivalence prevails, and the rationalizations and rituals so often surrounding the practice of hunting are all part of an overall effort to reconcile two conflicting tendencies.

Greenland native author Finn Lynge points out that according to ancient myths, respectful hunters will never go hungry: the animals will seek them out, as if asking to be taken. (In real life, however, animals often do resist their own killing.) Inuit hunters believe that disrespectful attitudes toward the animals will bring misfortune, mediated by the spirit powers of the animals concerned (Lynge, 1992). According to the Inuit, the animals depend on being hunted in order for their species to thrive, in other words, animals need killing. Thus the people take good care so as not to cause their prey animals unnecessary suffering, although many observers report that they do not extend this caring attitude to their sled dogs, who often are treated quite roughly (cf. personal communication, anthropologist A.J.F. Köbben).

A number of Australian Aboriginal tribes adhere to the notion that nature, including animals, is not self-managing but that it needs people to perform rituals. Lack of human ritual participation is thought to lead to a deterioration of the natural status quo. Here, too, animals are said to depend on human-performed acts (i.e., rites) for survival (Bennett, 1983; Morton, 1991).

Herding Animals

Nomadic hunters are not the only people famous for their close relationship with animals. The 1940 classic by Evans-Pritchard, *The Nuer,* is about African pastoralists who are—or at least were—profoundly influenced in their outlook by their dependence on and their love of cattle, a love that comes close to the affection people show toward pets. To the Nuer, cattle not only have instrumental value as sources of meat, blood, and milk but also are vital links in social relationships. However, although the Nuer spend their lives ensuring their animals' welfare, here too one may detect a certain ambivalence in their attitudes and practices, such as how they skillfully manipulate cow psychology and cow sociology.

Cattle are not primarily raised for slaughter: the oxen frequently play a sacrificial role in ceremonies. Although there are some special occasions when people gorge themselves on meat, generally it is thought that people ought not to kill an ox solely for food; the ox may even curse them. Eating an ox's flesh should only be done in severe famine. Nevertheless, any animal that dies a natural death is eaten. The Nuer are fond of meat and declare that on the death of a cow "The eyes and the heart are sad, but the teeth and the stomach are glad. A man's stomach prays to God, independently of his mind, for such gifts" (Evans-Pritchard, 1940).

In some pastoral cultures people traditionally believe that sanctions exist on failure to perform certain duties, sanctions not wielded by gods but by livestock. The Kel Ewey Tuareg people of northern Africa who keep goats and camels believe that their animals' social world is structured like their own and that animals relate to each other in the same way that humans do. According to these people, animals, like humans, have their own will, and human exploitation of animals is therefore restricted. The Tuareg typically do not control their camels' reproduction. Male and female animals mate far away in the desert and have to be chased and caught over and over again. Female goats may occasionally refuse to give milk, or so it is believed, when people have been known to sell that milk, which is considered immoral by traditional Tuareg.

Animal Suffering in Respectful Cultures

The bovine called "mithan" in tribal India serves as a beast of sacrifice, is worshipped, and is thought to be capable of all sorts of supernatural acts. However, this does not safeguard it from human cruelty, partly because the people do not perceive their treatment of the mithan as cruelty. For various hill tribes in India the mithan used to be important as a sacrificial animal, for instance, in marriage and death ceremonies, at times of illness and misfortune, in rites to maintain fertility and well-being, to mark important events, to seal friendship pacts, and in Feasts of Merit. The slaughtering and sacrificial methods (strangulation, random stabbing, or beating to death) are connected with religious beliefs about where the soul of the animal resides. (Simoons & Simoons, 1968). Various peoples of the Indonesian archipelago also venerate a bovine: the water buffalo. Here, too, the people's goodwill and respect for the buffalo is not always conducive to buffalo welfare. During the bloody sacrifice the buffalo is flattered and apologies are offered so as to get the victim to undergo its fate willingly and not revenge itself (Kreemer, 1956).

When a human moral and social system is projected onto the rest of nature, albeit in all sincerity, one runs the risk of losing sight of the animal as the Other. In that case animals are in danger of being deprived of their own domain and their own way of experiencing the world. Failure to acknowledge the animal's world as well as humanity's limitation in fathoming these worlds, results in a form of anthropocentrism.

The animal Other does not and cannot live in the same socio-moral domain as humans do. Animals live in their own societies and in their own ecosystems, of which humans may or may not be a part. Presumably animals could not care less what sort of perceptions their human killers carry around in their heads when performing the act of killing or sacrificing. The only real things to a wounded or dying animal are its own sensations and perceptions, the reality of pain and fear. To spear or knife an animal while at the same time humbly asking its forgiveness or uttering an incantation may be a sign of equality and respect on the part of the hunter, but it may be infinitely more horrifying to the animal than a quick bullet would be.

Although certain popular preconceptions prevail about indigenous peoples having a comprehensive and harmonious relationship with all that is natural, in native cosmologies there actually exists a high degree of differentiation in treatment and ideology regarding various animal species. Some species are almost semi-gods (the Rainbow Serpent in Aboriginal mythology), and others represent evil, such as the hyena in Bushman cosmology. Some animals, notably totem animals, may be treated well, whereas others not so well.

Further Resources

Bennett, D. H. (1983). Some aspects of Aboriginal and non-Aboriginal notions of responsibility to non-human animals. *Australian Aboriginal Studies, 2*, 19–24.

Elkin, A. P. (1967). Religion and philosophy of the Australian Aborigines. In E. C. B. McLaurin (Ed.). *Essays in honour of Griffithes Wheeler Thatcher 1863–1950*. Sydney: Sydney University Press.

Evans-Pritchard, E. E. (1940). *The Nuer, a description of the modes of livelihood and political institutions of a Nilotic people*. Oxford: Clarendon Press.

Kreemer, J. (1956). *De karbouw, zijn betekenis voor de volken van de Indonesische archipel*. Gravenhage/Bandung: Van Hoeve.

Lynge, F. (1992). *Arctic wars, animal rights, endangered peoples*. Hanover: University Press of New England.

Maddock, K. (1982). *The Australian Aborigines, a portrait of their society*. Ringwood: Penguin Australia.

Morton, J. (1991). Black and white totemism: Conservation, animal symbolism, and human identification in Australia. In D. B. Croft (Ed.), *Australian people and animals in today's dreamtime, the role of comparative psychology in the management of natural resources* (pp. 21–51). New York: Praeger.

Newsome, A. E. (December 1980). The eco-mythology of the red kangaroo in Central Australia. *Mankind, 12*(4).

Simoons, F. J., & Simoons, E. S. (1968). *A ceremonial ox of India, the mithan in nature, culture, and history*. Madison: The University of Wisconsin Press.

Barbara Noske

■ Culture, Religion, and Belief Systems
Indigenous Cultures, Hunting, and the Gifts of Prey

According to many indigenous traditions of North America, animals, rather than humans, control the successes of the hunt. Because the animals possess primary agency in the hunt, there persists a common belief that animals give or offer themselves to their human hunters. These beliefs are far from universal among North American indigenous traditions; therefore one must exercise caution when generalizing about such phenomena as animals with agency. Nonetheless, such perceptions offer intriguing doors for understanding relationships between humans and animals in hunting cultures.

Western biologists have long recognized that an animal such as elk, deer, and caribou will, at a critical moment when the animal becomes aware of its predator, stop and stare down its predator. According to some interpretations, this behavior is explainable as an evolutionary adaptation to predation, particularly by wolves. When the elk, for example, stops, the predator must also stop. On the one hand, this moment provides both animals with the opportunity to rest before resuming the chase. On the other hand, this moment of stopping provides the elk with an opportunity to gain a head start on the wolf by making the first move to flee. With such a head start any healthy elk or deer can easily outrun a wolf. However, when confronted by human predators, this tactic causes the prey to become particularly vulnerable to predation. By turning to face its predator, then, the animal provides the human hunter with the perfect opportunity to shoot.

Where Western science explains the phenomenon as an evolutionary maladaptation, a variety of indigenous cultures understand the action of the ungulate as the offering of the self to the hunter. In addition, indigenous cultures often believe that animals that require trapping rather than chasing, such as beavers, otters, or bears, equally possess the power to offer themselves to the hunter. By investing the animals with agency in the hunt, indigenous peoples believe that animals possess spirits, subjectivity, and awareness of the world around them.

Not all animals, however, possess the same degree of spirit or power. Among the Nutalo Koyukon Indians of the Arctic, moose only possess a spirit during the fall rutting season. Further, one species might possess more spirit than another animal. For example the Koyukon believe that bears possess greater spiritual power than moose or caribou.

Central to the spiritual power of animals is the belief that animals are as equally aware as humans of the world around them. Many indigenous cultures believe that animals are equally, if not more, aware of the actions, words, and thoughts of humans. This perception arises from the belief that humans and animals once shared culture and spoke the same language. If animals and humans once shared culture, then animals are cultural beings whose social groups overlap with those of humans in what some anthropologists call a "community of beings."

Within this community hunting operates as one way in which human and animal worlds interact in a reciprocal relationship. As actors in overlapping cultures, hunters must act with appropriate respect and right action when approaching the hunted animals. Failure to act with proper respect almost assuredly guarantees an unsuccessful hunt. Some cultures paint this relationship in terms of courtship. Therefore, if the hunter carries out proscribed courting rituals and actions, the game will submit to the hunter's will by offering its body.

Since the choice to offer itself to the hunter is in the hands of the animal, successful hunting never symbolizes skill on the part of the hunter. For the Cree, Koyukon, Ojibwa, and other Native Americans, a successful hunt can only be attributed to the graces of the animal. The hunter is successful only if the animal offers itself to the hunter. If the hunter can note any skill, it is the ability to act according to right praxis and ethics. Several of these practices and ethical ideas are common to many hunting cultures of North America.

Hunters should never boast of hunting skill nor should they speak ill of animals. The hunter should always speak as if he is the passive partner in the reciprocal relationship of hunting. Even everyday Koyukon conversation suggests that words must be chosen carefully when speaking of a hunt. Failure to show verbal and mental respect to the animal can result in failure while hunting. In addition, successful hunters are obligated to share their catches and kills with those less lucky in the hunt.

Just as the hunter should never boast about hunting skills, the hunter must always approach the hunt with equally appropriate respect. Before hunting, hunters should present offerings to the animals of prey. If the animals are "happy" with the gifts they receive, they will in turn offer their own flesh to the humans. Gifts can include utensils, clothing, body ornaments, or tobacco. Ojibwa tales depict beavers as particularly fond of tobacco. With tobacco, the pipe often acts as an intermediary between the human and beaver worlds. Anthropologists have referred to this reciprocal relationship as an "economy of sharing."

During the hunt, hunters should never kill wantonly or use oversized weapons. Hunters should only kill what they need for the sustenance of the community. From this perspective, it is not uncommon to hear criticisms of contemporary practices of trophy hunting. Further, hunters should ensure that the kill is quick, clean, and painless.

After the hunt, hunters must say prayers to the animal. Prayers not only function as tools for thanks but also aid the spirit of the animal in its transition from one material body to another. Failure to pray properly could result in the animal's avoidance of the hunters in future hunts.

Cree Indians also stress the transport of the animal's body after hunting. For example, a beaver is normally pulled on its back in the snow. With the black bear, the struggle of transportation, due to size of bear, continues to demonstrate to the hunter the power the animal has over him.

Once back in camp, the hunters perform rituals necessary for proper butchering of the prey. Many indigenous hunting cultures believe each particular animal requires unique and individual methods of butchering. Most importantly, however, one should butcher an animal so that no parts of the animal go to waste. Wasting the gifts of game ensures that an animal will not return to offer itself in the future because waste is one of the highest of transgressions in the economy of sharing.

As noted, sharing of game within the community is a crucial aspect of showing respect to the animal. According to the Cree, during the process of butchering, the bear decides and reveals to the hunter how he must divide the meat among the families. Just as one hunts with a concern for need over want, so the hunter must always think of himself in relation to the larger community.

Disposal of bones and nonusable remains of the kill are highly stressed elements of this reciprocal exchange. Cree, Ojibwa, and Koyukon narratives reveal the shared belief that the spirit of the animal resides in the bones of the animal. If the animal is to return to a new body the bones must be disposed of properly. Beyond the loss of future hunting success, improper disposal of bones could lead the spirit of the animal to haunt, molest, or bring ill fortune to the hunter, his family, and perhaps the community.

First, bones should never be broken, destroyed by fire, or consumed by another animal, such as a dog. Second, both Cree and Ojibwa believe that the unbroken bones must be returned to the natural habitat of that particular animal. For example, beaver bones need to return to the water for proper burial. Proper attention to the bones ensures a return on the part of the animal and therefore ensures hunting success in the future.

With proper treatment, action, and attitude throughout every stage of the hunt from the pre-hunt ritual to the disposal of bones, hunters ensure the willingness of animals to offer their bodies as sustaining gifts in future hunts. Throughout each stage the hunter must remain cognizant that it is the animals, not the people, who control the success of the hunt. From such detailed attention throughout the hunt to relations between human and animal communities, environmental ethicists seeking more sustainable ways of relating to the natural world frequently applaud the practices of indigenous peoples of North America. Broadly speaking, indigenous practices relating to the environment, such as these of hunting cultures, fall under the canopy of "Traditional Ecological Knowledge."

Traditional Ecological Knowledge stresses a tripartite plan of knowledge-practice-belief under which these elements of hunting fall. Central to this environmental ethic of belief and practice is the idea that humans and animals coexist in reciprocal communities. Proper respect and ethical practice, therefore, is always required for the continued survival of humans, animals, and the earth at large. Hunters must sustain the worlds of the animals so that the animals will sustain the worlds of humans.

Various efforts to celebrate the practices of indigenous peoples have been critically evaluated for a number of reasons. First, scholars warn against universalizing the practices and traditions of a large variety of peoples into monistic terms. Second, white scholarship on indigenous cultures often receives criticism as simply another form of colonial mining of traditional resources for the gain of another. Finally, various ethicists, particularly those concerned with animal rights, have argued that such celebrations of hunting cultures simply operate as false delusions to ameliorate the guilt associated with the killing of another life.

Noting the first criticism, traditional worldviews are diverse. However, many share the belief in a sacred, personal relationship between humans and other living beings. Because of this, some environmental ethicists argue that the belief in sacred relations between humans and the nonhuman world deserves celebration in contrast to Western conceptions of human domination over a soulless, impersonal natural world. Further, environmental ethicists counter the cynical voices of the third criticism by noting that at least these hunting practices work against the dominionistic, wasteful, and disrespectful trophy hunting practices of Western cultures. Moreover, these rites of ethics and praxis adhere to more holistic ethics, such as those of Aldo Leopold's celebrated "land ethic."

These worldviews teach human humility when relating to nature. More importantly, such practices and beliefs insist on an ethic that reduces humans from superiority to equality. In these hunting cultures, humans exist in a world of reciprocal relationship with animals. Both live and act in ways that mutually promote the well-being of the entire earthly community. Failure to do so results in tangible negative results and life lessons.

See also

Hunting, Fishing, and Trapping

Further Resources

Anderson, D. (2000). *Identity and ecology in arctic Siberia.* Oxford: Oxford University Press.

Bird-David, N. (1990). The giving environment: Another perspective on the economic system of gatherer-hunters. *Current Anthropology, 31,* 189–96.

Berkes, F. (1999). *Sacred ecology: Traditional ecological knowledge and resource management.* New York: Taylor and Francis.

Brightman, R. (1993). *Grateful prey: Rock Cree human-animal relationships.* Berkeley: University of California Press.

Callicott, J. B. (1994). *Earth's insights: A multicultural survey of ecological ethics from the Mediterranean Basin to the Australian Outback.* Berkeley: University of California Press.

Callicott, J. B., & Nelson, M. (2004). *American Indian environmental ethics: An Ojibwa case study.* Upper Saddle River, NJ: Pearson-Prentice Hall.

Feit, H. (1973). The ethnoecology of the Waswanapi Cree: Or how hunters can manage their resources. In B. Cox (Ed.), *Cultural ecology: Readings on the Canadian Indians and Eskimos* (pp. 115–25). Toronto: McClelland and Stewart.

Ingold, T. (2000). *Perceptions of the environment: Essays on livelihood, dwelling, and skill.* New York: Routledge.

Kerasote, T. (1993). *Bloodties: Nature, culture, and the hunt.* New York: Kodansha International Press.

Mech, L. D. (1970). *The wolf.* Garden City, NJ: Natural History Press.

Nelson, R. K. (1983). *Make prayers to the raven: A Koyukon view of the northern forest.* Chicago: University of Chicago Press.

Peterson, D. (2000). *Heartsblood: Hunting, spirituality, and wildness in America*. Boulder, CO: Johnson Books.

Scott, C. (1989). Knowledge construction among Cree hunters: Metaphors and literal understandings. *Journal de la Societe des Americanistes, 75*, 193–208.

Tanner, A. (1979). *Bringing home animals: Religious ideology and mode of production of the Missitani Cree Hunters*. New York: St. Martin's Press.

Samuel Snyder

■ Culture, Religion, and Belief Systems
Islam and Animals

The basis of Islamic belief is the Qur'an, an Arabic text believed by Muslims to have been revealed by God to the prophet Muhammad between 610 and 632 CE. Six of the Qur'an's 114 chapters (called *sūras*) are named for animals: the Cow (*sūra* 2), the Cattle (*sūra* 6), the Bee (*sūra* 16), the Ant (*sūra* 28), the Spider (*sūra* 29), and the Elephant (*sūra* 105). Among the animal species mentioned by name in the Qur'an are camels, cattle, horses, mules, donkeys, sheep, monkeys, dogs, pigs, snakes, worms, ants, bees, spiders, mosquitoes, and flies.

The Arabic word used in the Classical texts to refer to animals, including humans, is *hayawān*. In the Qur'an, however, the term used for nonhuman animals is *dābba* (pl. *dawābb*), often translated as "beasts," or *an'ām* when referring to livestock. The Qur'an states that all animals were created by Allah (2:159), "from water (24:44–45)," and in pairs (36:36; 51:49).

Human beings are often described in Arabic texts as "the speaking animal" (*al-hayawān al-nātiq*), despite the fact that the Qur'an itself acknowledges that nonhuman animals also have speech (27:16, 18). The Qur'an occasionally blurs the line between human and nonhuman animals, suggesting that it is possible for humans to be "demoted" to other species. One verse, for example, refers to "Those whom Allah has cursed, against whom He has been angry, of whom He has made monkeys and pigs because they worshipped the powers of evil" (5:60). Another verse refers to the Israelites of Moses' time who broke the Sabbath, to whom Allah declared, "Be despicable monkeys!" (2:65). Although it might be possible to read into these verses a metaphorical interpretation, Muslim commentaries have tended to take them literally.

Arabs in pre-Islamic times practiced animal cults, various meat taboos, sympathetic magic (*istimtar*), and possibly totemism. Some tribes had animal names, such as the Quraysh ("shark"), which was the tribe of the Prophet Muhammad, the Kalb ("dog"), and the Asad ("lion"). Certain animals, including camels, horses, bees, and others, were believed to carry blessing (*baraka*), whereas others, such as dogs and cats, were associated with the evil eye. Genies (*jinn*) were believed to sometimes take animal form. As in many cultures, pre-Islamic Arabs associated particular animals with human traits, such as the lion with bravery, the rooster with generosity, and the buzzard with stupidity.

The most important animal to the pre-Islamic Arabs was without question the camel, a species that provided them with food, shelter, clothing, and transportation. Due to its high value, the camel was considered the greatest sacrificial offering and was slaughtered at the time of the pilgrimage, to welcome honored guests, and often on the death of its owner (so as to serve him in the afterlife). The Bedouin believed that eating camel flesh was a religious act of devotion, and that the appearance of camels in dreams was an auspicious sign.

The Qur'an proscribed certain pre-Islamic practices related to animals, such as the consecration of animals to specific deities. The Qur'an clarifies that pre-Islamic taboos on the eating of cattle did not originate with Allah and should be abandoned, and in passing, it indicates that the Islamic requirement that an animal be slaughtered while pronouncing Allah's name had pre-Islamic origins (6:138–40).

The Arabs had many kinds of blood sacrifices, which mostly survived into the Islamic period though often in altered form. The most visible of the Islamic blood sacrifices, which is performed once a year by all Muslims able to afford it, is the Feast of Sacrifice ('Id al-Adha), which commemorates the prophet Abraham's willingness to sacrifice his son (Isma'il, not Isaac, in Islamic tradition). Many Muslims also make blood sacrifices in fulfillment of vows (nazr), seven days after the birth of a child (aqīqa), or on the tenth day of the lunar month of Dhu'l-hijja in atonement for transgressions committed during the pilgrimage to Mecca (hājj). The proper method for sacrificing an animal is called dhabh in Islamic law, the sacrificial victim is known as a dhabīha.

The only aspect of dhabh mentioned in the Qur'an is the saying of Allah's name at the time of sacrifice (5:4); the remainder comes from hadiths and the Islamic legal tradition. Thus, the Qur'an does not specifically require Muslims to sacrifice animals for food or for any other reason; it merely permits them to do so. And even so, the Qur'an reminds Muslims that if they do sacrifice animals, "neither their flesh nor their blood reaches Allah; it is only your righteousness that reaches Him" (22:37).

A belief in metamorphosis (maskh) also survived from pre-Islamic times; several examples occur in the Qur'an (5:65; 2:61, 65). Some heterodox Muslim groups even retained a belief in metempsychosis (tanasukh), but mainstream Islam considers such ideas to be heresy. Among the pre-Islamic Arab traditions that the Prophet Muhammad forbade to his followers are the practice of animal fights (though camel fighting remained popular, especially in Muslim India) and the cutting off of camel humps and sheep tails for food while leaving the animal alive.

Animals in the Hierarchy of Creation

Within the hierarchy of Creation, The Qur'an depicts humans as occupying a special and privileged status. According to the Qur'an, "Certainly, we have created Man in the best make" (95:4), and "Hast thou not seen how Allah has subjected (sakhkhara) to you all that is in the earth?" (22:65). The term khalīfa (lit., "successor"), which in the Qur'an is applied to humans, is generally defined by contemporary Muslims as "vice-regent," as in the following verses: "I am setting on the earth a vice-regent (khalīfa)," and "It is He who has made you his vice-regent on earth" (2:30; 6:165).

Humans as Conscientious Stewards

According to this view, although nonhuman Creation is subjugated to human needs, the proper human role is that of conscientious steward and not exploiter. "To Allah belong all things in the heavens and on earth" (4:131)—that is to say, not to humans. Moreover, the earth was not created for the sake of humans alone: "And the earth has He spread out for all living beings (anām)" (55:10). The Qur'an emphasizes that God takes care of the needs of all living things: "There is no moving creature on earth, but Allah provides for its sustenance" (11:6). Everything in Creation is a miraculous sign of God (āya), inviting Muslims to contemplate the Creator. Nonhuman animals are explicitly included among God's miraculous signs, both in general, "The beasts of all kinds that He scatters throughout the earth" (2:164), and in terms of specific species, as in the following verse:

> This she-camel of God is a sign to you; so leave her to graze in God's earth, and let her come to no harm, or you shall be seized with a grievous punishment (7:73).

Nevertheless, the Qur'an specifies that certain animals, such as cattle and beasts of burden, were created for the benefit of humans (6:142; 16:5–8). In fact, the Qur'anic "correction" of the pre-Islamic Arab practice of leaving certain "sacred" cattle unmolested (6:138) can be interpreted to mean that it is against God's will for humans *not* to use domestic animals for the purposes mentioned in the divine revelation. It is not only domestic creatures that are created for utilitarian ends. Marine animals, too, are said to exist so as to serve as food for humans (16:4).

Yet despite the clear interspecies hierarchy established in the Qur'an, humans are described as similar to nonhuman animals in almost all respects. The Qur'an states in several places that all creation praises God, even if this praise is not expressed in human language (17:44; 22:18; 24:41). In a verse that constitutes the very core of Islamic teaching on animal rights, the Qur'an further says that "There is not an animal (*dābba*) in the earth, nor a flying creature on two wings, but they are communities (*umam*, sg. *umma*) like you" (6:38). Nonhuman animals can even receive divine revelation, as in the following verse: "And your Lord revealed to the bee, saying: 'make hives in the mountains, and in the trees, and in [human] habitations'" (16:68).

Treatment of Animals

The Qur'an is supplemented as a source of guidance for Muslims by reports about the words and deeds of the prophet Muhammad, called *hadith*s. Many hadiths report Muhammad as reminding his companions to take the interests of nonhuman animals into consideration. In one famous hadith, a sinner is granted salvation for showing compassion to a thirsty dog; in another, a woman is condemned to hell for mistreating a cat.

Another hadith, which exists in many versions, reports Muhammad as saying, "There is none amongst the Muslims who plants a tree or sows seeds, and then a bird, or a person or an animal eats from it, but is regarded as a charitable gift for him."

Muhammad enjoined his followers to use animals only for necessary purposes, on one occasion reprimanding some men who were sitting idly on their camels in the marketplace, saying "Do not treat the backs of your animals as pulpits, for God Most High has made them subject to you only to convey you to a place which you could not otherwise have reached without much difficulty."

When traveling, Muhammad encouraged his followers to ride slowly if there was vegetation, so that their animals could graze, and quickly when in the desert; at night, they were to be protected from insects. Acts of cruelty such as branding or hitting an animal in the face were forbidden.

Muhammad seems to have recognized that animals have an emotional life that needs to be respected. Once a Companion of the Prophet was showing off some bird's eggs he had found while the mother bird fluttered about frantically. The Prophet is reported to have said, "Who has caused this bird distress by taking the eggs from her nest? Return them to her." In another instance, Muhammad spoke against the cutting of a horse's forelock, noting that "it is in the horse's seemliness or decorum." He once upbraided his favorite wife, 'Aisha, for being rough with her camel, telling her "It behooves you to treat the animals gently."

Killing Animals

Regarding wild animals, Muhammad forbade hunting for sport. Wild animal skins were not to be used as rugs or saddle covers, although the skins of domestic animals can be used for these purposes (presumably since they have already been killed for a "legitimate" purpose, namely, food). Muhammad is also reported to have said, "There is no man who kills [even] a sparrow or anything smaller, without its deserving it, but Allah will question him about it [on the Day of Judgment]."

The hadiths portray the Prophet as having insisted on the protection of some animal species and calling for others to be killed (though generally only when they pose some specific danger to humans or human interests). Muhammad disallowed the killing of frogs because he believed their croaking was in praise of Allah. Likewise he forbade Muslims to kill magpies because they were said to have been the first to perform the fast. Ants and bees were to be preserved because they were mentioned as the recipients of divine revelation. One well-known hadith mentions God reprimanding one of His prophets for needlessly destroying an ant colony, pointing out that they were "an entire community which sang My praise."

Other animals that Muslims are never to kill include hoopoes, swallows, and bats. On the other hand, Muhammad ordered the killing of mottled crows, dogs, mice, poisonous snakes, and scorpions. He permitted his followers to kill certain animals—including rats and mice, scorpions, crows, kites, wild dogs, lions, leopards, lynxes, and wolves—even when in a state of ritual purity (*ihrām*) during pilgrimage. Muhammad is also reported as having commanded that in cases of bestiality, both perpetrator and victim were to be killed, although other reports state that there is no punishment in such cases.

Ambiguity is often present in hadith accounts.

Shi'ite hadith collections differ considerably from those used by Sunnis in that Shi'ites admit different chains of transmitters, and because they include reports not only of the Prophet but also of his successors the Imams, they are more extensive. Moreover, because many of the Imams lived and traveled outside of Arabia, the contexts for Shi'ite hadith stories are more varied than in the case of Sunni hadiths, though many hadiths are also accepted by both sects (albeit through different transmitters).

One notable feature found in Shi'ite hadith accounts is that Muhammad and the Imams are portrayed as being able to converse with animals. Animals are sometimes reported as speaking in Sunni hadiths as well, but this is rare and presented in a less direct and matter-of-fact way as in Shi'ite stories. Numerous Shi'ite accounts have the Prophet or his descendants conversing nonchalantly with camels, birds, and other species, listening to their complaints and responding to them with compassion and understanding.

In one hadith, a pregnant lion asks Imam Musa al-Kazim to pray for her easy delivery. He does so, and in exchange, the lioness prays for the Imam, as well as for "his children, his partisans (lit., Shi'ites), and his friends."

Though the theme of compassion toward animals as a sign of piety is a familiar one, these stories emphasize several points more strongly than is found in the Qur'an or the Sunni hadiths. First, the attribution of language to nonhuman animals is very pronounced. Second, the Imams—like the Prophet himself but unlike ordinary humans— are able to speak animal languages. Third, and perhaps most important, animals pray, and their prayers are to be valued.

In the Shi'ite hadiths, at least, animals not only pray for the well-being of good humans but also call down God's wrath on bad ones, as in a hadith attributed to the eighth Imam, Reza, who warned his followers not to eat the lark or allow children to taunt it, for this species of bird prays repeatedly to God to curse the enemies of the Prophet's family. The Prophet is reported to have enjoined respect for the rooster, whose crowing signals the time for morning prayer; elsewhere, he suggests that the rooster's crowing is its form of prayer.

Islamic Dietary Laws

Islamic dietary laws as extrapolated in the legal tradition are based on the Qur'an and the hadiths. The overwhelming majority of Muslims eat meat; indeed, meat-eating is mentioned in the Qur'an as one of the pleasures of heaven (52:22, 56:21). The Qur'an appears to allow the eating of animal flesh, with certain exceptions, such as when one is in a state of pilgrimage (5:1). On the other hand, the Qur'an prohibits the eating of animals that have not been ritually slaughtered, as well as the eating of blood and pigs (5:3; 2:173; 6:145), except in cases of dire need (16:115; 2:173; 6:145). The pre-Islamic Arabs, who often had difficulty finding water while traveling in the desert, sometimes in desperation would slaughter a camel and drink its blood. Although the Qur'an prohibits this, the previous verses were sometimes invoked to justify the practice as a last resort.

Ritual Slaughter

Ritual slaughter (*dhabh*) is said to follow the principle of compassion for the animal being killed. According to a hadith, Muhammad enjoined his followers to "kill in a good way," stating that "every one of you should sharpen his knife, and let the slaughtered animal die comfortably." Yet, on another occasion, when Muhammad saw a man sharpening his knife while an animal waited nearby, he reprimanded him, "Do you wish to slaughter this animal twice, once by sharpening your blade in front of it and another time by cutting its throat?"

Ritual sacrifice, such as that customarily performed by Muslims on the occasion of 'Id al-Adha, is not prescribed as a duty in the Qur'an, but a hadith is sometimes cited to provide the sense that it is an obligation. Whether or not Muslims are obligated to perform a blood sacrifice during 'Id al-Adha has recently become a matter of debate.

The Qur'an and the hadiths are the main sources, along with analogical reasoning and consensus among scholars, for the body of Islamic law known as the *shari'a*. *Shari'a* law assumes without question that humans are going to make use of animals and to eat them. The legal questions therefore center on how to define and circumscribe the limits of these behaviors. The issues are *which* animals to eat, *how* to kill them properly in preparation for eating, and, to a lesser extent, *what responsibilities* humans have to the animals that serve them. Questions about *whether* humans have the innate right to do these things do not arise.

Islamic laws pertaining to animals are included under categories such as their treatment, their sale, how to include them in *zakāt* calculations, their lawfulness as food, prescriptions for slaughter, and restrictions on hunting. Thus, animals are discussed in terms of both their use by humans and, less extensively, the obligations of humans toward them.

The various schools of law each classified all known animals in terms of whether eating them was *halāl* (permissible), *harām* (forbidden), or *makrūh* (discouraged). All schools placed the vast majority of animals in the first, permitted category. Some animals presented special cases; frogs, for example, which would normally meet the conditions for a *halāl* designation, were determined to be *harām* on the basis of a hadith in which Muhammad forbade the eating of frogs.

Differences among the schools regarding these classifications occur mainly in cases of reasoning by analogy, such as whether or not to forbid the eating of animals that have similar names to those of forbidden animals, for example "dogfish." Another kind of ambiguity arises when an animal that would normally be considered *halāl* (such as an eel, which is a kind of fish) resembles an animal which is *harām* (for example, the snake, to which eels appear similar). The Maliki and Shafi'i schools allow the eating of fish found floating dead in the water, whereas other schools forbid it. Various schools disagree over the lawfulness of eating crustaceans and insects. Carnivores, which are *harām*, are identified in the legal tradition by their possession of fangs or claws; thus, there is disagreement over the lawfulness of eating elephants, because, although herbivores, their tusks resemble fangs.

Human Obligations to Animals

The Shafi'i jurist 'Izz al-din ibn 'Abd al-salam al-Sulami (d. 1262), in his legal treatise *Rules for Judgment in the Cases of Living Beings* (*Qawā'id al-ahkām fī masālih al-anām*), has the following to say about a person's obligations toward his domestic animals:

- He should spend [time, money or effort] on it, even if the animal is aged or diseased in such a way that no benefit is expected from it. His spending should be equal to that on a similar animal useful to him.
- He should not overburden it.
- He should not place with it anything that might cause it harm, whether of the same kind or a different species.
- He should kill it properly and with consideration; he should not cut its skin or bones until its body has become cold and its life has passed fully away.
- He should not kill an animal's young within its sight.
- He should give his animals different resting shelters and watering places, which should all be cleaned regularly.
- He should put the male and female in the same place during their mating season.
- He should not hunt a wild animal with a tool that breaks bones, which would render it unlawful for eating. (cited in Izzi Dien, 2000, pp. 45–46)

Although the rights of nonhuman animals are guaranteed in the legal tradition, their interests are ultimately subordinate to those of humans. As Sulami argues, "The unbeliever who prohibits the slaughtering of an animal [for no reason but] to achieve the interest of the animal is incorrect because in so doing he gives preference to a lower, *khasīs*, animal over a higher, *nafīs*, animal" (cited in Izzi Dien, 2000, p. 146).

Sport Hunting

Despite its prohibition in Islamic law, sport hunting remained a major form of entertainment in Muslim societies, especially among the elites. In Arabia the oryx was hunted to near extinction, and only recently have measures been taken to preserve the species. In Iran, species such as the lion, tiger, and cheetah were hunted into oblivion before modern times, and leopards have become exceedingly rare. Even gazelles, which were the favored game at royal hunting preserves right up until recently, are now generally found only on government lands where private individuals may not enter without special permission.

Historically the most egregious violations of the proscription against sport hunting were in India, where hundreds or thousands of creatures at a time would be indiscriminately slaughtered in bloody orgies of killing for the amusement of the rich and powerful. The favored method (a Central Asian technique called the *qamargha*) was to go out into the wilderness and create a wide circle of "beaters" who would make as much noise as possible as they slowly closed the circle, forcing huge numbers of terrified creatures toward the center. When the circle was almost closed, the royal hunters would fire at will into the throng of panic-stricken animals. So horrific was the resulting blood-bath that at one point the Mughal emperor Akbar the Great (r. 1555–1605) decided enough was enough and banned the sport, though apparently only for a time.

Preserves

The Islamic legal tradition contains two institutions that some contemporary scholars have argued could be considered as forms of wildlife preserves. They are the *himā*, "pro-tected area" or sanctuary, and the *harīm*, which was a "greenbelt" or easement around set-tled areas intended mainly to ensure a safe water supply. A related institution, the *harām*, refers to areas around the sacred cities of Mecca and Medina (called the *harāmayn*; "the two forbidden areas") where hunting is outlawed.

The *harāmayn* were apparently established in the Prophet Muhammad's time, when, according to the hadiths, he declared Mecca "sacred by virtue of the sanctity conferred on it by God until the day of resurrection. Its thorn trees shall not be cut down, its game shall not be disturbed." He also made a sanctuary of Medina, whose "trees shall not be cut and its game shall not be hunted."

The prohibition on hunting while on pilgrimage comes from the Qur'an, which states that the penalty for killing game is to offer a comparable domestic animal in sac-rifice (i.e., to God) by way of compensation (5:96). It would seem from this verse that killing wild animals when one is supposed to be in a state of purity is wrong because it is a crime *against God*, not against the animals in question. One must atone for this by paying the equivalent in one's own domestic livestock "back to God." This atonement for the killing of wild animals by killing yet more domestic animals can hardly be seen to benefit the animals themselves.

Some traditional *himā*s still exist in Saudi Arabia, but they are much diminished from former times and continue to disappear. Most of these preserves are aimed at excluding sheep and goats from grazing lands in preference to cattle, camels, and don-keys, but others exist to control the cutting of firewood or to keep flowering meadows intact for honeybees.

Even in the *harām*s around the holy cities, species such as the ibex and gazelle are no longer found. In fact the laws pertaining to these preserves have been generally ignored, on the basis that "development"—geared largely to servicing (and fleecing) the millions of pilgrims who now descend on the holy sites—is a need that overrides that of preserving nature.

What is important to note is that these areas were restricted primarily so that they might benefit humans. The *himā*, which in pre-Islamic times was an institution that allowed powerful landowners to keep others off their grazing lands, was transformed in the Prophet Muhammad's time into a means for preserving certain tracts of land for the public benefit. Significantly, the preserved areas were not to be too large, so as not to take too much land "out of circulation."

In short, the institutions of *himā*, *harīm*, and *harām* are all clearly meant to preserve resources for human needs, not those of animals. If animals are preserved, or if they ben-efit from the preservation of water and vegetation, this is a secondary benefit, because they themselves are seen in the law as existing for the good of humans.

Nonhuman animals are ubiquitous throughout the intellectual and artistic traditions of Muslim civilization. Almost invariably, however, animal figures are employed as symbols for particular human traits, or are entirely anthropomorphized actors in human-type dramas. In other words, even where nonhuman animals appear, the real message is about humans. This holds true in the realms of philosophy (e.g., Avicenna's *The Bird*) and mysticism (Attar's *Conference of the Birds,* or the many animal stories in Rumi's *Masnavi*), as well as in popular literature (*Kalila and Dimna, q.v.,* or the *Thousand and One Nights*) and the arts. Only in scientific works, such as treatises on zoology (e.g., al-Jahiz's *Book of Animals, q.v.*), are animals observed, discussed, and described in their own terms, and even then the emphasis is often on their uses or dangers for humans.

In conclusion, from this survey of animal-related material from the textual sources of Islam several points can be drawn. First, the tradition takes the relationship between humans and other animal species quite seriously. Second, animals are seen as having feelings and interests of their own. And third, the overriding ethos enjoined upon humans is one of compassionate consideration. Based on these sources it would seem that the Islamic ethical system extends moral considerability to nonhuman animals, although not on the same level as humans.

Further Resources

Foltz, R. C. (2005). *Animals in Islamic tradition and Muslim cultures.* Oxford: Oneworld.
Izzi Dien, M. (2000). *The environmental dimensions of Islam.* Cambridge: Lutterworth.
Masri, A. B. A. (1989). *Animals in Islam.* Petersfield, UK: The Athene Trust.
Pellat, C. (1971) Hayawān. In *Encyclopedia of Islam* (new ed.) (3, 305). Leiden: Brill.

Richard C. Foltz

■ Culture, Religion, and Belief Systems
Islam's Animal Fables: Kalila and Dimna

Perhaps the best-known animal stories in Muslim societies are found in the Iranian translator Ibn al-Muqaffa's ninth-century rendering from Middle Persian to Arabic of *Kalila and Dimna,* a collection of animal fables that came to pre-Islamic Iran from India. The general plot, connecting the various stories that occur as continuous digressions, follows the fortunes of two jackal brothers, Kalila and Dimna, at the court of the lion king. As in George Orwell's political allegory *Animal Farm,* the stories are really about people and politics and contain only the barest and most superficial observations about the traits of actual animals.

Dimna, the more ambitious of the two jackal brothers, decides to take advantage of his natural cunning and offers his services as advisor to the lion king, despite Kalila's warning to stay out of politics. Dimna is accepted at court and joins the king's inner circle. His first challenge is to advise the king on how to deal with the presence of a menacing-looking newcomer, a large bull grazing in a nearby field. Dimna offers his advice in the form of a fable, telling the king about a fox who comes upon a drum hanging in a tree and, not knowing what it is, rips it open and finds there is nothing inside. Likewise, he suggests, the bull may be nothing more than an empty threat.

With Dimna acting as go-between, the lion and the bull make contact and eventually become great friends. Dimna, dismayed at being displaced from the king's affections, begins to plot against the bull. In discussing his problem with Kalila, the two jackals exchange one tale after another to make their various points.

For example, Dimna illustrates the fact that the weak can conquer the strong through cleverness by telling the tale of a crow whose eggs are repeatedly eaten by a snake. At first the crow considers trying to peck the snake's eyes out while he is asleep, but her friend the jackal warns that this is risky. In his turn, he tells a story of some fish whose pond is about to be drained; they ask a crane to ferry them to safety, but the crane merely carts them off to a place where he can eat them at his leisure. Upon hearing this cautionary tale, the crow decides that instead of attacking the snake herself, she will get some humans to do it by stealing a necklace from them and dropping it down the snake's hole. Sure enough, the humans come out and finish off the snake, and the crow's eggs are safe.

Dimna's attempts to poison the ears of the lion king against his friend the bull also take the form of animal fables. The tale of the three fish and the fishermen, in which the wise fish wastes no time before escaping, moves the king to an attitude of prudence. Dimna continues with the story of the bedbug and the flea, in which a seemingly insignificant flea makes his more visible friend the bedbug pay the penalty for their joint crime of devouring a sleeping woman's luscious flesh. This tale persuades the king that his own friend, the peaceful, herbivorous bull, cannot be trusted.

Not content with turning the king against his dearest companion, Dimna then proceeds to incite the bull against the king, pointing out that lions are carnivores and eat herbivores like the bull. Recalling the story of the lion and the camel, in which a sickly camel takes refuge with a lion but at the instigation of the lion's fellow carnivores becomes a meal nevertheless, the bull falls into despair. Through a further set of tales, Dimna persuades the bull to launch a surprise attack on the lion king before the lion attacks him.

Of course, in the end the lion kills the bull, though immediately afterwards he is stricken with remorse for doing so. Kalila, disgusted by his brother's treachery, disowns Dimna forever, but that is the extent of the latter's punishment and Dimna regains his position as the lion king's advisor. The outcome is hardly morally satisfying. If there is an overriding theme in the Kalila and Dimna stories, it is that no animal can act contrary to its nature and that no one can really be trusted—the law of the jungle! Needless to say, this is not an Islamic teaching, but it surely represents the way many medieval Muslims viewed the world they lived in.

Further Resources

Wood, R. (1986) *Kalila and Dimna*. Rochester, VT: Inner Traditions.

Richard C. Foltz

■ Culture, Religion, and Belief Systems
Japan's Dolphins and the Human-Animal Relationship

Japan is an island country stretching 3,000 kilometers north to south, with a variety of marine animals along its coast. Dolphins are the marine animals that are most popular in Japan and that have had the most diverse relationship with humans. At the same time, there has historically been great change in that relationship, influenced by the Japanese government.

Dolphins have long been regarded as a marine resource in Japan, exemplified by dolphin hunts, small-scale coastal cetacean hunting in which dolphins are killed for food. Archeological digs show that such use of dolphins by the Japanese goes back to about 2500 BCE, which was the end of the Early Jomon period. The Chinese ideogram for "whale" means "huge fish," and there has been a tendency in Japan as well to regard dolphins as fish rather than mammals. With the issuing of the "Prohibition on Meat Eating" in 675, avoiding meat and eating fish became a general social trend, and dolphins and whales were considered fish and eaten. Fishermen, in particular, even now perceive dolphins and whales to be kinds of fish, as do government agencies because the agency in charge of cetacean matters is still the Fisheries Agency, not the Environment Ministry.

The relationship between humans and dolphins differs from one era to the next and also depends greatly on the geographical region. For example, there are a few places in Japan, such as Amakusa in Kyushu, where people have always coexisted with and left dolphins alone, regarding them as harmless wild animals. On the other hand, one can also see change over time in the view that dolphins and whales are marine resources, a view that has existed from prehistoric times. In times past, when capturing and killing large cetaceans was a serious life-and-death battle, people revered the animals they killed as formidable opponents, but at the same time they had compassion for the animals they killed.

Even for dolphins, which are small and easily killed, people felt remorse and for that purpose erected monuments to memorialize and still the souls of the dolphins and whales they killed. But in modern times, when capturing cetaceans no longer involves risk to one's life because of the use of advanced technology, economic efficiency reigns, and one no longer sees the awe and compassion of yesteryear for dolphins and whales. Buddhist rites for the killed whales and dolphins have become a annual ceremony, whereas dolphin hunts are now routine events, and some of the monuments erected in the past for the souls of cetaceans have been abandoned.

Japan's government encourages dolphin hunts and whaling, taking the position that eating dolphin and whale meat should be preserved because it is traditional food that supports Japan's culinary culture. The government even encourages schools to put whale meat in school lunches, and there are over 160 private-sector retail stores that sell dolphin and whale meat. However, strong criticism is directed at the Japanese government because eating dolphin and whale meat was originally a local phenomenon. Whereas the government spends tax money to promote whaling and to encourage people to consume whale meat, it has no interest in coexistence with dolphins and whales. Currently public opinion is so divided over whaling and dolphin hunts that the media tend to shy away from the topic due to its highly controversial nature. Yet, many people are uninterested in this issue because the amount of dolphin and whale meat consumed is less than 0.5 percent of beef, pork, and chicken consumption.

The polar opposite of killing dolphins is the type of involvement that regards dolphins and whales as wild animals that should coexist with humans in the limited environment of our earth. This philosophy appeared in Japan in the 1970s, but it was the 1980s and 1990s when it started gaining currency among the general public. At this time there were television programs and videos about dolphins, and many books about dolphins were published. In particular, *Messenger from the Sea* spoke out against dolphin hunts, revealed the differences in the Western and Japanese views on dolphins, and instilled new thinking into the Japanese view of dolphins, which had seen them only as marine resources. Meanwhile, diving became popular among the young, who found more opportunities to travel abroad and come into contact with wild dolphins. In the second half of the 1990s dolphins became one of the most popular marine animals among the Japanese, and there was dolphin-related merchandise everywhere.

Whale watching, which aims to achieve coexistence with dolphins and whales, was first carried out in Japan in 1988. Currently dolphin and whale watching are conducted in about thirty places in Japan, and over 1.5 million Japanese have participated. There was also a rapid change in how the Japanese responded to cetacean beachings. It was customary for Japanese who found cetaceans in shallows to bring them ashore and use them as "a blessing of the sea," but now large numbers of people will rush to the site of a cetacean beaching and try to push the animals back into the sea as long as they are still living. In 2002 when several melon-headed whales beached themselves in Ibaraki Prefecture, young people, surfers, and even children worked to save them. This one example shows that the general public's perception of dolphins and whales has changed much over the last several decades.

Aquariums in Japan are seen not only as recreational facilities but also as educational centers. Nearly all of Japan's aquariums display dolphins they have purchased from so-called "drive" fisheries, and they also put on shows featuring them. Dolphin shows started in 1957 in Japan, and they still have strong popularity. The number of facilities currently breeding dolphins numbers over fifty, including those that are not members of the Japanese Association of Zoos and Aquariums. Nearly all the facilities put on dolphin shows, and together they have almost 500 dolphins of various species (these figures are constantly changing due to dolphin deaths, replacement, resale abroad, and other factors). About ten facilities offer swimming with dolphins, and about seven conduct dolphin therapy. There are also people who have found spiritual value in dolphins and seek healing from their qi.

Since the mid-1990s there has been growing criticism of dolphin drive fisheries from nature, animal, and environmental protection organizations. Videos, photographs, and reports describing the dolphin hunts have been published. A growing number of people oppose the hunts, and criticism of live dolphin capture by aquariums is gradually beginning to build.

There are diverse forms of interaction between humans and cetaceans in Japan, and a wide range of thinking on the issue, although the Japanese government is pushing a hard-line policy that promotes the consumption of dolphins and whales. Meanwhile, the public is gaining knowledge of how dolphin and whale meat is contaminated with mercury and other organic chemical substances, and the international current of thought that wants wild animals to be left wild is taking hold in the country. The relationship between humans and dolphins in Japan will likely continue to change.

Further Resources

Elsa Nature Conservancy. (1997). *Wild orca capture: Right or wrong? A report on issues arising from the 1997 orca capture at Hatakejiri Bay, Taiji, Wakayama Prefecture, including an account of protest actions.*
———. (2001). *Dolphin hunting in Japan—How dolphins are captured for aquariums and food* [Videotape].
———. (2005). *Japan's dolphin drive fisheries: Propped up by the aquarium industry and 'scientific studies' the reopened dolphin hunts at Futo on the Izu Peninsula in Shizuoka Prefecture, and the dolphin export plan of Taiji Town in Wakayama Prefecture.*
Elsa Nature Conservancy and the Institute for Environmental Science and Culture. (1997). *A report on the 1996 dolphin catch-quota violation at Futo Fishing Harbor, Shizuoka Prefecture, & the issues it raised includes results of on-site investigation and report on protest actions.*
———. (2002). *New development at Futo: From dolphin hunting to dolphin watching—dolphin drives then and now on the eastern side of the Izu Peninsula, Shizuoka Prefecture, JAPAN.*
Hoyt, E. (2002). *Whale watching worldwide* (Japanese-English ed.). PACI.

Sakae Hemmi

■ Culture, Religion, and Belief Systems
Judaism and Animals

Judaism has developed across thousands of years and under a great variety of different cultural, social, geographical, political, and technological circumstances, each of which has left its mark on the role of animals in Jewish tradition and society. According to Jewish tradition, the Written Torah (the first five books of the Bible) may be understood as containing 613 commandments, which form the outline of Jewish law. The commandments are further expounded upon and extended by the Oral Torah, the living tradition of Jewish law that was first codified in the Mishnah (c. 200 CE) and further developed and expounded in the Talmud and many other works. According to one recent count, 138 of the commandments have some connection with animals.

Judaism has always valued the preservation of conflicting voices within the tradition, and countless references to animals are found throughout Jewish legal, philosophical, mystical, ethical, exegetical, liturgical, and homiletic literature. These two factors make it difficult to formulate statements that are universally true of Judaism in all of its varied manifestations. The goal of this article is merely to outline some of the major Jewish themes, ideas, and practices that directly involve animals.

Meat Eating and the Status of Animals

The biblical account of creation and other stories from the Book of Genesis may be read as laying out the foundations of human-animal relations in Judaism. According to the first chapter of Genesis, after creating the animals, God created a male and a female human in the divine image. They were meant to "rule the fish of the sea, the birds of the sky, the cattle, the whole earth, and all creeping things of the earth" (1:26) and they were told, "Be fruitful and increase, fill the earth and master it; and rule the fish of the sea, the birds of the sky, and all the living things that creep on the earth" (1:28). Some recent writers have claimed that these statements support the right of human beings to treat animals as they please. This impression is immediately tempered, however, by the next verse's call for vegetarianism: "God said, 'See, I give you every seed-bearing plant that is upon all the earth, and every tree that has seed-bearing fruit; they shall be yours for food'" (1:29).

Later stories in Genesis tell us that animals cannot fulfill the human need for human companionship (2:18–24), that Abel's sacrifice of sheep was more acceptable to God than was Cain's vegetable offering (4:3–5), and that Noah saved all of the land animals from extinction in the Flood by taking them aboard his Ark (6–8). The Flood story marks a major change in human-animals relations: after the Flood animals would fear humans (9:2) and humans would be allowed to eat flesh (9:3), but not flesh taken from animals while they are still alive (9:4). It should be pointed out that alongside the view of animals presented by Genesis, other parts of the Hebrew Scriptures (the Book of Job in particular) emphasize a rather different facet of animal-human relations. There, animals living in the wild are seen as exemplifying an aspect of God's creation that surpasses human understanding. By comparing God's intimate knowledge of the lives of wild animals with human ignorance of such matters, Scripture tries to make us aware of our human limitations.

The stories of Genesis have been the subject of massive commentary and analysis; perhaps their most systematic treatment in terms of animal-human relations may be found in Book III chapter 15 of Joseph Albo's (fifteenth century) *Sefer Ha'Ikkarim*. Albo suggests that the stories represent a dynamic process in which concern for the wellbeing of all living things is balanced against the need for people to understand the unique status of human beings; eventually God permits the eating of flesh as a concession to human weakness.

Some Jewish mystics claim that when properly executed, the slaughter and consumption of animals is actually beneficial to them as a way to free the "sparks of holiness" trapped within them. (The intricate rules of *shehitah*—Jewish ritual slaughter—are largely aimed toward minimizing the animal's suffering.) Other mystics support vegetarianism by claiming that practically no one achieves the degree of spiritual perfection required in those who can liberate "sparks" in this way. In the twentieth century, Rabbi A. I. Kook (1865–1935) argued that much of Jewish law points toward a rather strict vegetarianism and a radical concern for animal rights that are destined to become the accepted norm in the Messianic era. He also warned, however, that an overemphasis on animal rights and vegetarianism in our present unredeemed world would lead to the breakdown of respect for human beings rather than an increase of respect for animals.

Jewish attitudes toward vegetarianism reflect what might be called the basic parameters of Jewish attitudes toward animals in general. On the one hand, humans are created in the image of God and are fundamentally superior to other animals, which they are permitted to use for their own purposes. On the other hand, humans must take the well-being of animals into account, and the exploitation of animals for human ends must be regulated by moral considerations. Judaism's self-understanding of its concern for animals has developed in ways that parallel developments in Western moral philosophy. Some thinkers, such as Moses Maimonides (1135–1204), believe that the well-being of animals is of intrinsic moral importance, whereas others, such as Moses Nachmanides (1194–1270), believe that although only humans are intrinsically deserving of moral consideration, people must treat animals humanely in order to properly cultivate their own moral virtue.

Animal Welfare in Jewish Law: Working Animals

There is little mention of pets in Jewish law; it is usually assumed that animals kept in human possession provide goods or services to their owners. One might say that domesticated animals function as members of the household and are expected to pull their own weight. That, however, does not mean that the human masters are allowed to treat the animals as they please. Although wanton cruelty toward *any* animal is forbidden by Jewish law, someone who owns or works with an animal has many additional obligations toward it. For instance, Jews are required to make sure that their animals have been fed before sitting down to eat themselves.

The Torah contains a number of commandments that specifically deal with the working conditions of animals. According to Deuteronomy 25:4, one is not allowed to muzzle an ox while it is threshing grain. This commandment is understood to prohibit people from stopping any kind of animal from eating any kind of food with which it is presently working. One corollary of this rule is that a pack animal must be allowed to nibble from whatever it is carrying (Maimonides' *Mishneh Torah, Hilkhot Sehirut* 13:1–2). Deuteronomy 22:10 prohibits people from using a mixed team consisting of an ox and an ass to plow a field. Traditionally, this has been understood to imply a prohibition on making pure and impure (see below) animals work together to pull something, and further rabbinic legislation prohibits pulling something with any combination of animals from different species. One explanation for these laws is that animals often find it stressful to be forced into close contact with members of other species (*Sefer HaHinukh*, commandment 550); another possibility is that an animal from a weaker species will have trouble keeping up with a work-partner belonging to a stronger species. Other work-related laws include the obligation upon humans to assist in the unloading of a pack animal that has collapsed under its burden (Exod. 23:5) and the obligation to help a fallen animal get back on its feet (Deut. 22:4).

A vast section of Jewish law deals with the prohibition of work on the Sabbath and festivals. The Torah makes it clear that one's animals must also be allowed to rest on those days: "The seventh day is a Sabbath of the Lord your God; you shall not do any work—you, . . . your ox or your ass, or any of your animals" (Deut. 5:14); "On the seventh day you shall cease from labor, in order that your ox and your ass may rest" (Exod. 23:12). Many laws derive from these verses; for instance, an entire chapter of the Mishnah (Shabbat 5) is devoted to the question of which items one may have one's animal carry into a public area on the Sabbath. It is also prohibited for a Jew to lend or rent an animal to a gentile who might force it to work on the Sabbath (Maimonides Code Hilkhot Shabbat 20:3). A famous legend tells of a pious Jew who was forced by poverty to sell his faithful cow to a pagan gentile. The gentile later came to him to complain that the cow laid down and refused to do any work on the Sabbath. The Jew came over to the cow and whispered in its ear, "Oh cow, cow! When you belonged to me, you rested on the Sabbath but now that I suffer for my sins and had to sell you to this gentile, I implore you, stand up and do your master's will!" Immediately, the cow sprung to its feet, ready to work. Its new master was so impressed by the cow's theological acumen that he immediately converted to Judaism and became a scholar. He was known as "Rabbi Yohanan ben Torta ("Yohanan son of the Cow") (*Pesikta Rabbati* 14). Interestingly, Jewish law permits humans to perform certain kinds of work on the Sabbath that would otherwise be prohibited if it is necessary for their animals' well-being. For instance, Jews are allowed to milk cows on the Sabbath in order to alleviate the pain caused them by swollen udders. In Israel some milking parlors are fitted out with specially designed systems so that religiously observant Jewish dairy farmers can milk their herds on the Sabbath in a manner permitted by Jewish Law.

Laws Respecting the Parent-Child Relationship among Animals

Judaism places great stress on the importance of the human parent-child relationship, and this concern extends to parent-child relationships among animals as well. In his *Guide for the Perplexed,* which is usually considered to be the most important work of medieval Jewish philosophy, Maimonides writes that when animals see their offspring die, they "feel very great pain, there being no difference regarding this pain between man and the other animals. For the love and tenderness of a mother for her child is not consequent upon reason, but upon the activity of the imaginative faculty, which is found in most animals as it is found in man" (*Guide* III:48, Pines, 599).

Several laws reflect concern for the human-parent relationship among animals. Leviticus 22:28 prohibits the slaughter of an animal together with its offspring on the same day. Maimonides (loc. cit.) states that this law is intended to prevent situations in which the parent might witness the slaughter of its offspring. Similarly, Leviticus 22:27 states that a newborn animal "shall stay seven days with its mother, and from the eighth day on it shall acceptable as an offering by fire to the Lord." Deuteronomy 22:6–7 states that a mother bird and her eggs should not be taken together and that the mother bird must be shooed away before the eggs are taken from her nest.

"Clean" and "Unclean" Animals

The Hebrew scriptures (see especially Leviticus chapter 11) and Jewish law in general distinguish between animal species that are *tahor* ("clean" or "pure") and those that are *tamei* ("unclean" or "impure"). Among the mammals, only ruminants (animals that chew their cud) that have split hooves are *tahor,* such as cattle, goats, sheep, and deer. The rules for birds are trickier; suffice it to say that practically all of the fowl that people usually eat are *tahor,* including chickens, turkeys, geese and ducks, quail, pheasants, doves, pigeons, peacocks, and partridges. Fish with visible fins and scales are *tahor*, making catfish, sharks, and eels all *tamei*. All reptiles and amphibians are *tamei*, as are all invertebrates except for certain locusts (in practice, only Jews from Yemen still consider themselves permitted to eat locusts today).

The *tamei/tahor* split is not merely a matter of dietary law; *tahor* animals are often held to be somehow spiritually superior to the *tamei*. Even today, some Orthodox Jewish parents avoid having their children look at pictures of *tamei* animals or play with toy *tamei* animals. Some Medieval Jewish sages pointed out that ferocious beasts and birds of prey are all *tamei,* whereas the *tahor* animals are placid and have good dispositions. Given the common belief that "we are what we eat," it was suggested that by eating meat only from *tahor* animals, people can avoid becoming aggressive or cruel.

Attitudes toward Particular Species

Beyond the general preference for *tahor* animals over the *tamei,* more specific differential attitudes toward different kinds of animals can be found in the Jewish tradition. For instance, there is no doubt that Jewish tradition holds a very low opinion of pigs. The early rabbis decreed it illegal to raise pigs in the Land of Israel, and the authors of some classical Jewish texts were so disgusted by pigs that they avoided mentioning the species by name, referring to it as "another thing" or "that species." Today, many completely nonobservant Jews will still think twice before eating a ham sandwich. One reason why pigs are so unpopular in Judaism is that they symbolize hypocrisy. Viewed from the outside, a pig appears to be *tahor* because it has split hooves; however, it lacks the internal sign of purity—it does not chew its cud. The pig has also been associated with the enemies of the Jews. In the persecutions that led to the Hasmonean revolt in the second century BCE (whose victory is celebrated the festival of Hanukkah), Jews were forced by their Hellenistic oppressors to eat pork in order to challenge their loyalty to the faith. Later, the boar served as the emblem of the Roman military legion that occupied the Land of Israel. Eventually, disgruntled Jews came to use the Hebrew word for pig—*hazir*—to refer to the Roman Empire in general.

The attitude toward dogs in Jewish tradition is mixed. On the one hand, dogs are *tamei*, they are usually considered to be lowly, shameless creatures that hardly enjoy the honored status of "man's best friend." If someone barters a dog for another kind of animal, the purchased animal—even if it is *tahor*—is deemed unfit to be sacrificed in the Temple (Deut. 23:19). Rabbinical decrees require that dogs be leashed at all times and forbid people from raising "bad" dogs in their homes. On the other hand, classical Jewish texts do show appreciation for the loyalty and usefulness of dogs as guards and shepherds. Jewish law allows dogs to run free at night in frontier communities for security purposes. One Talmudic story tells of a man who was so appreciative of how his dog saved his wife from a rapist that he would allow the beast to eat at the table with human members of the household (Jerusalem Talmud, Terumot 3:9, 44b).

Horses played a particular military and political role in the historical period referred to by much of the Jewish Scriptures. In those days, horse-drawn war chariots were the hi-tech weapons used by powerful empires that threatened Israelite independence. Horses served as symbols of military might and political prestige in the same way as jet fighter aircraft do today. Often poorly equipped, the Israelites had no choice but to exclaim, "They [call upon] chariots, they [call upon] horses, but we call upon the name of the Lord our God." Accordingly, Israelite kings were warned not to aggrandize themselves by acquiring many horses (Deut. 17:16).

Donkeys seem to be more favored than horses by Judaism. The heroes of the bible are often depicted as riding on donkeys and showing concern for their well-being, and the biblical laws referring to animal rights often specifically mention donkeys. In a fascinating biblical episode, it is a donkey that is depicted as gaining the power of speech and describing the life of a loyal domesticated animal (see sidebar). In a way, the donkey almost manages to transcend the "dishonor" of being a *tamei* species. When the Temple in Jerusalem was still in existence, Jews would bring the firstborn of their *tahor* animals as sacrifices; that law applied only to donkeys of all the *tamei* species. (However, since a donkey is nevertheless *tamei,* it could not actually be sacrificed and eaten; a goat was offered as its substitute—see Exodus 13:13).

Moving on to the *tahor* animals, we find classical Jewish texts taking much more positive attitudes. The gazelle and the deer are wild *tahor* animals that figure prominently in biblical and later Jewish texts as figures of grace and beauty. The Hebrew word for gazelle—*tzvi*—actually serves as the word for "beauty" in biblical Hebrew. The author of Psalms (42:2) compares his longing for God to that of a deer crying out for water in a drought. The Book of Daniel and many later Jewish texts refer to the Land of Israel as the Land of the Gazelle.

Cattle figured prominently among the animals offered for sacrifice in the Temple. Both oxen and cows are frequently mentioned as draught animals in classical Jewish texts. As the story of Yohanan ben Torta cited previously demonstrates, bovines were valued as loyal workers; the prophet Isaiah (1:3) complained that whereas an ox can be trusted to recognize its master, the Jews had forgotten their God. Bovines do suffer one dishonor in the Jewish Scriptures: statues of calves were sometimes used as objects of prohibited idolatrous worship. Soon after leaving Egypt, the Israelites sinned by worshiping a golden calf (Exod. 32). They repeated the crime later by setting up two statues of calves for worship in Samaria (2 Kings 17:16).

Sheep and goats, usually mentioned together under the more general Hebrew term *tzon,* were also often offered in sacrifice. The relationship between a shepherd and his flock is an important metaphor both for the relationship between human leaders and their communities as well as for the relationship between God and humans. It is often pointed out in Jewish literature that the great leaders of the bible, such as Moses and David, were experienced shepherds and that the experience of caring for *tzon* taught them to take responsibility for their human "flocks." In the bible's most emotional description of a human's attachment to an animal, the prophet Nathan describes a poor man "who only had one little ewe lamb that he had bought. He tended it and it grew up together with him and his children: it used to share his morsel of bread, drink from his cup, and nestle in his bosom; it was like a daughter to him" (2 Sam. 12:3). Famously, the author of Psalms 23 proclaims that "The Lord is my shepherd; I shall not want." There is, however, a darker side to the metaphor. In the *U'Netaneh Tokef,* one of the central prayers of the Jewish High Holy day service, God is described as a shepherd culling his flock, deciding who shall live and who shall die in the course of the coming year. Despite the generally positive attitude toward *tzon,* for many years an ancient rabbinical decree forbade raising them in the populated areas of the Land of Israel because of the damage their grazing inflicted upon fields and gardens.

Animal Moral Psychology: The Law of Damages Inflicted by Animals

Sometimes animals damage human beings or human property—including other animals owned by humans. Because animals had a prominent presence in the societies in which the foundational works of Jewish law were complied and written down, it should hardly come as a surprise that Jewish law has quite a lot to say about destructive animal behavior. However, the concern that Jewish law demonstrates for the character and psychology of the guilty animal is remarkable.

Generally speaking, Jewish law holds the owner of an animal less accountable for the damage it does if the destructive behavior was unusual or out of character for the animal. (The present discussion is based upon Moses Maimonides' summary of the applicable laws as compiled in his *Mishneh Torah, Hilkhot Nizkei Mamon*). Some large predators, such as wolves, lions, and bears are never deemed fully tame. Human owners of such animals are always subject to the highest level of legal responsibility for their attacks. Domestic animals such as oxen present more complicated cases. An "innocent" (Hebrew: *tam*) ox is not prone to goring, whereas an ox with a history of aggressive behavior is called *mu'ad*—"prone" to violence. Following the principle explained previously, if a *mu'ad* ox gores another animal, its owner has to pay more in damages than if his ox had been *tam* and attacked another animal in an outburst of uncharacteristic aggression. An ox, however, is not simply *tam* or *mu'ad;* Jewish law recognizes a much more detailed categorization of animal behavior. An ox prone to attack other oxen may leave people alone. It may attack members of other animal species but steer clear of other oxen; it may get along with human adults but attack children; it may even be prone to attack on the Sabbath but not on workdays! Furthermore, if an aggressive ox becomes observably friendly, it may recover its status as a *tam*. Unless it can be shown that the ox was *mu'ad* to inflict the kind of damage in question, its owner cannot be sued for the highest level of compensation. On the other hand, because, for instance, all oxen enjoy eating carrots, they are all considered to be *mu'ad* for damages caused by their eating carrots.

The case of an animal that has killed a human invites deeper consideration of its moral psychology. In theory, an animal that kills a human being can stand trial and be killed if found guilty (Exod. 21:28). (Jewish courts have denied themselves the authority to mete out capital punishment to humans or animals for almost two thousand years. Even the ancient courts that did try capital cases rarely executed anyone because of the extremely strict rules of evidence and testimony that applied.) This conceptual possibility allows for the development of some very interesting ideas about the moral responsibility of animals. For instance, in order for such an animal to be found guilty, it has to be established that the animal *intentionally* killed the person. If an ox accidentally trampled someone while running to get some food, or if it accidentally gored a human being while trying to attack an animal, it would not be subject to the death penalty. Even more remarkably, even if the animal intentionally killed a human being, it could only be killed by the court if the attack sprang from *its own character*. Thus, if a person incited a dog to kill someone, the dog could not be killed; it had not acted of its own natural volition. Similarly, if a usually friendly animal (such as an ox) had been deliberately trained to be aggressive—perhaps in order to fight people or other animals for the "amusement" of human beings—that animal would not be held accountable for killing a person. The aggressiveness that led to the attack was not really part of the animal's nature; its original good character had been artificially distorted by human intervention.

This article has only described "the tip of the iceberg" of the ideas and practices relating to animals that can be found in Jewish tradition. The following references should help interested readers begin to gain a more complete grasp of the topic.

Further Resources

Bleich, J. D. (1989). Animal experimentation (pp. 194–236) and Vegetarianism and Judaism (pp. 237–250b). In *Contemporary Halakhic Problems*, Volume III. New York: Ktav Publishing/ Yeshiva University Press.

Cohen, N. J. (1976). *Tsa'ar Ba'ale Hayim: The prevention of cruelty to animals, its bases, development and legislation in Hebrew literature*. Jerusalem and New York: Feldheim Publishers.

Kalechofsky, R. (Ed.). (1992). *Judaism and animal rights: Classical and contemporary responses*. Marblehead, MA: Micah Publications.

Pines, S. (Trans.) (1963). *Moses Maimonides'* The Guide of the Perplexed. Chicago: University of Chicago Press.

Schochet, E. J. (1984). *Animal life in Jewish tradition: Attitudes and relationships.* New York: Ktav Publishing.

Toperoff, S. P. (1995). *The animal kingdom in Jewish thought.* Northvale: Aronson.

Berel Dov Lerner

Balaam's Ass

Berel Dov Lerner

Chapters 22–24 of the biblical Book of Numbers tell the story of Balaam, a wizard hired by the Moabites and the Midianites to curse the Israelites. While on his way to do their bidding, an angel of the Lord blocks Balaam's path, standing with a drawn sword. Balaam cannot see the fearsome angel, but his ass [Hebrew: *aton*; a female donkey] on which he was riding can. The ass tries to get out of the angel's way by moving off into a field, prompting Balaam to hit it. Next the angel traps them in a narrow passage through vineyards. The ass again tries to avoid the angel, pressing itself against a fence and squeezing Balaam's leg in the process. In response, Balaam hits it again. The angel repositions himself leaving the ass no avenue of escape; in desperation, the ass crouches down on the spot and refuses to move. All this time Balaam has no idea what is going on. He begins beating the beast with a stick. At that point God miraculously "opens the ass's mouth," allowing it to speak in its own defense: "What have I done to you that you beaten me these three times?" Balaam is disgusted with the ass, and threatens that if he only had a sword he would kill it. At that point the ass makes an impassioned speech: "Am I not your ass upon which you have ridden all along until this day? Have I ever risked doing something like this to you in the past?" Balaam accepts the ass's complaint, after which the angel finally reveals himself to the now terrified wizard. The angel immediately begins berating Balaam for beating the animal that had actually saved his life; if the ass had not avoided the angel, Balaam would have been killed and the ass spared!

This unusual story has invited a wealth of interpretations. Some classical Jewish commentators think that it is meant to recount an actual miraculous event; others, such as Maimonides, believe that it describes a dream that Balaam dreamt. In any case, the story also offers a fascinating way to think about the rights of domesticated animals.

■ Culture, Religion, and Belief Systems
Mongolia's Tsaatan Reindeer Herders

Forgotten Lessons of Human-Animal Systems

For millennia, humans have depended on animals for resources necessary for survival. Through domestication, man has sought to ensure these resources by creating systems that bring humans and animals together into arrangements of coexistence. In the developed world, as social, economic, and environmental factors have grown increasingly complex, systems of domestication have often tended toward mass production, resulting in pollution, cruelty, and imbalance. Few models remain that remind us how humans can utilize animals for resources while still maintaining balance, respect, and sustainability in the face of an ever-changing world.

In a remote corner of Central Asia, one community is struggling to do just that. The *Tsaatan* people of northern Mongolia are a nomadic people who depend on reindeer for nearly all aspects of survival, as well as cultural, spiritual, linguistic, and economic identity. Originating from the Sayan mountain region in Russian Siberia and Mongolia's northernmost province of Hovsgol, the Tsaatan, or ethnic *Dukha* people, are credited as one of the world's earliest domesticators of any animal. Written records by a traveling Chinese monk in 499 CE and 3,000-year-old stone carvings of reindeer in the area are evidence of the Tsaatan's ancient relationship with their reindeer (Donahoe, 2003; Vitebsky, 2005). For generations, the group practiced nomadic reindeer husbandry throughout the Sayan region until border closures in the 1920s restricted their movements to within Mongolia. Now isolated from their ancestors and other reindeer herding peoples of the region, such as the Evenki, Tofa, Tozhu, and Soyots, the Tsaatan reside between 51 and 52 degrees north latitude and are Mongolia's only reindeer herders (Jernsletten & Klokov, 2002). Today just over 200 Tsaatan individuals live in the Mongolian taiga, an ecosystem characterized by larch trees, high moisture, and subarctic conditions. They move regularly between two main camp areas, referred to as *Barone* (West) and *Zuun* (East) Taiga, respectively, seeking pasture for their deer. Tsaatan maintain small herds of between 7 and 160 reindeer per family, utilizing the domesticated deer (*Rangifer tarandus*) for renewable resources and imbuing them with value as essential members of the community.

Reindeer are milked daily, providing the staple component of the Tsaatan diet, which largely consists of reindeer dairy products. Naturally shed antlers are used to carve tools and make handicrafts that are sold to tourists, a recent but increasingly significant shadow economy for the community. Nutrient-rich velvet antler is periodically harvested in early summer for international medicinal markets, but the practice is declining due to controversy over cutting the still-living structures (Haigh & Keay, 2006).

Perhaps the most significant use of deer in the Tsaatan system is for transport. Reindeer are ridden and used as pack animals during nomadic moves that occur every 2–10 weeks. They are an essential mode of transportation in the high mountain, roadless taiga. A single reindeer can accommodate up to 65 kilos of weight, consisting of household items, personal belongings, and of course riders (Donahoe, 2003). Sometimes logs used in the construction of the Tsaatan's tipi dwellings, or *ortz*, will also be loaded in order to transport them from one camp to the next if wood sources are scarce. The docile animals are ridden by young and old alike and are often used to travel 30 kilometers or more in a single day (Keay, 2002).

Unlike many reindeer-herding peoples in the circumpolar region of the world, such as the Evenki of Siberia or Saami of Scandinavia, meat harvest is not a predominant feature in the Tsaatan system (Donahoe, 2003; Vitebsky, 2005). Deer are more valuable as transport animals and milk providers than as a source of meat. In fact, the community's ancient tradition of shamanism largely precludes the slaughter of reindeer by placing sacred value on the animals. Instead, hunting wild game such as elk, moose, bear, sable, and boar have historically provided a consistent protein source.

In exchange for renewable resources and reliable transportation the Tsaatan select optimal pasture for the herds, provide protection from predators, and offer their reindeer salt treats. In essence, both humans and animals are linked through a system of interdependence that enables both to survive in the harsh landscape of the northern taiga. Without the Tsaatan, reindeer would be at the mercy of wolves, left to compete against wild caribou for resources, with natural selection inevitably taking its toll. As for the Tsaatan, one herder succinctly stated: "If our reindeer die, we die. They are not just our living but our life," (Keay, Sanjim interview, 2002). This unique relationship is apparent in daily life and is chronicled in multiple Tsaatan tales of origin, such as the following:

> A poor woman in ancient times was wandering in the mountains without any animals. She came upon a deer and was careful not to disturb it. For three days she returned to the same spot and found the deer waiting curiously for her. On the third day, there were two deer, and she called to them, "Goo goo goo," and said, "If you come home with me, we can take care of each other. I will protect you from wolves and give you salt to eat, and you can give my people food and a way to travel." The deer followed the woman home, and as it was fall, the deer mated, and the first reindeer was soon born. (Keay, Erdenshimik interview, 2002)

This story illustrates reciprocity and reverence uncommon to most human-animal systems in the developed world. In the Tsaatan model, people's very identity and origin is linked with that of reindeer. This fosters an unspoken commitment to the animals, making mistreatment and disrespect of reindeer a cultural taboo. Even children are expected to uphold these standards, which are learned through stories, songs, and living examples. As adults, Tsaatan people express a kind of gratitude and appreciation for reindeer that might seem trite or out of place in a western model of domestication. But in the Tsaatan community, reindeer evoke an ethical responsibility characterized by respect and reverence. This is indicated by the tradition of having a shaman in the community identify one sacred, or protected reindeer per family that is permitted to wander freely much like a member of the family. Cultural beliefs such as this offer guidelines for maintaining balance, sustainability, and respect between humans and animals, particularly in the face of change.

At present, the Tsaatan system of reindeer husbandry is in a state of transition, challenged by modernity, development, and an ongoing period of rapid change. Forced relocation, slaughters of the herd, and upheavals of the Tsaatan's socioeconomic situation during Mongolia's socialist era (1921–91) placed pressure on herders to radically alter or abandon their lifestyle. The political transition in 1991 from socialism to a free-market democracy further affected the community, creating a novel environment in which to adapt. In recent decades, many families have tended toward other pursuits, leaving the reduced number of herders in the taiga in a position to face challenges never before encountered.

Today, strict hunting laws and expensive permits along with mounting concerns over wildlife populations have made hunting costly and controversial resulting in an unfamiliar dependence on reindeer as a meat source. Traditional knowledge, such as the use of herbs for treatment of wounds and ailments, has been forgotten or is no longer considered relevant. The emergence of tourism and mixed use of the taiga (i.e., mineral exploration and nonsubsistence hunting by various individuals) has escalated traffic and influence within the community, and a growing need for cash in order to secure relatively new commodities such as antibiotics, flour, school supplies, and two-way radios all threaten the lifestyle of the Tsaatan.

As a result, change and adaptation are constant forces in the community. In response to changing food needs, herders are beginning to adjust management and breeding strategies of their animals in order to provide a sustainable meat source while still honoring ancient values associated with the use of reindeer. Elders have begun to revitalize traditional healing techniques, whereas the incorporation of modern pharmaceuticals has supplemented changing health needs. New activity and outside visitors to the taiga are often perceived as welcomed opportunities, and herders have begun to pursue shadow economic activities that offer cash income while continuing to herd reindeer.

Some new factors however, have created complex scenarios of conflicting interest that seem like roadblocks to the Tsaatan's ultimate survival. One example is the face-off between nomadism and sedentarism, each offering distinct benefits to the community. Herders' desires to access schools, town centers, outside visitors, and economic opportunities outside the taiga have stimulated some families to settle in semi-permanent camps close to the *ecotone,* or ecosystem edge, between taiga and steppe biomes. Although the socioeconomic benefits of this trend are apparent, the lack of movement and removal from the reindeer's natural taiga habitat disallows for proper forage access, thereby compromising the nutrition and health of the herds. Furthermore, exposure to livestock in settled areas increases the risk of disease for reindeer, a reality that may be responsible for the epidemic proportions of Brucellosis and other ailments affecting the reindeer and people of the community. Finding solutions to dilemmas such as this are indeed complex endeavors.

Understandably, it is daunting for the Tsaatan to develop approaches that sustain reindeer husbandry and fulfill new economic and social needs. The extent of change and development seems to undermine the very fundamentals of the community, leaving the culture's eventual extinction a threatening possibility. In fact, few societies in the world have managed to maintain the integrity of subsistence-based human-animal systems in the face of development and modernity. But for the Tsaatan, cultural values and traditional practices continue to be informants of daily choices and visions of the future. The Tsaatan identify as reindeer herders above all else, frequently asserting that reindeer are still the most valuable of all possessions and are the central feature of life. Reciprocity, respect, and sustainability remain core values, as does the desire to maintain balance in the face of adaptation, modernity, and development. Such statements might echo latent sentiments felt by western cattle ranchers, for example, but for the Tsaatan these words are enduring and motivating realities. This has likely played a role in helping make the Tsaatan people among the last remaining animal-dependent and predominantly subsistence peoples in the world, and it may be the critical element needed to achieve balance

In a unique way, both the Tsaatan and Reindeer are linked through a system of inter-dependence that enables both to survive in the harsh landscape of the northern taiga. Courtesy of Morgan G. Keay.

The Tsaatan people of northern Mongolia are a nomadic people who depend on reindeer for nearly all aspects of survival, as well as cultural, spiritual, linguistic, and economic identity. Courtesy of Morgan G. Keay.

in human-animal systems. Perhaps our own society can take a lesson from this in order to reconsider systems of domestication as opportunities to foster productivity, sustainability, and compassion simultaneously.

Further Resources

Clutton-Brock J. A. (1987). *Natural history of domestic animals.* Austin: University of Texas Press.

Donahoe, B. (Spring 2003). The troubled taiga. *Cultural Survival Quarterly, 27*(1); http://www.cs.org/publications/csq/csq-article.cfm?id=624

Haag, A. L. (2004, December 21). Future of ancient culture rides on herd's little hoofbeats. *The New York Times, Science Times.*

Haigh, J. C., & Keay, M. G. (in press). The Tsaatan culture and the management of Mongolian reindeer. IVth World Deer Congress. Melbourne, Australia.

Hudson, R. J., Drew K. R., & Baskin, L. M. (1989). *Wildlife production systems: Economic utilization of wild ungulates.* Cambridge: Cambridge University Press.

Jernsletten, J. L., & Klokov, K. (2002) *Sustainable reindeer husbandry. Project of the sustainable development working group, Arctic Council.* Norway: Centre for Sami Studies, University of Tromsø. http://www.reindeer-husbandry.uit.no.

Keay, M. G. (2002). *Creatures of culture: An investigation of the semi-domestic Mongolian reindeer.* Brattleboro, VT: School for International Training.

———. (Spring 2002). Interview with Erdenshimik, female Tsaatan herder.

———. (Spring 2002). Interview with Sanjim, male Tsaatan elder.

Vitebsky, P. (2005) *The reindeer people: Living with animals and spirits in Siberia.* New York: Houghton Mifflin.

Morgan G. Keay

■ Culture, Religion, and Belief Systems
Monkeys and Humans in India

The relationship between people and monkeys dates back to the beginning of human existence. Influenced by Hindu mythology, people in India revere monkeys as helpers of God. They allow them to freely inhabit cultivated fields, temples, villages, towns, and cities.

India is one of twelve mega-biodiversity countries with unique fauna and flora; it harbors a total of fifteen species of nonhuman primates, which range from the small lesser ape, Hoolock gibbon, to the large lion-tail macaque. Only three species of monkeys, the Hanuman langur (*Semnopithecus entellus*), bonnet macaque (*Macaca radiata*), and rhesus macaque (*Macaca mullatta*), live in human habitats and interact with people on a regular basis.

The Hanuman langur is frequently mentioned in the Hindu epic, the Ramayana. These leaf-eating monkeys live throughout the Indian subcontinent. The rhesus macaque inhabits northern India, whereas the bonnet macaque is found only in southern India. Macaques have cheek pouches to temporarily store food, and they can grab and store an entire meal in a hurry. Humans feed the langurs and macaques regularly in many parts of India, indirectly helping to keep the monkey population somewhat stable over the years.

During the 1950s and 1960s, monkeys were periodically harvested and exported to the United States for biomedical research, but the practice was banned in 1978. All primate species in India are now protected by law. During the 1980s the population of rhesus macaques was believed to be around 200,000; the current population is estimated at half a million individuals, with about 50 percent of them living in human-dominated landscapes. As a result, human-monkey conflict in towns and cities intensifies daily, and in the absence of an effective management plan of forests and monkeys, this conflict may worsen in the future. Due to the lack of natural food sources in urban areas, the monkeys have learned the art of living with people and finding human food for their survival. The urban monkeys know how to beg for food and also steal food from people. Despite their religious beliefs, urban dwellers are slowly losing their goodwill toward monkeys and are gradually showing animosity toward them.

In India's capital city of New Delhi, animals on the streets are a familiar sight. Although elephants, cows, dogs, and pigs often wander through the city, the red-face rhesus macaques are the ones causing the most trouble. Over 5000 of them live inside the city limits, and the efforts to get rid of the roguish monkeys have failed so far. Capturing the monkeys, to release them in other villages and forests, is not welcomed by local governments because they don't want them in their own "backyard." Monkey bites are also increasing; it is estimated that 100 people are bitten per day in urban areas across India. During observations of rhesus and bonnet macaques, it was found that women and children were harassed by monkeys more often than men. Monkeys have learned that it is easier to scare women and children than adult men.

One can avoid getting bitten by these urban monkeys by knowing more facts about their behavior. The best way to avoid being harassed by monkeys is not to feed them or get close to them. While carrying food, people should not show it, so that the monkeys will not follow. If monkeys approach people, direct eye-to-eye contact should be avoided because

A bonnet macaque sits comfortably on a bus to receive food from passengers. Courtesy of G. Agoramoorthy.

it provokes aggression; the monkeys may chase or attack humans. If monkeys grab the food out of people's hands, it is better to leave the food with the monkey and walk away. Chasing, yelling, and throwing sticks or stones, for example, will further aggravate the monkeys.

It is possible to solve the man-monkey conflict in India only if a successful management plan is implemented by the respective local, state, and federal government agencies. The first step is to ban all trappings of monkeys from forest areas. Monkeys need natural food sources, so planting of wild fruit trees should be carried out in areas where monkeys are known to live. Adequate waterholes in natural habitats should be provided for the macaques; they need water to drink, and they also love to swim. Urban monkeys with no natural food sources can be captured, screened for contagious diseases, and the healthy animals can be sterilized before relocating them to forest areas with plenty of natural food, water, and shelter. In order for those procedures to take effect, well-trained field biologists, effective government officials, and workable management strategies and polices are urgently needed in India.

Further Resources

Agoramoorthy, G., Smallegange, I., Spruit, I., & Hsu, M. J. (2000). Swimming behavior among bonnet macaques in Tamil Nadu: A preliminary description of a new phenomenon in India. *Folia Primatology, 71,* 152–153.

Pirta, R. S., Gadgil, M., & Kharshikar, A. V. (1997). Management of the rhesus monkey, *Macaca mulatta* and Hanuman langur, *Presbytis entellus* in Himachal Pradesh, India. *Biological Conservation, 79,* 97–106.

Seth, P. K., & Seth, S. (1992). Long-term studies of free-ranging rhesus monkeys in India. *Primate Report, 32,* 49–60.

Govindasamy Agoramoorthy

The Monkey God Hanuman

Govindasamy Agoramoorthy

The Hanuman langur (*Semnopithecus entellus*), the most widespread among all nonhuman primates in South Asia is named for the Hindu monkey god Hanuman. According to the Hindu epic Ramayana, the wife of Prince Rama, Sita, was abducted by the demon Ravana. Hanuman led his army of monkeys to build a bridge across the Indian Ocean, from the southern tip of India to the island of Sri Lanka, where Sita was being held. As Hanuman's army attacked, Rama rode on Hanuman's shoulders and fired an arrow into Ravana, killing the demon and liberating Sita. Thus Hanuman became one of the most loved of all Hindu deities in India—twice a week, Hindu faithful flock to Hanuman temples to make offerings and seek divine blessings. People also regularly feed monkeys as part of the ritual to get blessings from the monkey god.

The black-faced, silver gray–bodied Hanuman langurs have long limbs and tails that are longer than their head-and-body length combined. These highly adaptable monkeys survive the cold in altitudes up to 4,000 meters in the Himalayas of Nepal and the heat of Jodhpur on the fringe of the Great Indian Desert. They feed on leaves and buds, flowers, fruit, seeds, and occasionally other cultivated crops. They also feed on insects, fungi, and gum. The stomach of these fascinating Colobines is divided into four sections to aid in the digestion of cellulose. The first two compartments serve as fermentation chambers, in which the stomach contents are blended and decomposed by cellulose-degrading bacteria.

Adult female hanuman langur opens her mouth as a threat towards a human intruder. Courtesy of G. Agoramoorthy.

Hanuman langurs live in well-organized social groups with an adult male and several adult females and their offspring. There are also all-male groups of varying ages. As with humans, langur mothers bear one infant at a time, with twin births occurring rarely—only one case of quadruplets has been recorded. Within hours of the birth, other group members, excluding the adult males, help care for the infant; this behavior is known as "allomothering." Among adult males, there is intense competition for the leaderships of groups, and soon after a new male takes over a harem, all weaned male youngsters are shunted out. Infanticide, or brutal killing of infants, also occurs when a male becomes the new leader of a group to induce sexual activities in females.

Wild hanuman langurs receive food from a human. Courtesy of G. Agoramoorthy.

The Hanuman langurs are considered endangered, and they are facing difficulties surviving in the wild due to ongoing habitat fragmentation, human population growth, agricultural expansion, and poaching. Although they seldom bite people intentionally, langurs living in urban environments have frequent contact with humans, sometimes raiding houses in search of food and occasionally destroying television antennas and other delicate equipment on rooftops. Due to the Hindu religious faith, they enjoy a higher level of cultural esteem in India than the rhesus and bonnet macaques. They are also thought to be more elegant and less aggressive. They are common in cities, towns, villages, forests, and tourist sites, and people interact with them frequently. Since they rely more on natural vegetation and less on agricultural crops, they are seldom considered pests, and they pose no threat or economic loss to people. In fact, people in India tolerate Hanuman langurs more than the problematic macaques.

Further Resources

Agoramoorthy, G. (1994). Adult male replacement and social change in Hanuman langurs of Jodhpur, India. *International Journal of Primatology*, *15*, 225–39.
———. (2005). Hanuman langur. In A. J. T. Johnsingh (Ed.), *The mammals of South Asia*. Delhi: Oxford University Press (in press).
Hrdy, S. B. (1977). *The Langurs of Abu: Female and male strategies of reproduction*. Cambridge, MA: Harvard University Press.

■ Culture, Religion, and Belief Systems
Mustang: A Personal Essay

I sit here on a cold day, a Chickasaw woman, covered with a nineteenth-century painted buffalo hide, listening to the mustang's hooves run, pounding the earth outside. Mystery is the true name of my horse because that's what she is in this world, a mystery of survival. That's what all the wild horses are.

Like all true wild horses, she has fewer vertebrae than any of the domesticated, human-bred horses. She was not created for appearance but for sturdiness and wilderness survival, short and muscular. She was rounded up when she was young, but her memory is very old.

There is a relationship between us, and when we walk together, there is a bond, a feeling that surpasses language. She feels this, too, because somehow horses *know,* and the mustang knows even more than those who have had a human hand in their creation. This capacity to know, this unity, is something that intrigues me.

Horses have always been valued by Native peoples. The relationship has been a long one, and it has resulted in immense cultural change for tribes. No matter where they came from—the Spanish horses from DeSoto in the Southeast, those coming upward into Texas, New Mexico, and California, the Russian horses from the North—their presence was an agent of cultural change so powerful that the U. S. government eventually tried to kill all those that became Indian horses in order to win its war against tribes.

It was said in history books that Indians first thought the human and the horse were one animal, but Native people are intelligent and practical people and could tell the difference between muscle and armor, especially when the armored men fell off in the heat and couldn't get up because of the weight of armor. However, horses became so significant a part of Native cultures that they almost are one animal, the unity between them is so strong. It is as if, with horses, a new way of being, a new relationship, was created.

Horses quickly became a part of religions, brought into traditional realms. Gifts were given to them, and they were decorated with eagle feathers, paint, cloth, and beadwork. Called Mysterious Dog, Elk Wolf, and Sacred Dog, horses are the one thing from the European continent that has been incorporated into Native mythologies as special beings. Their names alone reveal this: Sky Gods, Medicine Dogs, Wolf Dogs. Across the United States and Canada, and probably in Latin America as well, are horse songs. From the Piegan and Bloods, to the Nez Perce who bred the Appaloosa, all the way down into Texas, there are beautiful songs for these animals who have shared not only our lives, but our history. The Aztecs laid down red cloth for those maned, long-tailed creatures to walk upon.

The mustangs' presence opened a world for Indians across the continent, and for the gift of them, the people have been grateful. Ceremonies revolved around the horse, including Black Elk's Horse Dance, which appeared to him in a dream and then was taken up by his Lakota nation. They are not just animals once used for hunting and warfare, but as "special beings," treated with reverence, considered to be sacred animals. The Navajo use them in the Enemy Way, where a horse carries a medicine bundle as part of the remarkable and serious long-lived ceremony.

Mystery is the epitome of the Indian horse in every way. She is short-necked, sturdy, with feet like "agate," as a Navajo song says. The feet of the wild horses are studied at veterinary schools and by breeders who want to strengthen the hooves of the other breeds. Misty, as I call her for short, was injured in the process of being rounded up. A rifle I once carried frightened her, and I believe they used rifles to create noise to move all the horses together into their roundup pen. She tells me her story in pieces this way. No one is allowed to touch her left ear, and so I realize she was "broken" in a manner that uses the ear twitch to make the horse submissive. Through our years together, I have slowly tried to get us to the point where I can at least scratch behind her ear. And she no longer tries to kick me when I clean her back hooves. She has given in to me in so many ways. Riding, working on the ground, she is happy. She loves order. She is happy to stop and be obedient when I say "Ho," or when I simply stop walking. It is part of being a herd again, part of doing what the others do:

> Child, you've said it yourself,
> the way a horse thinks,
> the way a gathering of fish swim.
> One turns, they all turn.
>
> Tsimmu

We know each other as if we shine from the same star, as if we rode the continent together. We breathe each other in, through nostrils, the smell of grass. The feet of stone, black rock, the history of a world I come from, except she was another birth. We will live the same number of years, our life spans the same, together. Horse woman, I put my arms around her. At night, out the window we talk, borderless the way it is with horse and the one who loves her. We have a clarity together. I sit with her and watch the meteor shower as we pass through their orbit. Her colors are constantly changing. She is a blue corn horse, sometimes a pale blue, gray, then turning black, then the color of water willows when they become red just before the break of spring, the long hairs on her stomach.

I can only imagine the relationship between those who loved and whose lives depended on their horses, the ones now called mustangs, meaning "untamed."

Her hunger tells another story. It is great, always frenzied, tearing at the grasses as if she were hungry before, and she probably was. Survivor of hard men, perhaps of waterless places.

Mystery. There will always be an aspect to her that she must keep, the untamed part, the wildness, the memory of what she used to be, where she used to be. She is constantly on guard, on the lookout, her head turning toward a sound, her eyes always searching for predators. Still, she is vulnerable at times, sleeping in the moonlight, large, breathing widely. Yet, if a sound, an ear moves. Snap of a twig and she rises up to protect. Studying the history of Chickasaw horses, I found a drawing that looked just like Misty. It is believed by many that the journey of horses across the country first began in this Southeast territory. The Chickasaw horses in the Southeast were so famous that all horses at one time were called Chickasaws. They were traded along the Mississippi, for we lived in an important trade center with other tribes and then with European traders. I have wondered if my ancestors knew her ancestors, if we have a history together. The Cherokee, leaving first along the Trail of Tears, took these ponies with them and they disappeared somewhere in time. We took them with us also along the Trail Where People Cried, and thieves who followed us stole them at night. Many were interbred and are now said to have become the American Quarter Horse. A taller horse was in demand. Some, however, escaped into the wild herds and traveled to places unknown. Because of their size it would have been hard to breed with larger horses.

The history of mustangs is not without its tragedy. Killed in great numbers by the Americans during what they called The Indian Wars, their losses were great, their suffering greater. General George Custer killed thousands to keep the people of the continent from escaping. General Nelson Appleton Miles and other military men did the same. Because these mustangs' structure lacked European concepts of beauty, they were considered less worthy than the horses the army kept. Nevertheless, many of the horses they killed must have been their own, for the Indians became famous for horse theft. It was a necessity for survival, as was their close relationship, for they depended upon the horse to work with them in ways the military did not. They developed ways of riding, on the side of the horse and even underneath, against the belly.

The mustangs and horses became a sign of wealth for tribes. They also wrote history on this land. Goodland, Kansas, was eroded because of the cavalry chasing Indian people back and forth into Oklahoma, where they were to build a wall around Indian Territory to keep the unwanted people held inside. Black Kettle was chased by the Americans so often that one of his ponies escaped and was simply called Black Kettle's pony. Black and shiny, it had a mane and tail that touched the ground. For thirteen years people tried to round up this horse that Black Kettle had so cared for. It was finally caught by the same man that bought the skull of Joseph, the Nez Perce chief, and used it for an ashtray. The pony, it is certain, had a miserable outcome.

As for Mystery, we have the same life spans. She has U.S. branded on her neck, just below the mane, to remind me daily of our shared history. We were *removed* into Oklahoma. I think sometimes of my grandfather and his favorite short, black horse that he brushed until it gleamed like silver, according to my father who, as a boy, tried to ride this one-man horse and, to his dismay, ended up being ridden against trees, into limbs and branches, until his father saddled up another horse and rode up beside them to rescue the young man. My grandfather loved Shorty because he so resembled the Chickasaw pony that disappeared around the time of his birth but was spoken about so much.

Misty and I share a life. At night I speak to her from the window in bed and she talks back. If a predator has been near, I leave a light on for comfort. She watches me move about the room as much as I watch her in daylight. I hear her run in the morning, back and forth, in the summer when it is still cool, in fall when the leaves have fallen, in the snow of winter, her hooves pounding earth. I imagine what her herd must have sounded like, running. I think, perhaps I know her ancestors and what they did for us, how they ran for us to save our lives, trudged alongside us on the Trail of Tears before disappearing into the larger continent. I grieve for the mustangs, many now being sent to other countries as food, killed as dog food, their land taken over for cattle.

Recently a herd of mustangs was released on Lakota land. As the people watched, feeling something returned, a herd of buffalo also came over the hill to watch the horses run. It would seem the relationships are more than just a bond between the human and animal, but the animal and the land, including the others that dwell there.

Further Resources

Dobie, J. F. (2005) *The mustangs*. Lincoln, NE: Bison Books.

Linda Hogan

■ Culture, Religion, and Belief Systems
Native American Vision Quests

What is a Vision Quest?

The vision quest is a ritual practice in which an individual spends time, often several days, alone in a wild or natural place seeking communication with the spirit world, especially with a spirit helper. The spirit helper might come in a vision or a dream, often appearing as an animal. This vision may provide vital information, direction, or wisdom of some kind.

The kind of animal met in a vision is important because help is provided in ways characteristic of the species involved. Also, a quester may meet more than one animal. According to Black Elk of the Lakota, for example, the famous warrior chief and priest Crazy Horse received most of his power through this practice, "which he did many times a year, and even in the winter when it was very cold and very difficult." He received visions of the Badger, a prancing horse (the source of his name), and the Spotted Eagle, as well as nonanimal helpers, the Rock and the Shadow (Joseph Epes Brown, 1953, p. 45).

The vision quest is widespread among Native Americans both historically and at present. Beyond this basic outline, the practice varied substantially from place to place. Sometimes it was essentially a coming-of-age rite. For other groups it could be practiced at any time and possibly often, as with Crazy Horse or Black Elk himself. In some traditions those who had visions stood out as potential shamans or healers; in others it was a universal expectation and thus a kind of disgrace if you did not quest or if you went on a quest and did not receive a vision. Sometimes only the men were expected to engage in a quest; for others, women sought visions in this way, as well.

As with all customs, a vision quest can be expected to include features fitting with other local practices and beliefs. Among peoples of the Northwest Coast of North America, for example, there was strong sense of inheritance. People inherited not just physical property but also their guardian spirits. The shaman inherited his powers. But as Robert Torrance (1994) observes, it might be clear which spirit you will inherit, but "at least a pro-forma" quest is still needed. A long solitary quest was essential for personal encounter, at least a person's first encounter, with empowering spirits. Among the Coast Salish of British Columbia, the two were combined, in that "ritual transfer of shamanic power from father to son was preliminary to the boy's personal spirit encounter" that came about through the vision quest (Torrance, p. 179).

Among many Northwest Coast peoples the vision quest was not just for shamans but for everyone, roughly corresponding to the tribal initiations found among other peoples. It was still important that a person obtain his or her own spirit guardian by their own efforts. We are told that among some groups, children who balked at the task were whipped. Preparation could be extensive, and the dangers, both physical and mental, were real, and not everyone succeeded.

For some people the quest was practiced, but it was not central to their culture. Thus, among the Northern Paite, the dream is the central means of acquiring spiritual power. Dreams may come unsolicited or be sought via a quest ("most commonly in a mountain cave," says Torrance). And some Native American peoples do not engage in the vision quest at all. Among sedentary Pueblo agriculturalists, the wild, uncontrolled visionary experience was rare or at least unimportant, while the seminomadic Apache bands valued both the visionary shamanistic experience and ritualized ceremonies. The religious leader "in training" would carefully study and memorize many ceremonies, yet would value his own personal encounter with supernatural powers even more (Torrance, p. 192).

Conversely, the vision quest was not limited to the Americas. To refer again to Robert Torrance's rich survey of the spirit quest, among the peoples of the Malay peninsula and Indonesia the shaman would often engage in a kind of battle with his spiritual foes. Communication might involve a "solitary vigil beside an open grave or in the dark forest" (p. 145). Among the Iban of Sarawak, an individual might engage in a vision quest for his or her own healing. Often, communication with spirits was via dead relatives and other humans rather than animals.

Animals as Spirit Helpers: An Example

Because these matters are complex, difficult to generalize, and varied from people to people, a concrete example will do most to convey a sense of the animal-human relationships involved in the vision quest. Thanks in part to the memories of Lakota holy man Black Elk, compellingly told by John Neihardt (1932) and Joseph Epes Brown (1953; 1992), the rites of the Lakota (referred to as the Oglala Sioux in many older publications) are among the most accessible and widely known. Called *Hanblecheyapi*, or "Crying for a Vision," the vision quest among the Lakota was practiced throughout one's lifetime, not just as a coming-of-age rite, and this "lamenting," as it was also known, might have any of a number of goals, from seeking a lifelong spirit helper to seeking a solution to a specific problem.

Among the Lakota, the animal that is met would not be *just* a flesh and blood owl or bison such as you might encounter at any time, but one bearing a spirit with power. As for Plains Indians generally, "each animal was really a crystallized projection of the abstract spirit" (Brown, 1992, p. 37). Yet the animal was not just incidental either, a kind of arbitrary messenger or random form taken by a spirit. The kind of animal involved in a given vision mattered a great deal for the kind of help and power received (Brown, 63–64).

Not surprisingly, the bear is associated with fierceness and unpredictable forces and was appreciated for its lack of fear of animal or human. But bears are also associated with knowledge of underground or earth forces. This includes roots and herbs, and because of this, visions of the bear were often associated with healing powers. Thus Black Elk, who was both a medicine man and a holy man, said that he received many of his curative powers from the bear. He knew some 200 distinct herbs, the understanding of which he received along with other "bear powers" through dreams and during vision quests.

For all the strength and fearlessness of the bear, it is the wolf that captured the attention of the warrior. Visions involving wolves would be related to strength, swiftness, and success in warfare. Members of the Lakota Wolf Society took the wolf pack as a model for the strength and cooperation of the human society. The fox, on the other hand, is thought of as gentler and less aggressive, and in its gentleness is strength and courage of a different kind.

The vision quest ends with a sweat lodge that is prepared for the returning seeker. Here the vision is described for and interpreted by an expert. Later, the vision might be incorporated into songs, medicine-bundles (made up, in part, of the fur or other elements of the spirit-helper animals), and a range of rituals. One's name might be changed, the animal's form might be painted on clothing, tepees, or shields, and a person might join a particular society, such as the Kit-fox society, each of which originated from someone's earlier vision of the animal. These societies and their character relate as much to the character of their animal as to the specifics of any one individual's vision.

What Really Happens During a Vision Quest?

As Philip Jenkins (2004) points out, the vision quest, along with other elements of Native American culture, is being discovered by many non-Native Americans. As a result, people want to know if these visions are real. In one sense they are certainly real, often life-changing experiences. Visions can be vivid and tangible, and they are easily distinguished from ordinary experience. This is almost by definition: the experience would have to differ from the ordinary or it wouldn't qualify as a vision. For example, a vision might involve an animal talking. It would be silly to imagine that Native Americans are unaware of the fact that animals do not normally talk. Indeed this is a key—if you thought ordinary animals could speak, you would hardly treat a message from one as a transforming event.

But behind the basic question is often one far more difficult to answer, such as, "was an eagle really there or is it all in their heads?" Anthropologists usually reject the visionaries' explanations of their own experiences, especially the fundamental idea that there are spirits out there who communicate with us. This is not a prejudice against Native American tradition; it is how any religion is studied, a perspective with a long tradition in anthropology (Wason, 2004). If we don't accept the informant's view as authoritative, then what is going on? Robert Torrance offers one typical perspective.

> Clearly mastery of such powerful spirits presupposed the difficult mastery of self, a principal goal of the arduous quest . . . for the guardian spirit was an alter ego resulting from purposeful self-transformation, an expanded self acquiring transcendent power through communication with a larger than human world, and therefore a self not given but created and found. (1994, p. 180)

This seems plausible and could well be part of what is going on. But it is hard to imagine Black Elk being satisfied with such an explanation. Native American cosmological and metaphysical thought is rich and complex, and varied in place and time, but many of those who practice the vision quest believe that reality consists of more than molecules and forces and natural laws. They believe often in a creator-God, in a range of active spiritual beings, and they understand the categories of human, animal, and spirit as fluid rather than fixed and separate. The quester, then, would not believe an ordinary animal was speaking, but that the quest was a transcendent, spiritual experience. If one's vision consists of a wolf walking out of the woods to face you and deliver a message, this might be a *living Canis lupus,* and he may have the bad breath to prove it, but it is no *ordinary* animal. Some descriptions give the impression that the vision occurred in an altered, perhaps dream-like state in which a different dimension of reality opened up. In other cases it sounds much more physical. But for many engaged in the vision quest, separating "physical" from "spiritual" might be no easier than imagining an unbridgeable gap between human, wolf, and spirit.

Further Resources

Brown, J. E. (1953). *The sacred pipe: Black Elk's account of the seven rites of the Oglala Sioux, recorded and edited by Joseph Epes Brown.* New York: MJF Books.
———. (1992). *Animals of the soul: Sacred animals of the Oglala Sioux.* Rockport, MA: Element, Inc.
Harrod, H. I. (2000). *The animals came dancing: Native American sacred ecology and animal kinship.* Tucson: University of Arizona Press.
Hirschfelder, A., & Molin, P. (2001). *Encyclopedia of Native American religion* (updated ed.). New York: Checkmark Books/Facts on File.
Jenkins, P. (2004). *Dream catchers: How mainstream America discovered Native spirituality.* Oxford: Oxford University Press.
Neihardt, J. G. (1932; 1972). *Black Elk speaks: Being the life story of a holy man of the Oglala Sioux, as told to John G. Neihardt (Flaming Rainbow).* New York: MJF Books.
Tooker, E. (Ed.). (1979). *Native North American spirituality of the eastern woodlands: Sacred myths, dreams, visions, speeches, healing formulas, rituals and ceremonies.* Mahwah, NJ: Paulist Press.
Torrance, R. M. (1994). *The spiritual quest: Transcendence in myth, religion and science.* Berkeley: University of California Press.

Wason, P. K. (2004). Naturalism vs. science in the anthropological study of religion. *Omega: Indian Journal of Science and Religion, 3*(2), 27–58.

Wiebe, P. H. (2004). *God and other spirits: Intimations of transcendence in Christian experience.* Oxford: Oxford University Press.

Paul K. Wason

■ Culture, Religion, and Belief Systems
Noah's Ark

The biblical story of the Flood—"Noah's Ark," found in Genesis 6:5–9:19—is arguably one of the most widely recounted and popular stories from the Hebrew Bible (to Christians, "The Old Testament"). It is depicted in religious visual arts (particularly artwork for children who identify with the animals and the rainbow). It is referenced in sacred music, told in Christian and Jewish religious settings globally, and Christian theologians from Thomas Aquinas to Martin Luther to Andrew Linzey analyze its meaning. Twentieth-century zoos sometimes self-designated as "arks," attempting to save a remnant animal population in the midst of species extinction.

Noah's Ark is not the only account of a great flood. Similar stories of catastrophe emerge in many of the world's religions and from geographically and historically distinct cultural groups. But the Hebrew account mirrors closely those of two other ancient Mesopotamian epics, *The Epic of Gilgamesh* and the *Atrahasis Epic. Gilgamesh* even includes a group of animals aboard a ship. Myriad interpretations of the flood story focus on elements such as the re-creation or the new covenant. But in almost every case, the role of animals is ignored, or mentioned only briefly. While they abound in artwork, other than human animals are quite marginal in most exegetical works. If they are included in these interpretive and substantive pieces, animals seem to fare rather poorly. A brief outline of the story and overview of analyses of it provide a glimpse into its impact on human-animal relations.

The Hebrew account, as told in the book of Genesis, combines several sources into one story. In short, it describes the process through which God, angered at the "wickedness of humankind," destroys and then re-creates the world, establishing a new covenantal relationship with the humans and other animals on the earth. God does this by flooding the entire creation, but forewarns one faithful person thus allowing a remnant to survive and repopulate the world. The one person, Noah, is deemed worthy: he "found favor in the sight of the Lord" (Gen. 6:8). So God provided Noah with very specific instructions to build an ark. This is where animals enter the story.

Although the biblical account explicitly states that the Lord "was sorry that he had made humankind on earth" (Gen. 6:6), a flood that covers the entire world would, obviously, kill all the land animals, not just the humans. So God tells Noah to bring "two of every kind into the ark, to keep them alive with you" (Gen. 6:19). Later in the text, biblical scholars will explain that this is a point where different oral versions are blended—rather awkwardly in this case; God's instructions shifted slightly, and the Lord tells Noah to bring "seven pairs of all clean animals" and "a pair of the animals that are not clean" (Gen. 7:2). The significance of this shift is addressed later. After seven days the rain begins to fall, and it continues for forty days. As a result "all flesh died that moved on the earth, birds, domestic animals, wild animals, all swarming creatures that swarm on the earth, and all human beings" (Gen. 7:21).

Eventually God makes the waters subside and Noah sends out various birds to look for dry land. Finally, a dove returns with an olive branch. The story also indicates that the ark comes to rest on the top of a mountain (specified as Ararat). After months, God tells Noah to leave the ark and to bring "every living thing that is with you of all flesh—birds and animals and every creeping thing that creeps on the earth—so that they may abound on the earth, and be fruitful and multiply on the earth" (Gen. 8:17). This is the moment of re-creation, with similar instructions regarding filling the earth, reflecting back to the first two chapters of Genesis (the biblical Creation story).

The return to dry land was not, however, a blessing for all of the animals. God instructs Noah to build an altar and sacrifice one of "every clean animal and of every clean bird" as a burnt offering (Gen. 8:20). God finds the odor "pleasing" and in response to the sacrifice of animals declares, "I will never again curse the ground because of humankind . . . nor will I ever again destroy every living creature as I have done" (Gen. 8:21). It must have been knowledge of this sacrifice that required extra pairs of "clean animals" when Noah was initially loading the ark. But, here again, a blending of sources means that different pairings of animals are reported.

After the command to be fruitful and multiply, God announces that the relationship between humans and animals is changed forever. In the creation accounts (specifically Gen. 1:29–30), God gives "every plant yielding seed that is upon the face of all the earth, and every tree with seed in its fruit" to humans for food. Thus, the reader functions under the assumption that all humans are vegetarians at this point. The change in Genesis 9 carries a huge impact as God declares to Noah (and thus to all humans) that the "fear and dread of you shall rest on every animal of the earth, and on every bird of the air, on everything that creeps on the ground, and on all the fish of the sea; into your hand they are delivered. Every moving thing that live shall be food for you; and just as I gave you the green plants, I give you everything" (Gen. 9:2–3). This momentous and devastating change reads almost as a curse. Most likely this is an acknowledgment of humans eating other animals and of different sources telling the story.

God then establishes a covenant with both humans and "every living creature that is with you" (Gen. 9:10). God promises never to destroy the earth with a flood again and places a rainbow in the clouds as a sign of this promise.

Noah's Ark has been interpreted variously. From Martin Luther, whose exegesis of Genesis 9:3 states that God "set himself up as a butcher" and allowed Noah to be provided for "more sumptuously now" (Luther, *Lectures on Genesis*), to John Olley, who focuses on the "covenant" made equally with all living beings. Andrew Linzey, a Christian theologian, and Dan Cohn-Sherbok, a Jewish theologian, coauthored one of the first books on animals and liberation theology in these traditions—*After Noah: Animals and the Liberation of Theology*. The title is telling and reveals the considerable impact of the Noah story on Western culture's perception of human-animal relations. It is not a straightforward account, and interpretations are not only complex, but also problematic on many levels. But its impact on the understanding of the relationship between humans and other animals, particularly in the history of Western civilization, is undeniable.

See also

Culture, Religion, and Belief Systems—Christian Theology, Ethics, and Animals
Culture, Religion, and Belief Systems—Judaism and Animals

Further Resources

Alter, R. (1996). *Genesis*. New York: W. W. Norton.

Brueggemann, W. (1982). *Genesis*. Atlanta: John Knox.

Habel, N., & Wurst, S. (Eds.). (2000). *The earth story in Genesis*. Sheffield, UK: Sheffield Academic Press.

Linzey, A., & Cohn-Sherbok, D. (1997). *After Noah: Animals and the liberation of theology*. London: Mowbray.

Watermann, C. (1984). *Genesis 1–11: A commentary*. Minneapolis, MN: Augsburg.

Laura Hobgood-Oster

■ Culture, Religion, and Belief Systems
Pantheism

Pantheism is a spiritual belief in which the divine and the natural world are understood to be one and the same. The word *pantheism* stems from two Greek words, *pantos* ("all") and *theos* ("God"). Pantheists believe that whatever exists is God, and God *is* all that exists. The pantheist's world is divine, from spiders to chickens, from rocks to fish; God is all, and all is God.

"Panentheism" is a related belief whereby the divine permeates the natural world but is not exhausted by what we see and experience. Panentheists view the Ultimate Reality as *more than* the natural world, whereas pantheists identify the Ultimate Reality directly with the physical world.

Pantheism exists in nearly every religious tradition, especially among mystics, who aspire to unity with the divine. Hinduism, the dominant faith of India, has particularly strong pantheistic elements. Important Hindu sacred writings, such as the *Upanishads* and the *Mahabharata,* teach of pantheism.

For Hindus, *Brahman* is the highest principle of the Universe, the substratum underlying the universe, the unknowable, indefinable power behind and within all that exists. (Some authors translate *Brahman* as "God.") The *Upanishads,* composed about 2,500 years ago, teach that each individual is *Brahman*: "This Great Being . . . forever dwells in the heart of all creatures as their innermost Self . . . [and] pervades everything in the universe" (*Svetasvatara* 122–23). *Brahman* is identified with nature and nonhuman animals:

> Thou art the fire,
> Thou art the sun,
> Thou art the air . . .

> Thou art the dark butterfly,
> Thou art the green parrot with red eyes,
> Thou art the thunder cloud, the seasons, the seas. (*Svetasvatara* 123–24)

The holy essence of *Brahman* pervades every living being, and every creature shares the essence of this ultimate reality. The ground of each individual's being is identical with that of *Brahman*. Piglet, sparrow, human being—Hindu pantheists find "in all creation the presence of God" (Dwivedi, 2000, p. 5).

To understand what it means to be human in the Hindu worldview is to understand what is important in the existence of salmon or sheep because the foundation and fundamental core of each is *Brahman.* "[A]s by one clod of clay all that is made of clay is known," so all things are one in essence, and that essence is sacred (*Chandogya* 92). As a pinch of salt dissolved in a glass of water cannot be seen or touched, but changes the contents of the glass from fresh water into salt water, so the subtle essence of *Brahman* runs through all beings, yet cannot be perceived or touched. This subtle essence makes each living being holy. As all rivers are temporarily distinct but ultimately join one great sea, so do all living beings appear in separate bodies but are ultimately united by *Brahman.*

The sacred *Mahabharata* teaches that those who are spiritually learned will behold all beings in Self, Self in all beings, and *Brahman* in both. Hindus understand all living beings to have *atman,* usually translated as "souls," and that *Brahman,* or God, is that soul. In the most famous portion of the *Mahabharata,* the *Bhagavad Gita,* the beloved God, Krishna, explains what it means to be divine: "I am the life of all living beings. . . . All beings have their rest in me. . . . In all living beings I am the light of consciousness" (*Bhagavad* 74, 80, 86). The divine is not viewed as separate from the natural world of

cows and rabbits, but as indwelling. Krishna is often pictured as standing with cattle or as playing in the rural countryside of Northern India. The *Bhagavad Gita* presents the divine as so close to us that divinity is an essential part of who we are, and an essential part of all other aspects of creation: "I am not lost to one who sees me in all things and sees all things in me."

The *Bhagavad Gita* also notes that we exist in the heart of all other beings, and the heart of all other beings dwells within our own self. Not only is the divine in all beings, but we, as part of the divine, are also part of all other living entities. If we understand the world in this spiritual light, the divine is in our hearts, and we have the spiritual understanding to acknowledge the divine in all other beings. In this way Hindus come to love *all* beings, and the pleasure and pain of other creatures becomes personal (*Bhagavad* 71–72).

Given that Hinduism has rich pantheistic teachings, it is not surprising that *ahimsa* is central to Hindu morality. *Ahimsa,* often translated as "nonviolence," is more literally translated as "not to harm." This moral ideal requires Hindus to avoid bringing suffering to any living creature whenever possible. It is small wonder that most Hindus have, for centuries, been vegetarians.

Ahimsa is the cornerstone of the Jain faith, a religion that emerged in Northern India around 500 BCE. Buddhism arose at

An image by a Kalighat artist of the Hindu God Krishna as a cowherd milking a cow, ca. 1890. ©The Art Archive/British Library.

roughly the same time in the same area of India. The Jain and Buddhist religious traditions have much in common, including an emphasis on *ahimsa* inherited from the earlier Hindu religious tradition. Jains do not eat either the flesh or the eggs of other creatures.

Pantheism remains central to Hindu belief and practice. Hindu spirituality requires that people strive not to harm the natural world because the divine *is all* that exists. Devout Hindus must extend love not only to other human beings, but to dogs and fish, chickens and hogs.

Pantheistic spiritual beliefs, beliefs that see the divine in the world around us, discourage human arrogance and pride that might otherwise lead people to believe that we are superior to other animals, that we are separate and more important than the rest of the universe. Pantheism, such as that seen in Hinduism, teaches that every aspect of the natural world is divine, and that the spiritual life therefore requires us to show respect and reverence toward all living beings and the natural world.

Further Resources

Dwivedi, O. P. (2000). Dharmic ecology. In C. K. Chapple & M. E. Tucker (Eds.), *Hinduism and ecology: The intersection of earth, sky, and water* (pp. 3–22). Cambridge: Harvard University Press.

Embree, A. T. (Ed.). (1972). *The Hindu tradition: Readings in Oriental thought.* New York: Vintage.

Harrison, P. A. (1973). *Elements of pantheism.* Tamarac, FL: Llumina Press, 2004.

Mahabharata. (1973). (W. Buck, Trans.). Berkeley: University of California Press.

Nelson, L. E. (2000). Reading the *Bhagavadgita* from an ecological perspective. In C. K. Chapple & M. E. Tucker (Eds.), *Hinduism and ecology: The intersection of earth, sky, and water* (pp. 127–64). Cambridge: Harvard University Press.

Svetasvatara Upanishad. (1948). (Swami Prabhavananda and F. Manchester, Trans.). *The wisdom of the Hindu mystics: The Upanishads: Breath of the eternal.* New York: Mentor.

Lisa Kemmerer

■ Culture, Religion, and Belief Systems
The Parrot Book

Like the tales in *Kalila and Dimna* (q.v.), *The Parrot Book* stories are said to have originated in India and to have come to their compiler, Zia al-din Nakhshabi (d. 1350), in the form of a Persian translation, which he then revised and supplemented with quotations from the Qur'an and Persian literature. Indeed, ancient Sanskrit texts (such as the *Panchatantra,* the *Sukasaptati,* and others) seem to be the major source for this genre of literature. As the translator of *The Parrot Book* notes, these works were mainly intended as entertainment for the ruling classes, and animal characters were often used as a way of describing social injustice and oppression.

Nakhshabi's *Parrot Book* consists of the usual Russian-doll arrangement of tales within tales, told against the backdrop of a marital drama. A traveling businessman leaves at home his young wife, who (naturally!) is tempted to infidelity during his absence. She confides in their pet myna bird, who tries to dissuade her from her illicit desires, but the sexually frustrated wife refuses to listen and cruelly kills the bird. Next she turns to their pet parrot, who, being somewhat more clever than the myna, tries to avoid the latter's fate by responding more subtly, through stories.

Similar to the plot in *One Thousand and One Nights,* where Scheherazade keeps her husband from committing his crimes by enthralling him with tales until dawn, the parrot prevents its mistress from going out to visit her lover by recounting tales (mainly of other women's immoral behavior) through the night. This goes on for fifty-two nights, until the husband finally returns. As is generally the case in Muslim literature, the morals issues addressed in *The Parrot Book* are really human ones and from a decidedly male (not to say misogynistic) point of view. Animals exist as stand-ins or fly-on-the-wall observers of human behavior, rather than as beings in their own right.

The narrator of *The Parrot Book* is an impossibly humanlike parrot that can recite the Qur'an and is valued at a thousand dinars. Though it is not hard to imagine some actual Muslim parrot-owner training his bird to recite the Qur'an, the description of this particular parrot is obviously exaggerated. His role, perhaps, mimics that of a faithful servant (who would have had to be a eunuch under the circumstances)—but, of course, parrots are more colorful.

The first tale told by the parrot narrator is of a parrot very like himself, forced to play the role of mediator in a dysfunctional human marriage. In a status-conscious traditional society with little means for upward mobility, the reversal of power relationships can be a powerful literary tool, whether it be lower-class humans miraculously becoming royalty, as in certain *One Thousand and One Nights* stories, or the orchestration of human events by animals in *The Parrot Book.*

Another tale seems to offer a similar reversal: that of a dog who "was accorded more esteem than anyone in the village." This is no ordinary dog, of course: it is the dog of Layla, the young beauty for love of whom the madman Majnun condemned himself to a life of solitude in the desert. If the dog possesses more importance than most humans, it is nevertheless from a human, his mistress Layla, that this importance derives. Once again, the elevation of a lowly animal to an exalted station seems to be primarily a narrative device.

Occasionally, animal examples are used to illustrate the "laws of nature," to which humans, too, are presumably bound. For example, in the fifth night's story, a mother parrot instructs her young that there can be no friendship between birds and other species, since their interests are fundamentally at odds. The message here corresponds neither to interspecies relations in nature nor to human communities, where cooperation between differing communities is vital to the survival of all. Rather, it would appear to be a somewhat feeble attempt to justify xenophobia and class divisions.

"The Story of the Lion and the Cat" (the fifteenth night) is an example of political allegory illustrating the strategies of those who wish to remain in royal service by keeping themselves indispensable. The lion (who is, as always, the King) has a mouse problem, and hires the cat to handle it. The cat knows he will hold onto his job only as long as he allows some mice to survive; one night, however, his overzealous son exterminates all the mice, and, sure enough, once the lion realizes mice are no longer a problem, he fires the cat. Real lions and cats, of course, have no such relationship.

We have already seen that what appear to be animals can sometimes be other beings—such as humans or genies—which have taken animal form. In the story of the eighteenth night, Moses saves a pigeon about to be eaten by an eagle, offering the eagle a pigeon's weight of his own flesh in exchange. It turns out the birds are actually the angels Michael and Gabriel, come to test his generosity and selflessness. A story such as this is more likely intended to highlight the exceptional qualities of an ancient prophet than to recommend anyone actually offer his own flesh to birds.

An interesting twist on the Moses episode occurs in the story of the thirtieth night, when a nasty woman thrown out of the house by her husband finds herself in the jungle and in danger of being eaten by a leopard. Like Moses, she offers herself as a sacrificial meal, but in her case, it is for the purpose of tricking the leopard. In *The Parrot Book,* all women are conniving liars, and in this story, the hungry leopard is merely her buffoon.

The irremediable infidelity and deceitfulness of women is the thematic thread tying together the parrot's fifty-two stories, not the lives or qualities of the many animals that adorn the tales. The parrot narrator, having won his freedom by informing on his mistress, announces that if all humans are like her, he wants nothing to do with them. The woman's husband, after killing her to save his honor, decides the same and ends his days as a pious hermit.

Further Resources

Nakhshabi, Z. (1978). *Tales of a parrot: The Cleveland Museum of Art's Tūtī-nāma.* M. A. Simsar (Trans.). Cleveland: The Museum.

Richard C. Foltz

■ Culture, Religion, and Belief Systems
Potcake Mongrels in the Bahamas

Local mongrels, or strays, in the greater Bahamian archipelago (i.e., the Bahamas and the Turks and Caicos Islands) are called "potcakes," which is the same name given to burnt food at the bottom of a cooking pot. Traditionally, these scraps were fed to dogs, and so the name has been applied to the dogs. Similarities in this name can be seen in the use of "pothound" in Grenada and "rice dog" in Guyana. The preference of potcakes for cooked food was demonstrated in a trapping exercise in Providenciales, where cooked bones, rather than raw bones, were effective in enticing dogs into traps.

The Name *Potcake*

The use of the term *potcake,* as applied to dogs, appears to have a long oral tradition going back to—at least—the end of the 1800s. The word *potcake* used in reference to dogs does not appear to have appeared in print until the 1970s, although veterinarians used the term in the 1960s to indicate a local dog of no defined breed. We surmise that "potcake" may have been considered a vulgar or slang word which would not have been used by the educated classes or the newspapers. In the 1960s and before, newspapers commonly used the word "cur" for a mongrel dog (almost certainly a potcake). In the 1960s, *cur* was clearly a very derogatory word, and its use in parliament at that time caused an outrage. An alternative explanation for the rise in the use of the term *potcake* suggests that it was increasingly used in the late 1960s and early 1970s as a way of distinguishing between local dogs and the purebred dogs which were then becoming increasingly popular.

With the advent of independence and a greater awareness of indigenous art and culture, the potcake, too, has received more prominence. The Bahamas Kennel Club has set a standard for potcakes to encourage their care and increase their prominence, and attempts have been made to "rebrand" potcakes as something desirable by calling them "Royal Bahamian Potcakes." They have even been used as mascots in antidrug campaigns and to promote improved pet welfare.

The potcake has now been integrated into popular culture. "The Cry of the Potcake," a song by Phil Stubbs, and pictures and cartoons by Eddie Minnis have brought the potcake to the attention of a wide audience. In the Turks and Caicos Islands, a Potcake Foundation has been established to promote animal welfare in general—and potcakes in particular.

Usage of the Term *Potcake* in Language

The word *potcake* has now developed an even wider usage, but one related to its canine use. In a small island community, family relations are important and can influence how a person is received. Surnames are often classified as being Bahamian or non-Bahamian and then sometimes qualified by which island a person was born on. Family connections are used both to make opinions of a person and to exert or gain influence. Because potcakes have unknown lineage, they are typically considered "low" dogs. Hence, potcakes are considered as being outcasts, and this is reinforced by the fact that they are typically kept outside the home, given little health care, and may still be fed household leftovers. These nuances of outcast, low breeding, being common, irresponsible, or not belonging have passed into a usage that can be applied to humans or even culture— "potcake culture." To say "You do not have the breeding to be a potcake" would be to attempt to thoroughly insult someone. The potcake is typically considered a prolific breeder or of great prowess, and these perceptions have developed into sexual connections when "potcake" is applied to humans. The tenacious aspect of a potcake, which makes a good watchdog, however, can be considered a trait of which to be proud.

Potcakes in the Bahamas

Roaming dogs have officially been "a nuisance" since 1841, when legislation was passed admitting that Nassau, the capital, had too many dogs. Despite numerous acts designed to curb the dog population, islands such as New Providence, Grand Bahama, and certain parts of Abaco still suffer from an overpopulation of dogs.

The first detailed description of potcakes is thought to be that of L. D. Powels in 1888:

> The origin of the extraordinary collection of mongrels that inhabit this city and its suburbs and pass for dogs must ever remain a puzzle. Mr. Drysdale says they are "the most fearful and wonderful productions of nature." Like the majority of living things in Nassau they are half-starved, and spend their nights wandering about the wealthier parts of the city, trying to pick-up scraps. Their howlings, and the crowing of the cocks, which invariably commence at 11 p.m., and continue for several hours without ceasing, make nights hideous. Some time ago a dog tax was imposed by the Legislature, but it became so unpopular, and so extremely difficult to collect that it had to be ignominiously abandoned. Wherever you go in coloured settlements, dogs run out every minute to bark at you, but I never heard of their biting anyone, and they run away if you merely turn and look at them.

This description could have been written more recently than 1888. It captures the key features of potcakes seen today, except that the roaming dogs are not so starved, but owners continue to be reluctant to license their dogs. Their skittish behavior, streetwise ways, and ability to blend into the general environment of men and cars makes them tolerated by passersby, who are usually untroubled by them. However, their behavior exhibits marked, localized differences. In residential areas, roaming dogs are wary and often very shy, but can respond affectionately if encouraged. In tourist areas, their behavior is respectful but less obviously wary. Typically, they will approach and sit by a car or person and patiently wait to be fed. Potcakes are considered to make good watchdogs—but poor *guard* dogs. Barking has long been a feature of potcakes, which endears them to owners, but for others their barking at night has been a source of complaint since at least the 1880s.

Despite the negative image associated with potcakes, they can make wonderful companions; this is recognized by both residents and tourists, some of whom adopt roaming potcakes and take them back to their homelands.

Roaming Dogs

As previously noted, potcakes are the most common type of dog seen roaming, but purebred dogs also roam. Despite it being common for roaming dogs to be viewed as only potcakes or "the potcake problem," it is best not to equate potcakes with roaming dogs, as not all potcakes roam nor are all roaming dogs potcakes.

Residents have an ambivalent attitude toward roaming dogs. While the dogs are considered a nuisance, many people feel sorry for them and feed them. In addition, generally—but particularly in Providenciales—garbage, either from bins that are incorrectly stored or that are thrown on the ground, ensures that there is a large amount of food accessible to roaming dogs. In tourist areas, restaurants and hotels are also sources of large quantities of food.

While food does not appear to limit the population, other factors do. The subtropical environment of the Bahamas is hostile to the roaming dog population. Heartworm is ever present, and venereal tumors and prolonged droughts—among other influences—constrain the life expectancy of roaming dogs and limit their reproductive ability. Occasionally, distemper outbreaks occur, so the population is never static. Data from the Animal Control Unit indicate only one peak in the number of puppies per year. This suggests that dog are breeding at intervals greater than six months (otherwise there should be two peaks per year, as the animals that bred in the spring would be responsible for a second peak later in the year). We believe that on average, often less than one puppy per litter survives to breeding age.

Little is known about the sex ratio, age distribution, and weights of roaming dogs, but they are considered large for roaming dogs. This may be due to the availability of food in garbage or from handouts, or it may reflect the fact that about 40 percent of owned dogs are allowed to roam. Some people let their dogs roam because they feel that it is "cruel" to confine dogs. The most common nuisances of roaming dogs reported are those of tipping over garbage bins and barking at night. However, this second complaint may be a misattribution, as 70 percent of owned dogs are kept "for protection" and so would be expected to bark.

Roaming dogs are scarce in the hotter hours of the day. Their activity is crepuscular. For example, our observations have demonstrated that their activity is closely aligned to sunrise; consequently, they are more visible in the cooler winter months, when sunrise is later and there is an overlap with peaks in human activity (travel to work). This can give the impression that the dog population increases in the winter months.

Owned Dogs

Between 30 and 40 percent of households own dogs, and the average number of dogs per household is about 2.5. These Bahamian dog-owning households have more dogs per household than in many other countries. Potcakes make up at least 70 percent of the owned dog population, and maybe as much as 90 percent. The average age of owned dogs is three years. The lack of confinement and the low neutering rate result in many unwanted litters. One study in Marsh Harbour, Abaco, found that all the litters in the previous twelve months had been born to dogs that were able to roam. This suggests that

Bella is a young female potcake dog (sometimes called pothound), typical of the species. Courtesy of Tamara L. Kerner.

all these litters were unplanned and so may have resulted in unwanted puppies. Long-term observation of one group of four cared-for potcakes, which almost certainly received no health care, showed that they were unable to increase their numbers over a five-year period, despite having several litters. These observations contradict the perception that potcakes are "always breeding." While potcakes do breed all year round, an individual dog does not appear to breed often. Abandonment of dogs is considered to be the engine that maintains the roaming dog population.

Potcakes are considered "survivors" who do not need health care, and this may account for the fact that only 30 percent of dog owners routinely take their dogs to the veterinarian. Potcakes generally receive little or no health care, whereas purebred dogs are more likely to do so. Despite this difference in healthcare, potcakes have the same average age as purebred dogs. Potcakes are valueless (no one buys a potcake), and this may explain the reluctance of owners to spend money on them. If the animal dies, it is easily replaced by adopting another roaming dog. Because few dogs have collars or are licensed, establishing formal ownership is not easy.

It is clear that potcakes are, in many ways, typical of subtropical mongrel dogs, but their long association with humans has given potcakes a special place in Bahamian society. This has led the potcake to be regarded as an icon—but also as a fallen pet. While most Bahamians feel sorry for the roaming potcake, they also consider him a nuisance who they do not want harmed. This contrasts with many owners who still keep their dogs outside, free to roam and feed on scraps. It is hoped that, as owners are made more aware of their responsibilities toward their pets, potcakes will be better cared for and will be elevated from watchdog to companion.

Further Resources

Allsopp, R. (1996). *Dictionary of Caribbean English usage*. Oxford: Oxford University Press.

Fielding, W. J., Mather, J., & Isaacs, M. (2005). *Potcakes: Dog ownership in New Providence, The Bahamas*. West Lafayette, IN: Purdue University Press.

Humane Society International. (2001). *Dogs on Abaco Island, the Bahamas: A case study*. Washington, DC: Humane Society International.

The Nassau Guardian. (2004, June 11). *Miller ignores LNG critic.* Retrieved March 3, 2005, from http://archive.nassauguardian.net/archive_detail.php?archiveFile=./pubfiles/nas/archive/2004/June/11/NationalNews/104814.xml&start=40&numPer=20&keyword=potcake&sectionSearch=&begindate=1%2F1%2F1999&enddate=12%2F31%2F2010&authorSearch=&IncludeStories=1&pubsection=&page=&IncludePages=1&IncludeImages=1&mode=allwords&archive_pubname=Nassau+Guardian%0A%09%09%09

Powles, L. D. (1888). *The land of the pink pearl: Recollections of life in the Bahamas.* Reprinted 1996. Nassau: Media Publishing Ltd.

William Fielding

■ Culture, Religion, and Belief Systems
The Pothounds and Pompeks of Grenada

The "ordinary"—or "regular"—dog in Grenada, West Indies, is called a "pothound." One is told that the dogs are called pothounds because "dey is as many as potholes in de road." Some say it is because the local dogs are fed whatever is left in the pot or get to lick the cooking pot, which is more likely. But with Grenadians' having healthy appetites, not much would be left in the pot, and a Grenadian woman would *never* allow a dog to lick from her cooking pot! But, more seriously, the name does come from the common way of cooking. As with the Grenadian national dish, "oildown," which is cooked in a big pot, the same goes for the local dogs: "dey all mix up in one big pot and out comes de pothound!"

The pothound is a medium-size dog weighing between twenty-five and thirty-five pounds, depending on how well fed or how good at scrounging they are. Pothounds, if owned, are fed on chicken back, chicken neck, and rice, all boiled together. Any vegetable scraps, such as figs, bluggoes (a type of banana), or breadfruit guts go in the cooking pot for the dogs. In fishing villages, the smaller fish, called "jacks", and any fish scraps are put in the dogs' cooking pot. Grenadians rarely feed their dogs raw scraps, as they claim "dey jus' vomit it up." By cooking the food, the bones are softened and are more readily chewable. That is not to say the pothounds don't have problems with chicken and fish bones. Veterinarians and volunteers often have to pull such bones out of pothounds' throats and out of the rectums of small puppies that cannot digest them.

Pothounds are often observed as a pack, lounging for long hours outside a window of a kitchen where food is being prepared. There they patiently wait for the scraps or leftovers from the pot. Unfortunately, when the boiling water is thrown out, the poor dogs are not always quick enough to get out of the way and can get severely burned. It rarely happens to the same dog twice, but it is not uncommon for dogs to be treated for serious burns from boiling water having been accidentally poured on them.

Unowned, stray pothounds wander and scrounge for food, their numbers determined by its availability, as with stray dog populations in other parts of the world. The more aggressive and stronger dogs survive; the weaker dogs do not. Many dogs succumb to either or both sarcoptic and demodectic mange, further debilitating already impaired immune systems. Scratching, hairless dogs can be seen roaming restaurants and roadside chicken barbecue stands. These dogs scratch open lesions in their skin, which is then secondarily colonized by bacteria causing chronic skin infections. Unless someone takes pity on these dogs, they have a high mortality rate at an early age.

The female pothound will breed every six months, giving birth to six or eight puppies. On average, two of these puppies will be male, although there are rare exceptions. Pothounds make good mothers, moving their puppies for safety. A female with puppies can become very aggressive and will attack humans it does not know if they come too close. When the mother is off scrounging for food, her puppies are usually quite safe from predators; the only predator in Grenada is the mongoose. The mother keeps her litter hidden beneath any strewn housing materials (like boards or concrete). The mother will make every effort to raise her young, and most litters survive. Occasionally, it is reported that a mother has eaten her young. The only other way a litter may die is if the mother is poisoned by the reckless use of weed killer or through the consumption of deliberately deposited rat poison.

Pothounds have shorthaired coats and come in a multitude of colors. The most common is a fairly uniform light brown or tan coloration. The most desirable color is white with tan, brown, or black spots or markings; the least desirable pothounds are black with pointed ears. These are often parented by feral dogs and do not make good pets or good watchdogs, but they are great hunters. Increasingly, more brindle-colored (locally known as "tiger") pothounds are being seen. Most pothounds have dark brown eyes, but occasionally a rare and beautiful golden eye color, which is quite desirable, is seen. Many pothounds are born with short tails. The short tail ranges from a little stub to about four or five inches. Many people feel this is a desirable trait, as they believe short-tailed dogs do not get worms. Some owners even have the tails of their dogs removed because of this belief.

The two males out of a litter will be kept by the owner, given away, or sold first. Male dogs are much more popular than female dogs. Pothounds are used as pets or as watchdogs, and most homes keep several in their yards or gardens. Pothounds are very protective of their territory and can be quite aggressive. They are very loyal to their owners and will follow them everywhere they go. Even if they get no attention from their owners, they instinctively know who they belong to and will protect the entire family. They will not allow any other dog to trespass on their boundary and will often chase and fight with wandering dogs that are looking for food. Human trespassers, including children, must be wary of the protective pothound.

The pothound is a predictable dog and will give a fair warning to trespassers. The dog will growl first, then bark, and, if that does not work, it will remonstrate with lunging motions accompanied by a range of growls. If, at that point, the intruder turns and runs, the pothound will give chase and even bite the person on the ankles. The pothound will normally stop this attack at its boundary. If stood up to, the pothound will usually back off and slink under the house or to the back of the yard. However, it will continue to bark until the owner is alerted or the trespasser moves on. Pothounds have never been known to actually attack a human, other than to chase them off the property. They rarely protect themselves and prefer to run and hide when threatened.

Pothounds are great hunting dogs and are valued for this role by local hunters. A hunter usually selects the more feral type of pothound for hunting. These dogs are fed only after the hunt and live on the very fringe of the hunter's property. The hunter will generally go out to hunt with four or five dogs. *Manicou* (opossum), iguana, and *tatoo* (armadillo) are hunted in the forested habitat preferred by these animals. The dogs naturally hunt and will corner the hunted animal in a tree so that the hunter can then capture or kill the animal. These dogs are very devoted to the hunter and rarely hunt alone.

Pothounds are generally sweet-natured, tolerant, and loving dogs and make great family pets. Children can climb all over them, drag them around, and dress them up. A well-fed and well-cared-for pothound is one of the easiest dogs to own. They prefer living outside, under the house in the shade, or on the front porch or veranda. They do not require much grooming, having short hair. Once they are spayed or neutered, they rarely wander off their property, except to follow their owner. They don't eat much and can survive on little food, and they will eat almost anything.

Pothounds can be trained to do almost anything and are outstanding on an obstacle course. The only thing the pothound is not too tolerant of is water. Many do not like going into the sea, especially if that is where they are bathed. It is believed that seawater is good for dogs, and pothounds are often forced into the sea to be washed. It is also believed that salt water kills fleas and ticks, and the dogs are sometimes held under the water against their will. Many people use seawater in an attempt to treat mange and wounds on their dogs. One will often see pothounds lounging on the beaches, but rarely will one see them willingly go into the sea.

A very enduring trait that pothounds have, not often seen in any other breed, is smiling. Most dogs tend to yawn or make yawning motions when they are not sure what is expected of them, but a pothound will actually *smile*. When a person makes small talk to a pothound that is not used to being spoken to in a kindly manner, it will often curl its upper lip, show its teeth, and look away and smile. This is an expression of wariness but acceptance. Not all pothounds smile, though, and, like people, some do it more often than others. Because of this practice, many pothounds are named Smiley.

There are several local organic prescriptions and treatments for parasites and mange. For worms, local people commonly use the leaves of the soursop tree. Soursop, or "guanabana," have shiny, dark, leathery leaves that display a pungent odor when crushed. The leaves of the soursop are crushed and boiled to make a "tea" that is drunk by the dogs; it is believed that this kills parasites.

To treat mange on dogs, it is a custom to use the fruit from the calabash tree, locally called "boley"—so named because a bowl can be made out of the dried fruit, which is quite large. The pulp is scraped out, and the pulp and seeds of the calabash fruit are crushed and spread on the dogs' skin. Another treatment for mange is to pour used car oil all over the dog's skin, in an attempt to smother the mites. The dogs then lick the oil and become quite ill; this treatment is not recommended, but one does, occasionally, see "oily" dogs walking around.

To elicit vomiting after a dog has ingested poison, Grenadians commonly use sugar water or warm stout and force the substance down the dog's throat. Both are quite effective emetics.

The diseases that inflict pothounds are mostly infectious diseases that can be prevented. Pothounds do not seem to have a propensity for noncommunicable diseases, including inherited diseases, such as hip dysplasia, that often come with highly bred dogs. Extra toes and dewclaws are often seen on pothounds. The condition is not debilitating and is ignored. If kept clean and free of fleas, ticks, and mites, pothounds rarely fall ill. The average lifespan of a pothound, if not treated with any preventative medicines, may be between four and six years. Pothounds treated with preventative medicines can live up to fifteen years, and sometimes even longer. They are usually quite healthy in their old age. An elderly pothound is rarely seen with arthritis and can be very active in its senior years.

Poor nutrition and being hit by cars are common causes of death in pothounds. Another cause of premature death is heartworm (*Dirofilaria immitis*). Serological and necropsy studies have shown that infection increases with age, and almost a third of dogs over six years old have heartworm. Carval syndrome, the most severe form of heartworm disease, was seen in 13 percent of just over four hundred dogs examined. The parasite has a ubiquitous distribution throughout Grenada. Heartworm is likely to be transmitted all year round, with a peak in the rainy season (from June to early March) corresponding with the mosquito population density changes. Because so many different genera of mosquitoes naturally transmit heartworm to dogs in Grenada breeding in various water sources, it is difficult to envisage how this zoonotic infection can be reduced in its importance in the pothound population.

Since early 2000, cases of "tick fever" (*Ehrlichia canis*) have been recorded in pothounds. This blood-borne disease is believed to have come from Trinidad, where Grenadians purchased pit bulls and brought them back to Grenada. The pit bulls came with either the disease or the infected vector ticks (*Rhipicephalus sanguineus*). Whatever the origin, the introduction of tick fever spread quite rapidly and is now common among pothounds.

At the beginning of the rainy season, in June or July, parvovirus commonly occurs in puppies—perhaps it is the heavy rains that help spread the infection. With mothers giving birth outside and with infected puppies excreting the virus in their diarrhea, with Grenada's hilly landscape, the heavy rains could spread infection. Parvo lasts for the first two or three months of the rainy season every year then abates. Another common infection seen in pothounds, mostly during the rainy season, is leptospirosis, which is caused by species of the genus *Leptospira,* a spirochetal zoonosis that can, theoretically, infect all species. Detailed studies have not been conducted, but dogs are known to have a specific serovar (*L. canicola*). Transmission and spatial distribution of leptospirosis is facilitated by water, and, for this reason, the bacterial infection may be more common in the rainy season.

Along with the Grenada pothound is another local dog, called the "pompek" or "puddle" (poodle). These are small dogs, weighing anywhere from eight to twenty pounds. These little dogs are highly prized and very sought after. It is believed they originated from a cross between a Pomeranian and a Pekinese from England many years ago because pompeks have many of the characteristics of both. They have thick, fluffy fur and full, curled-up tails. The two most common colors of pompeks are white and reddish brown. They often have an underbite and bad knees from being poorly bred.

It is rare to see pompeks as street dogs, like the pothound. However, over the years, several have been turned in to the shelter when found wandering the streets on their own. Pompeks do not do well as street dogs and can be killed by packs of loose pothounds following a female in heat. Pompeks cannot compete with the pothounds for scrounging food but can hunt lizards and mice to survive.

Pompeks have very thick fur. Because Grenada's climate is quite damp and hot, they often suffer from fungal disease. The little pompeks also suffer from all of the same endo- and ectoparasites that afflict pothounds. They are often covered in ticks—which can become numerous before the owner spots the problem, as the ticks can hide in their long coat. The coat of the pompek can become very matted and requires constant grooming.

Many Grenadans believe that the pothound and the pompek are both delightful dogs with great personalities and are proud of both these local dog types.

Peggy Cattan and Calum Macpherson

■ Culture, Religion, and Belief Systems
The Pure Brethren of Basra

The so-called "Pure Brethren" (*Ikhwān al-safā'*) were a group of radical Muslim philosophers who lived and wrote in the southern Iraqi city of Basra during the latter third of the tenth century. Their very name contains an animal reference, as it is borrowed from a fable about a group of doves in the *Kalila and Dimna* stories (*q.v.*). The Brethren wrote their treatises collectively and—due probably to the unorthodox nature of many of their positions—anonymously. The best-known of their fifty-one works, titled *The Case of the Animals versus Man before the King of the Jinn,* is probably the most extensive critique of mainstream human attitudes toward animals in the entire vast corpus of Muslim literature.

In this unusual book, representatives from the animal kingdom bring a court case against the human race, whom they accuse of abusing their superior position. The animals point out that before the creation of man they roamed the earth in peace and harmony— what might be called in contemporary language "natural balance"—until the arrival of humans, who do nothing but exploit and destroy and who lack any sense of justice.

The trial proceeds, and the humans present one argument after another to support their claims of uniqueness and entitlement. Though the humans insist that "philosophical and rational proofs" will suffice to establish their case, the evidence they bring is anything but. "Our beautiful form, the erect construction of our bodies, our upright carriage, our keen senses, the subtlety of our discrimination, our keen minds and superior intellects all prove that we are masters and they slaves to us," they say (Goodman, p. 56). Each of these subjective assessments is demolished in turn through counterexamples from the animal world, but the humans persist in providing more of the same. "We buy and sell them, give them their feed and water, clothe and shelter them from heat and cold" (p. 60). This justification, too, the animals eloquently refute.

The humans then turn to character assassination, criticizing the qualities of various species such as the rabbit, the pig, and the horse. Each of these maligned species speaks up in its own defense, richly describing the special qualities and merits of its kind. In every case, the animals are the ones providing the rational arguments, in contrast to the humans' arrogant and self-serving claims.

Fearing they may be losing the case, the humans begin to contemplate bribing the judges, as well as other sneaky behavior. The animals, meanwhile, hold councils according to their kind—predators, birds, sea creatures, crawling things—and consult with one another respectfully and democratically to elect a representative from each animal category to present their case to the King of the Jinn.

In the discussion among things that crawl, the Brethren go so far as to challenge the example of the Prophet, in defense of animals' right to live. Where Muhammad included poisonous snakes among creatures that Muslims should kill, the Brethren's snake spokesman points out that snakes, too, play a vital role in the natural community, both as predators keeping other species in check and in producing poisons that not only kill but have medicinal uses as well.

The Brethren recognized that in the world of creation, every species plays its assigned role and knows its proper place, with the sole exception of humans. In fact, the Brethren's understanding of natural processes and relationships was sophisticated enough that they are sometimes claimed to have foreshadowed Darwin. Yet, while one can see in their cosmology an ecological vision that is, in some respects, strikingly modern, they were creationists, not evolutionists; animals occupy specific vital niches, not because they have evolved into them through natural selection but because God, in the perfection of his creative plan, has distributed them that way.

On another level, it could be argued that *The Case of the Animals versus Man* is an almost postmodern work, in that nonhuman animals are presented as subjects of their own experience, not merely as objects observed by humans; at least, that is the apparent reading up until the last page of the book. But the persuasiveness through which the reader is made sympathetic to the animals' view only makes the culminating scene more shocking: the King of the Jinn, in the end, decides in favor of the humans, basing his judgment on nothing more than the capricious and unproven premise that humans alone can have eternal life.

This unexpected, abrupt, and, from an animal rights perspective, highly unsatisfying conclusion leaves one wondering just what point the Brethren were trying to make. Is their treatise intended to awaken the reader to a non-anthropocentric reality? If so, the ending is clearly unacceptable. But if the intention is to reassert the view of human uniqueness, why so convincingly make the case on the animals' behalf? Or is the reader's frustration meant to be turned against God, for having established the hierarchy of creation on the basis of such unfair and arbitrary principles? The question is not easily resolved.

Whatever the ultimate intentions of the authors, *The Case of the Animals versus Man* remains unique in the context of Muslim society, today as much as during the century in which it was written. It is important to note that the views of the Pure Brethren were never accepted into mainstream Islamic thought, and, in subsequent centuries, only the heterodox Ismaili Shii sect, identified today with the Aga Khan, adopted their writings as authoritative. Yet, it may be that in regard to animal rights, the Pure Brethren (like St. Francis in Catholicism) were simply ahead of their time, and, as such, they may have more to teach us in the twenty-first century than they did to the Muslims of their own era.

Further Resources

Goodman, L. E. (Trans.) (1978). *The case of the animals versus man before the king of the jinn.* Boston: Twayne.

Richard C. Foltz

■ Culture, Religion, and Belief Systems
Religion and Animals

Religion has long been one of the most powerful forces shaping our beliefs about human-animal relationships. From ancient times, religious traditions have regularly sought to provide perspectives on humans' possibilities with other living beings. Some attitudes advanced by religious traditions have been very respectful of other animals, holding them to be bringers of blessings and, sometimes, even divinities. Decidedly more negative attitudes toward animals have also been advanced by religious traditions—in this article, a range of attitudes will be discussed as they have been advanced by religious traditions other than Christianity.

Today, some people dismiss religious attitudes as irrelevant to human-animal interactions, but several factors make religious attitudes extremely important when assessing our current human-animal relationships. First, religion continues to be one of the principal factors, around the world, affecting people's attitudes and actions toward other animals. Second, even though in some ways there have been important shifts in attitudes toward religion in the Western world—for example, in *Irrational Man* (1962) William Barrett argued that the decline of religion was "the central fact of modern history in the West" (p. 24)—the attitudes that prevail in secularized spheres of today's world had their origin in religious traditions. In other words, those who hold religion irrelevant often advance attitudes that were first formulated and then made popular in the culture by some now forgotten religion. To understand why people think as they do about the rest of the world's living beings, then, it is necessary to engage perspectives developed within the world's diverse religious traditions.

For a second kind of reason, religion plays another major part in how we think about these issues. Because humans are, in modern biological terms, members of the animal family, one could say that religion was both developed for and practiced by animals. But in this article we ask, what about all of the *other* animals? Although the first-century Roman writer Pliny the Elder asserted that an elephant had been seen "worshipping the sun and stars, and purifying itself at the new moon, bathing in the river, and invoking the heavens," the most relevant issue is not whether other animals are themselves religious, but, instead, how religions have helped or hindered people in thinking about and treating Earth's other living beings.

What makes this inquiry complicated is that nonhuman animals are very diverse—they range from extremely simple creatures to far more complicated domesticated work partners. Some have the largest brains on Earth (sperm whales), and others are free-living, intelligent creatures who exhibit rich emotional lives worked out in their own communities.

What part have these many different kinds of beings had in human religious life and belief? As the twenty-first century begins, scholarship on this subject is only beginning to reveal how rich the intersection of "religion" and "animals" is. This modern scholarship is intensely interdisciplinary; that is, it calls regularly upon many approaches from both the sciences and the humanities to help discern what happened in the past, what is happening now, and what might happen in the future in our relationships with other living beings.

It surprises some that nonhumans have had a remarkable presence in religious beliefs, practices, symbols, and ethics, from ancient times. Such presence is diverse and complex—sometimes religious believers have had a very positive attitude toward other life, while at other times they have thought of other-than-human animals as mere objects created expressly for humans' use.

In the most ancient religions, other animals' presence in the religious imagination was pervasive. Hunting societies characteristically thought of the game animals as holding special power and holiness. But, as noted by many scholars, the development of agriculture and animal domestication diminished humans' encounters with game. Before this development, other animals were often seen as individuals in every sense that humans are individuals—more importantly, they were often seen as ancestors, clan members, fellow travelers, separate nations, and, often, intermediaries between the physical world and the supernatural realm.

For many native peoples, relationship with other life forms was maintained through dreams and waking visions. Ritual ceremonies often honored interspecies bonds as central to these believers' lives. As such, other animals often energized religious sensibilities dramatically.

The altered status of nonhuman animals that followed the development of agriculture and domestication of some animals was far less spiritual than the ancient vision of connection and community. A modern statement of the reduced status of other animals appeared in 1994, in the revised *Catechism of the Catholic Church* (Paragraph 2415): "Animals, like plants and inanimate things, are by nature destined for the common good of past, present, and future humanity."

Thus, over time, many humans' interest in nonhuman animals narrowed. The development of some new religious traditions caused many believers to shed an already minimal appreciation of nonhuman creatures. In the Western cultural tradition, for example, studies of animals by Greek, Roman, Jewish, and Islamic theologians and interested believers declined as the centuries passed.

The cumulative result of many religious thinkers failing to notice other animals, or to take them seriously, has had some complicated byproducts. In the seventeenth century, for example, the philosopher René Descartes thought of other living beings as being more like clocks than like humans—he paid attention to the side of religion that dismissed living beings, while ignoring the side of religion that recognized other animals' sentience, intelligence, and connection to all life and its source. Scientific experimentation on other animals became particularly prevalent in scientific circles from the late-nineteenth through the twentieth centuries, and religious institutions offered no substantial opposition to this trend.

The upshot has been that nonhuman animals have been marginalized by many who follow, or were impacted by, the influential cultural traditions of the Greeks and Romans, and by the Abrahamic religions (Judaism, Christianity, and Islam). The prevalence in westernized secular circles of ignorance about other animals, which resulted in their radical dismissal in morals and ethics, thus has its roots, as it were, in the failure of these cultures and religions to notice other animals and take them seriously.

Some forms of each of these traditions—Greek, Roman, Jewish, Islamic, and, of course, Christian—continued to assert the importance of "animals," but the human-centeredness of the mainline versions of these religious and cultural traditions obscured this important option. Today, these traditions are sometimes wrongfully characterized as completely "bad news" for other animals—in reality, each of these religious and cultural traditions has many features that can be seen as animal-friendly and promoting rich, healthy human-animal relationships.

Hinduism

Of great importance, as well, are some fascinating features of India's major religious traditions. Hinduism offers an astonishing range of views about the living beings that share our ecological community. Humans are recognized in Hinduism to be in a continuum with all other life, for Hindus believe in reincarnation—this belief constantly promotes awareness of humans' connections with and obligations to other living beings. Yet, humans are also considered the highest form of what biological life should be. One thus often finds the hierarchical claim that human status is *above* that of any other animal.

The Hindu tradition, then, also exhibits a mix of views about other living beings. The belief that nonhuman animals are inferior to any human sometimes produces scorn for other animals (there can also be scorn for lower-caste humans). But arrayed against the negative side is Hinduism's deep ethical sensitivity to other animals as beings who should not be killed. This sensitivity is sustained, in part, because Hindus understand all animals to have souls—this alone makes them worthy of ethical consideration, such that Hindus commonly assert the importance of *ahimsa*, or "non-harming."

Further, many Hindu scriptures hold that the earth was not created for humans alone, but for other creatures as well. If one visits India today, one still sees many examples, in daily life, of coexistence with other animals, the best known of which is the sacred cow.

Buddhism

Buddhist attitudes are somewhat similar—all animals, human and otherwise, are viewed as fellow voyagers in this world's perpetual repetition of birth, death, and rebirth. This tradition, like Hinduism and other influential Indian religions (such as Jainism), produces much compassion toward other animals. Buddhist literature has countless expressions of concern for other living beings—this is one reason many scholars have claimed that Buddhism takes a kind, sympathetic view toward nonhuman lives.

But alongside the high-profile positive Buddhist attitudes, however, are complicating features. Buddhists group all nonhuman animals into a single realm, which is inferior to the human realm. This realm is often viewed as an unhappy place—as the historical Buddha said, "many are the anguishes of animal birth." Being born a nonhuman is, in some ways, seen as a punishment, and some Buddhist scriptures characterize animals as pests. But, even if these factors lead to occasional negative descriptions of nonhuman animals, Buddhist views always exhibit central concerns for compassion, the most note-worthy of which is monks' and ardent laypersons' daily recitation of a commitment to refrain from killing any life forms.

Judaism, Christianity, and Islam

As Buddhism, Hinduism, and other India-based religions share features in how they view nonhumans, so do Judaism, Christianity, and Islam. The latter trio features a promi-nent tendency to focus on the human species as if humans alone should be the principal concern of moral protections. Each of these religious traditions holds that our world was created by a divinity that elevated the human species above all other forms of life. Consequently, one finds many attempts to justify practices that harm other animals. In Judaism, ideas about nonhuman animals are, as with the other religions, rich, diverse, and ultimately in tension with one another. One reason for this is that the Hebrew Bible contains some contradictory views about which beings really matter. Sometimes, the Jewish scriptures focus on keeping humans safe from dangerous animals, while at other times, the Hebrew Bible features a utopian vision of peace with wild animals. The first of these views is more common, which is not surprising, given that human interests are characteristically seen in Judaism as far more important than the interests of any other biological beings.

Sometimes the view of other animals is quite negative—Philo Judaeus, a first-century Jewish historian, employed an image of continuous warfare by the animals against humankind. But one does not have to look far in Jewish sources to find positive images as well, because the tradition holds that other animals were created by God, who is proud of them (as expressed in various passages in Job) and daily feeds them. So one finds provisions, such as the law codes (Exod. 22–23 and 34; Lev. 22 and 25; and Deut. 14–26) that recognize, to some extent, the welfare of other animals.

Such provisions that notice and take the welfare of other animals seriously can have their limits, however. Some are concerned only, or primarily, with the welfare of domes-tic animals—that is, those that work for or produce benefits for humans. Others are related only to how sacrificial animals can be killed. Such protections may or may not produce benefits for the many kinds of nonhumans that are not domesticated or used for sacrifice.

Another important connection to other animals exists because of the number of specific animals mentioned in the Hebrew Bible. Psalm 104, for example, mentions many different kinds of animals. The breadth and frequency of such references, as well as observations about many varieties of life found on Earth, suggest that the early Hebrews appreciated the interconnectedness of human and nonhuman beings. Overall, then, Jewish attitudes provide many bases for arguing that this tradition, like Islam and Christianity, has core values and insights that can be cited when protecting other animals' welfare. In similar ways, Islamic views reflect both an extraordinary emphasis on humans as the centerpiece of a created universe *and* recognition of the moral importance of other animals. The Qur'an, Islam's all-important revealed scripture, frequently asserts that nonhuman animals exist for humans' benefit, but the issue of how humans treat other animals, deemed creatures of Allah, is also given emphasis in the tradition. The Qur'an's Surah 6, Verse 38, notes: "There is not an animal in the earth, nor a flying creature on two wings, but they are communities like unto you." The revered sayings of Muhammad also include references such as the following to the importance of kindness to animals: "Whoever is kind to the creatures of Allah, is kind to himself."

Although humans are divinely appointed representatives of Allah, slaughter of animals for food is ritualized in ways that include concerns for humane slaughter. Such regulation is often cited as reflecting the complex Islamic tradition's core commitment to the view that nonhuman animals have, as creatures of Allah, an integrity and inherent value of their own.

Native and Other Religious Traditions

Outside these five major religious traditions, often called "the world religions," views about and relationships with nonhuman animals remain diverse and profound. Native traditions around the world characteristically view humans as having a spiritual kinship with many kinds of nonhuman living beings. Many feature origin stories that hold all living beings to be the children of one mother and father. For example, the famous *Black Elk Speaks: Being the Life Story of a Holy Man of the Oglala Sioux* begins with this kinship image: "It is the story of all life that is holy and is good to tell, and of us two-leggeds sharing in it with the four-leggeds and the wings of the air and all green things; for these are children of one mother and their father is one Spirit."

Indigenous believers often speak of communication with nonhuman animals, and of the importance to human spiritual development of studying and understanding animal behavior. This kind of taking nonhuman animals seriously is evident in many other religious traditions, including Chinese folk religions, Confucianism, Daoism, Japanese Shinto, and the Jain and Sikh traditions of India—these and many other religious traditions offer important insights into human connections with other natural beings. One of the most important of such insights appears in an observation by the Catholic theologian Thomas Berry: "Indeed we cannot be truly ourselves in any adequate manner without all our companion beings throughout the earth. This larger community constitutes our greater self" (Waldau and Patton, 2006).

As enduring cultural and ethical traditions, religious institutions often have been the primary source of answers to one of humanity's most fundamental questions: "Which living beings really should matter to me and my community?" The religions of the world will no doubt continue to have major impacts on how believers, as well as the secular world, look at our own moral development and the related responsibilities to these nonhuman beings. Religions can contribute constructively to helping us notice and take seriously the actual realities of other animals, even as we shed our biases and fantasies about the diverse beings "out there" in the other-than-human world.

See also

Culture, Religion, and Belief Systems—Christian Theology, Ethics, and Animals
Culture, Religion, and Belief Systems—Islam and Animals
Culture, Religion, and Belief Systems—Judaism and Animals
Culture, Religion, and Belief Systems—Religion and Human-Animal Bonds

Further Resources

Barrett, W. (1962). *Irrational man: A study in existential philosophy*. New York: Anchor Books.

Black Elk & Neihardt, J. G. (1932). *Black Elk speaks*. New York: W. Morrow.

Waldau, P. (2001). *The specter of speciesism: Buddhist and Christian views of animals*. New York: Oxford University Press.

Waldau, P., & Patton, K. C. (2006). *A communion of subjects: Animals in religion, science, and ethics*. New York: Columbia University Press.

Paul Waldau

■ Culture, Religion, and Belief Systems
Religion and Human-Animal Bonds

Religion influences our understanding of human and animal relations in three principal ways, through our perceptions, our values, and our practices. The first is the contribution that religion makes to our perception. People sometimes refer to "religious vision," meaning there are ways of seeing that are deeply rooted in religious traditions that can enrich our perspective. The way we view the world is indebted to a range of influences, and religion is one of them.

On the animal-human bond, the influences are both negative and positive. Negatively, some religions tend to exalt human power over animals and exclude animals from the bonds of friendship with humans. Perhaps the most extreme version of this tendency can be found in St. Thomas Aquinas (1225–74), who held that friendship with animals was impossible because they are not rational. Because friendship was only possible, according to Aquinas, with rational creatures, animals were deemed incapable of "fellowship with man in the rational life" (*Summa Theologica*, Part 1, Question 65.3). This strong emphasis on rationality, which in Western religious traditions was denied to animals, has meant variously that they were largely perceived as being without a mind or an immortal soul, and incapable of having a relationship with God.

Although Judaism, Christianity, and Islam all recognize that animals are creatures of God, that their life belongs to God, and even that God loves all creatures, it remains true that all three have given animals a low status in comparison with that given human beings. There is little in their religious literature that champions relations with animals. Only, perhaps, in hagiography are relations with animals recognized and celebrated. St. Francis of Assisi is the obvious example, but there are countless other Christian saints of the East and West, including St. David of Garesja, St. Anthony of Padua, St. Catherine of Siena, St. Guthlac of Crowland, St. Werburgh of Chester, and St. Columba of Iona, who befriended animals and had friendly relations with them. St. Francis's idea that animals are our "brothers and sisters" has had great symbolic power within the Christian tradition, though it appears to have influenced behavior very little.

On the positive side, Eastern religious traditions have envisaged a much stronger bond between animals and humans. Jainism, Hinduism, and Buddhism all offer cosmologies that explicitly link humans and animals. Chief among the animal-inclusive concepts is the notion of *samsara* (cycle of death and rebirth), which expresses a radical continuity between all living beings. All life exists as in a chain, and all are linked together. From this perspective, animals and humans are not creatures but *subjects*—all life is in a state of progress or regress determined by *karma*, understood as a moral law of cause and effect. Animals and humans, thus conceived, are obviously interrelated: each individual soul has not just a biography, but also an ancestry.

The second contribution that religion makes concerns values. In the West, the predominant view of animals is that they exist to serve human interests. The originator of this view, or at least its earliest philosophical exponent, was Aristotle (384–322 BCE). He maintained that because "nature makes nothing without some end in view, nothing to no purpose, it must be that nature has made them [animals and plants] for the sake of man" (*The Politics*). Although not specifically religious, it became the predominant lens through which later religious thinkers, including Augustine, Aquinas, Luther, and Calvin, interpreted the place of animals.

What Aristotle held to be the "end" (*telos*) of animals became in later Jewish and Christian thought the God-given purpose of animals as well. Even the Hebrew and Christian scriptures were subsequently interpreted in terms of this "instrumentalist" model; thus, for example, "dominion" in Genesis 1:28 comes to be seen as God's validation of human supremacy. The irony is that, in its original context, dominion (*radah*) meant something quite different, namely God's commission to humans to care for the rest of creation. Proof that this is the correct reading is given in the subsequent verses (29–30) in which humans, like animals, are given a vegetarian diet—a situation that is only reversed after the Fall and the Flood (Gen. 9:3ff). God's original will in Genesis 1 was therefore for a peaceful, nonviolent creation. But the idea that animals are given for our use—either through the designs of nature or divine providence—has so caught a hold that Western society still principally regards animals as tools, machines, commodities, and resources for human use.

In Eastern religion, the idea of *ahimsa* (noninjury, nonviolence) has a long provenance. Arguably, Jainism taught the concept of nonviolence to the world; it has certainly influenced Hinduism and Buddhism and perhaps far wider. It is the noblest of all Indian ethical injunctions, expressed in the incomparable words of the venerable Mahavir:

> For there is nothing inaccessible for death. All beings are fond of life, hate pain, like pleasure, shun destruction. To all life is dear. (*Acharanga Sutra*)

These words are the result of a simple but profound spiritual discovery: all life is holy, sacred, or God-given. Life, therefore, has intrinsic value, and all that lives has an interest in living. It does not follow of course that all life is accorded the same value. *Samsara* is not an egalitarian doctrine; on the contrary, those who commit misdeeds (or rather those with "bad" karma) are sent back to live as one of the "lower" forms of life. While life is an interconnecting chain, humans still represent the apex of the moral hierarchy.

The third contribution that religion makes is in terms of practice. How people perceive the world obviously affects what they do. Religious practices can therefore be seen as the embodiment of religious perceptions of animal-human relations. The obvious example is animal sacrifice. It has been said that the most usual characterization of animals in the Hebrew scripture is as objects for sacrifice. In fact, there are a wide variety of characterizations—for example, as creatures, as covenant partners, or as possessors of *nephesh* (God-given life)—but it is the case that animals and birds are most regularly used throughout the Hebrew scripture as means of sacrificial offering.

Interpreting what this practice means is less than straightforward. As one might expect of any practice lasting more than 1,000 years, various interpretations are possible. Negatively, it can be seen (as is most usually seen) in terms of using animals as a means of reparation for human sin or appeasing the divine.

But it is worth pointing out that this is only one view of many. For example, another view is that sacrifice is to be understood as the returning of an animal to its Creator who made it, so that, far from involving the gratuitous destruction of a creature, the practice paradoxically involves its liberation—its final union with God. Whatever interpretation is given, it is significant that within the Hebrew Bible itself there is a developing criticism of the practice as inefficacious or immoral. Psalm 50 describes the Lord opposing sacrifice on the grounds that creatures belong to him:

> I do not reprove you for your sacrifices;
> your burnt offerings are continually before me.
> I will accept no bull from your house,
> nor he-goat from your folds.
> For every beast of the forest is mine,
> the cattle on a thousand hills.
> I know all the birds of the air,
> and that moves in the field is mine. (7–11, *RSV*)

The logic of this protest appears to be that humans should not appropriate what in fact belongs to God—not only are animals his, but he also knows them individually and cares for them.

Eastern religious traditions have, however, firmly set themselves against animal sacrifice, though it is true that Islam retains animal sacrifice for major festivals. And of course, both Judaism and Islam maintain the practice of "religious slaughter," called *shehita* and *halal,* respectively. Again, Jainism led the way in rejecting animal sacrifice and in commending the way of peaceable living with all nonhuman creatures. In Mahayana Buddhism, the Bodhisattva postpones his own enlightenment in order to save all living things from the cycle of misery and death:

> I have made a vow to save all living beings . . .. The whole world of living beings I must rescue from the terrors of birth, old age, of sickness, of death and rebirth . . . I must ferry them across the stream of *samsara* . . . I will help all beings to freedom. (*The Bodhisattva's vow of universal redemption.*)

This vision of humanity using its power to save other living creatures—and doing so sacrificially—is characteristic of Jain and Buddhist thought, which seeks *moska,* or liberation, for all. But it is not completely unknown in other religious traditions. In Christianity, the redemptive effects of the death of Christ are understood as inclusive of all beings, as for example, in Colossians, in which Christ is described as "the first born of all creation." Through Christ, God has determined "to reconcile to himself all things, whether on earth or in heaven, making peace by the blood of the cross" (1:15–19). In Judaism, there is the vision of a future heaven and earth in which the lion lies down with the lamb, where there is universal peace, and "they shall not hurt nor destroy in all my holy mountain" (Isa. 11: 6–9).

Oddly, there are no religious rites in Eastern traditions that unite concern for animals and humans or specifically celebrate the animal-human bond. It may be that because *ahimsa* is such a widely accepted practice that no need was felt for any specific rites. In Western traditions, there are likewise no specific rites, except that Catholicism has always accepted the appropriateness of blessings for animals, presumably mirroring God's own blessing of the creatures recorded in Genesis 1: 20–22. These appear in the *Romanum Rituale,* the priest's service manual, first written in 1614 and left virtually untouched until

1952. This provision has enabled animal blessing services and latter animal welfare services arranged by all Christian denominations in the West. These are usually held on St. Francis Day, October 4, which is now designated "World Day for Animals," and the first Sunday of each October, which is designated as "Animal Welfare Sunday."

One of the new, unofficial rites specially concerns the celebration of human relations with companion animals. The service involves the bringing of animals to the front of the church, where their human companion publicly promises to be faithful in care and love—mirroring God's own covenantal care shown in Genesis. The priest then says "May the God of the new covenant of Jesus Christ grant you grace to fulfill your promise and to show mercy to other creatures, as God has shown mercy to you" (*Animal Rites*, 1999). Services of celebration and blessing have been held in many cathedrals in Britain and America.

There are resources within almost all the religious traditions of the world for a celebration of the animal-human bond. But it must be said that many of the more positive ideas have been obscured by "instrumentalist" elements that present animals as wholly separate from humans or that suppose that animals exist only to serve us. There is a need for religious traditions to respond creatively to the new voices of ethical sensitivity to animals that are now increasingly heard in Western society in particular. At the heart of this sensitivity needs to be a reevaluation of human relations with animals—from one of crude dominance to friendship and respect. Ironically, although religion is often seen as an antiprogressive force because of its social conservatism, it contains many subtraditions that offer precisely that vision of filial relations with animals.

Baptist preacher Charles Spurgeon once recounted the view of Rowland Hill that a person "was not a true Christian if his [or her] dog or cat were not the better off for it." And Surgeon commented "That witness is true" (*First Things First*, 1885). The same should also be said of all world religions.

See also

Culture, Religion, and Belief Systems—*Animal Immortality*
Culture, Religion, and Belief Systems—*Christian Theology, Ethics, and Animals*
Culture, Religion, and Belief Systems—*Religion and Animals*
Culture, Religion, and Belief Systems—*Religion and Predators*
Culture, Religion, and Belief Systems—*Religion's Origins and Animals*

Further Resources

Birch, C., & Vischer, L. (1997). *Living with the animals: The community of God's creatures*. Geneva: WCC Publications.

Chapple, C. K. (1993). *Nonviolence to animals, earth, and self in Asian traditions*. Albany: State University of New York Press.

Kapleau, P. (1981). *To cherish all life: A Buddhist case for becoming vegetarian*. Rochester, NY: The Zen Center.

Linzey, A. (1994). *Animal theology*. London: SCM Press.

———. (1999a). *Animal gospel: Christian faith as if animals mattered*. London: Hodder.

———. (1999b). *Animal rites: Liturgies of animal care*. London: SCM Press.

Linzey, A., & Yamamoto, D. (Eds.). (1998). *Animals on the agenda: Questions about animals for theology and ethics*. London: SCM Press.

McDaniel, J. B. (1989). *Of God and pelicans: A theology of reverence for life*. Louisville, KY: Westminster John Knox Press.

Phelps, N. (2002). *The dominion of love: Animal rights according to the Bible*. New York: Lantern Books.

Sorrell, R. D. (1988). *St Francis of Assisi and nature: Tradition and innovation in western Christian attitudes toward the environment*. New York: Oxford University Press.

Walters, K. S., & Portmess, L. (Eds.). (2001). *Religious vegetarianism from Hesiod to the Dalai Lama*. Albany, NY: State University of New York Press.

Webb, S. H. (1998). *On God and dogs: A Christian theology of compassion for animals*. New York: Oxford University Press.

Andrew Linzey

■ Culture, Religion, and Belief Systems
Religion and Predators

For many millennia humans knew themselves as both predator and prey, one species competing and cooperating in relation to many other species. This interaction is discernable in religious systems of various cultures, wherein nonhuman animals are symbolically and physically woven into ritual practice, and are prominent actors in religious cosmologies (myths dealing with the structure and nature of the universe). Perhaps because their presence in the physical landscape demands attention, predator animals have attracted a great deal of religious attention as well, oftentimes disproportionate to their numbers. Religious attitudes toward predators thus provide a particularly powerful lens through which to view the larger network of relationships that humans share, or hope to share, with the natural world.

The Spiritual Power of Predators

In the myths and rituals of hunter-gatherer peoples, religious specialists frequently call upon predator animals for spiritual power, guidance, and provision. Stories from various cultures tell of humans who are singled out by animal spirits to act as mediators between human and animal worlds; it is not uncommon for these religious specialists to gain their abilities to cure through dream states in which they are dismembered and then put back together through the power of predator animals. Bears have figured prominently in this capacity, leading human ecologist Paul Shepard (1996) to comment that the bear "is the most significant animal in the history of metaphysics in the northern hemisphere." From the circumpolar regions, to the North American Pacific Northwest, to ancient Greece, bears have also been featured in ceremonies designed to expose and initiate young people into the mysteries and reciprocities of human and nonhuman animal relationships.

Other predator animals have special significance in the religious lifeways of tribal peoples. For example, Tukano Indian religious specialists of eastern Columbia utilize the power of thunder, which is believed to be related to the jaguar spirit, even appearing as a roaring jaguar on occasion. On the Great Plains of North America, prior to extensive contacts with European settlers, many Indian tribes revered and imitated wolves as master hunters, and individuals often took wolf names as a mark of their special relationship with them. In the Russian Far East, the Tungus peoples still make offerings at local shrines and observe special proscriptions in relation to the Siberian tiger, believing it to be a powerful spiritual being. On the islands of the Pacific Ocean,

Solomon Islanders, among other peoples, worshipped sharks. Such examples could be multiplied and are commonly found among peoples who view animals as able to communicate, instruct, and assist humans who show them due respect. Hunter-gatherer peoples generally view the world as a place consisting of a finite number of spirits, and therefore, humans and nonhuman animals are implicated in a cosmic process of spiritual exchange; especially when life is taken, proper conduct is required to ensure that the cosmic cycles in general, and predator-prey relations in particular, continue in a favorable way.

The Impacts of Domestication

Many scholars, most notably historians, cultural geographers, and archaeologists, claim that a dramatic shift in religious attitudes toward nonhuman animals in general, and predators in particular, occurred with the social changes produced by the domestication of plants and animals approximately 10,000 years ago in the ancient Near East (similar patterns of domestication occurred later in other parts of the world). Archaeological and literary evidence suggests that agricultural production led to population increases, which demanded increasingly hierarchical social arrangements, including a ruling class to regulate food surpluses and a specialized priestly class responsible for appeasing deities associated with weather phenomena and seasonal changes.

As hunting for food became secondary or unnecessary, predator animals diminished in immediate religious significance, eventually becoming confined to metaphorical symbols of ruling power, exampled by the association of lions with royalty in ancient Sumeria and Egypt. The domestication of animals such as sheep and cattle also resulted in alternate social and religious configurations. Even today, pastoral cultures—Sami reindeer herders in Sweden, for example—typically harbor ambivalent if not hostile attitudes toward predator animals that threaten the domesticated animals on which they depend.

While humans benefited in many ways from domesticating animals and plants, the process of domestication also led to an increasing physical separation between humans and nondomesticated animals, with a corresponding symbolic impact: the line between culture (associated with human technologies, sedentary populations, and domesticated animals) and nature (associated with uncultivated lands and non-domesticated animals) became more rigid. Religious systems and symbols reflected such social changes, and as time passed, animal deities in many cultures gradually came to be represented by hybridized human-animal forms or by strictly human physical characteristics.

Conquering the Beasts Within and Without

Despite the gradual increase of social and physical separation between nonhuman predators and humans that occurred over time, however, animal predators continue to loom large in religious art, literature, and symbol. In her 1997 book *Blood Rites*, Barbara Ehrenreich writes that "Probably the single most universal theme of mythology is that of the hero's encounter with the monster that is ravaging the land or threatening the very foundations of the universe: Marduk battles the monster Tiamat; Perseus slays the sea monster before it can devour Andromeda; Beowulf takes on the loathsome, night-feeding Grendel." Though such monsters may seem to be fantastical projections of the human imagination, Ehrenreich states that "it might be simpler, and humbler, on our part to take these monsters more literally: as exaggerated forms of a very real Other, the predator beast which would at times eat human flesh."

This seems to be reflected in texts from the ancient Near East, in which predator animals often serve as metaphors of destruction and potency. In the prophetic writings of the Hebrew Bible, God's judgment is occasionally compared to the ferocity of a lion (Isa. 31:4, Jer. 49:19, Hos. 5:14); in other passages, God is depicted as using predators to enact punishment upon people: "Therefore a lion from the forest will attack them, a wolf from the desert will ravage them, a leopard will lie in wait near their towns to tear to pieces any who venture out, for their rebellion is great and their backslidings many" (Jer. 5:6; see also 2 Kings 17:25).

Undoubtedly such proclamations were intended to scare wayward Israelites to rededicate themselves to their covenantal relationship with God, underscoring how predators can be an effective tool to invoke humility. The book of Job in the Hebrew Bible stands out in this respect (see especially Job 38–42:6). In one memorable passage, God reminds Job of the deity's power by invoking Leviathan, a creature that the religionist Rudolph Otto names as the ultimate representation of the monstrous wholly Other, and what writer David Quamman refers to as "the archetype of alpha predators" (2003).

Humility may be one reaction to predator symbols, but it is certainly not the only possible response. Predators have been esteemed in some religious systems, representing the more powerful forces of the natural world, yet they have been feared for similar reasons. This is particularly evident in instances in which humans define themselves as unique (or superior) in relation to other animals, and in such cases predators are often cast as symbolically embodying the "bestial" or "savage" side of human nature.

Finding it difficult or impossible to tame large predators, humans have often resorted to projecting their greatest fears upon them, casting them as villains in a sacred drama and as representatives of all that is uncivilized and debased. In the following passage, the Roman poet Ovid (43 BCE–17 CE) expresses such themes, contrasting the longing for the "golden age" of times past with the debauchery of the present, which he believed was distinguished by its predatory tendencies:

> Flesh is the wild beasts' wherewith they appease their hunger, and yet not all, since the horse, the sheep, and cattle live on grass; but those whose nature is savage and untamed, Armenian tigers, raging lions, bears and wolves, all these delight in bloody food. Oh, how criminal it is for flesh to be stored away in flesh, for one greedy body to grow fat with food gained from another, for one live creature to go on living though the destruction of another living thing!

Ovid's moral misgivings are not exceptional. Predator animals throughout history have often been the scapegoats of humanity's worst tendencies. The associations of predator animals with violence and bloodshed remains prominent among many traditional Christian theologians, who assert that the natural world is corrupted by sin and who look forward to a redeemed earth in which the "lion will lie down with the lamb" (Isa. 11:6). For example, Andrew Linzey (1995), a contemporary Anglican theologian and strong advocate of vegetarianism, asserts in his book *Animal Theology* that the point of Isaiah's vision "is not that animality will be destroyed by divine love but rather that animal nature is in bondage to violence and predation." Humans should seek to free themselves of such animality, Linzey asserts, in order to achieve a "higher order of existence." In this type of theological schematic (often referred to as the "peaceable kingdom"), animal predation is considered an anomaly (or atrocity) awaiting harmonious correction by God. Though this represents only one possible Christian theological interpretation among many, the notion of a fallen world subject to decay until humans (and/or God) restore it is a common theme in Western history in general and Christian theology in particular. Literal attempts to realize visions of paradise do not often account for, or simply negate, the gustatory needs of meat-eaters.

Yet, non-Western religious beliefs and narratives are not immune from hierarchical worldviews or from privileging human beings over other animals. Despite sometimes being viewed as alternative sources of wisdom that are more ecologically sensitive than Western traditions, close study of Asian religious traditions reveals frequent categorical distinctions between humans and other animals. In text, if not in practice, animals— especially predators—are understood as signifying a "lower" order of existence.

Buddhism, often simplistically hailed as an environmentally beneficent religion, invokes many of the same prejudices against nonhuman animals found in other religious traditions. Scholar of religion Paul Waldau explains in his 2002 book *The Specter of Speciesism* that in ancient Buddhist texts (specifically, the Pali canon), "Even though it is clearly assumed that all life has value, it is *also* assumed that human life is of a qualitatively *better* kind than is the life of any other animal." Though a general continuity between humans and animals is affirmed in the karmic cycle (the understanding that all beings are subject to rebirth), the dominant view remains one of discontinuity, in which nonhuman animals must first achieve human status before they can reach enlightenment. Similarly, "regression" to animal status once one is a human is considered punishment for one's failings in life.

Furthermore, nondomesticated animals are frequently depicted in Buddhist texts as representatives of undesirable wildness and carnivorous appetites. Predator animals are thus moral guides of what *not* to be. To be born as a tiger, for instance, after once being in human form, is considered a punishment because animals do not typically possess the spiritual faculties to achieve liberating insight. As the Buddha warns in the Lankavatara Sutra: "Such ones [human meat-eaters] will fall into the wombs of such excessive flesh-devouring creatures as the lion, tiger, panther, wolf, hyena, wild-cat, jackal, owl, etc. . . . Falling into such, it will be with difficulty that they can ever obtain a human womb; how much more difficult attaining *Nirvana!*"

Perhaps it is unsurprising that predator species often fill analogical roles in religious systems, serving as cautionary object lessons of human depravity, for one of the functions of religion is to instruct humans, not nonhuman animals, how to live. Yet, it is important to recognize that religious narratives inform human behavior toward actual, not just imaginary, animals. As human population and competition for habitat has increased over time, negative views of predator animals have led to drastic reductions for many species, and in some cases, extinction.

Predator Values and Ecology

Since at least the mid-twentieth century, ecological understandings that highlight the vital function that predators play in ecosystems have provided some persons with a renewed sense of the value of predators. Among the earliest of predator advocates in North America, Sierra Club cofounder John Muir—whose religious views have been described as pantheistic (God-in-everything) or animistic (animals and plants contain a spirit)—wrote in *A Thousand Mile Walk to the Gulf* (1998) that he considered all creatures to be "earth-born companions" no matter how "noxious and insignificant" they might seem. He even quipped that human arrogance led him to consider that "if a war of races should occur between the wild beasts and Lord Man, I would be tempted to sympathize with the bears." Recognition of the important role that predation plays in evolutionary processes has also contributed to positive reevaluations of predator animals. For example, early twentieth-century poet Robinson Jeffers, whose religiosity has also been described as pantheistic, often wrote of the beauty he perceived in the natural (and oftentimes violent) cycles of life and death, as in this excerpt from his poem "The Bloody Sire": "What but the wolf's tooth chiseled so fine/The fleet limbs of the antelope?/What but fear winged the birds and hunger/Gemmed with such eyes the great goshawk's head?/ Violence has been the sire of all the world's values."

For wildlife organizations, environmental activists, conservation biologists, and others in the twentieth and twenty-first centuries, predator animals have increasingly taken on a symbolically salutary role, representing the "wild" evolutionary processes that are threatened by human abuse. Indeed, much of the urgency expressed by environmental activists derives from grave concerns related to species extinction and ecosystem fragmentation. Because of predators' general dependence on large food bases and habitat, images of predator animals are often utilized by environmental groups to promote ethical consideration for these animals and their habitats. In this capacity, predator animals sometimes act as sacred icons of ecological integrity and authentic wildness.

Religion can be considered a negotiation between the human and the nonhuman, and religious myths and ethical norms can either subvert or reinforce cultural understandings of human dominance. The manner in which predator animals are regarded reflects important aspects of how different peoples "map" both spiritual and physical terrain. Because humans are able to recognize similarities between themselves and other predators, as well as exaggerations of their own strength, intelligence, and dexterity, predator animals are excellent candidates for religious symbols and metaphors, often providing access to mysterious worlds that are beyond human control.

Further Resources

Burkert, W. (1998). *The creation of the sacred: Tracks of biology in early religions.* Cambridge, MA: Harvard University Press.

Cauvin, J. (2000). *The birth of the gods and the origins of agriculture* (T. Watkins, Trans.). New York: Cambridge University Press.

Diamond, J. (1997). *Guns, germs, and steel: The fates of human societies.* New York: W.W. Norton.

Ehrenreich, B. (1997). *Blood rites: Origins and history of the passions of war.* New York: Henry Holt.

Glacken, C. J. (1967). *Traces on the Rhodian shore: Nature and culture in western thought from ancient times to the end of the eighteenth century.* Berkeley: University of California Press.

Jeffers, R. (2003). *The wild god of the world: An anthology of Robinson Jeffers.* Stanford, CA: Stanford University Press.

Linzey, A. (1995). *Animal theology.* Champaign: University of Illinois Press.

Midgley, M. (1995). *Beast and man: The roots of human nature.* New York: Routledge.

Mighetto, L. (1993). *Wild animals and environmental ethics.* Tucson: University of Arizona Press.

Muir, J. (1998). *A thousand-mile walk to the Gulf.* Boston: Mariner Books.

Quammen, D. (2003). *Monster of god: The man-eating predator in the jungles of history and the mind.* New York: W.W. Norton.

Reichel-Dolmatoff, G. (1975). *The shaman and the jaguar.* Philadelphia: Temple University Press.

Shepard, P. (1996). *The others: How animals made us human.* Washington, DC: Island Press.

Sideris, L. (2003). *Environmental ethics, ecological theology, and natural selection.* New York: Columbia University Press.

Waldau, P. (2002). *The specter of speciesism: Buddhist and Christian views of animals.* New York: Oxford University Press.

Walters, K. S., & Portmess, L. (2001). *Religious vegetarianism: From Hesiod to the Dalai Lama.* Albany: State University of New York Press.

Worster, D. (1994). *Nature's economy: A history of ecological ideas.* New York: Cambridge University Press.

Gavin Van Horn

■ Culture, Religion, and Belief Systems
Religion's Origins and Animals

Among features said to distinguish humans from other animals, religion has had an especially secure place. Many other aspects of the traditional human-animal distinction have been undermined by ongoing discoveries of similarities and continuities between humans and nonhuman animals. We now know that chimpanzees, for example, resemble humans in being self-aware, having complex communications that include a degree of symbolism, and having cultures that include tool-making traditions. Nonetheless, few writers attribute anything like religion to them, and most accounts of religion still assume that it is uniquely human. Overlooking nonhuman animals, however, may mean overlooking valuable clues about the origins of religion.

Several assumptions about religion and about animals underlie the view that religion is peculiar to humans. Adherents of religions, for example, usually assume that religions are primarily social relationships between persons and a deity or deities. Judaism, Christianity, and Islam (the prototypes of "religion," a Western concept) assume further that animals are not persons, and so cannot have such relationships.

Many nonbelievers, as well, assume that animals cannot be religious, though often for different reasons. One view (held by Freud among others) is that religion is wishful thinking, primarily wishing for immortality and for postmortem retribution. Hence it is a kind of self-delusion. Because animals presumably do not worry about the inevitability of death, they do not delude themselves with dreams of immortality. Another skeptical view (held for example by David Hume and E. B. Tylor) is that religion is an explicit attempt to explain the world. Again, nonhuman animals evidently do not engage in explicit explanation and thus do not have religion.

A third, again largely nonbelieving, view is that religion is symbolism and is not meant literally. In this view, prayer and sacrifice, for example, nominally may be directed to a deity but actually are covert statements, directed to one's fellow humans, about human social relations. Whether or not religion is symbolic in this broad sense, symbolism more narrowly is a requisite of language and hence is required for prayer, as well as for other ritual communication. Once again, nonhuman animals are thought not to symbolize and hence to have nothing like religion.

One might question these assumptions about religion and about animals, asserting that each is either mistaken or correct only to a degree. One thus could reopen (as have A. F. C. Wallace and Walter Burkert) the possibility of religion, or its precursor, in animals. An alternate account of religion itself, however, together with new observations about animals, reduces the importance of the distinction between humans and nonhumans.

This alternate account, which emphasizes cognition, holds that religion is not sharply different from other human thought and action but is continuous with them and, like them, is a product of natural selection. In this account, religion, like other thought and action, is mainly concerned with interpreting and influencing the world in general. What distinguishes it is only that its central assumptions are anthropomorphic and animistic: they attribute qualities of humans and of life to things and events that lack them. At the same time, recent observations suggest that not only humans but also other animals display animism, in one or more senses of this term.

Animism and Anthropomorphism

Animism has a range of meanings, but two are salient. One meaning is a belief, sometimes termed "vitalism," that the cause of life in living organisms is a vital principle (e.g., a spirit) distinct from the sum of physical organic processes. The other common meaning is the attribution of characteristics of life to things and events that are biologically inanimate. These two senses of the term may overlap and also may be combined.

The vital principle involved in the first sense of animism may be conceived as material, as immaterial, or as somewhere in between, and either as separable from the body or as inseparable. Conceptions of it as more or less immaterial and as separable from the body occur cross-culturally, including in religion, where the principle may be called by such terms as spirit or soul. Whether any nonhuman animals have such a conception is unknown, although there is suggestive evidence.

The second kind of animism salient here, the attribution of life to the inanimate, also appears universal among humans. For example, we may sense a loose thread tickling our skin as a crawling insect, or feel that a tangled cord's resistance to us is perverse. Animism in this sense appears widespread in other animals as well, as when a predator pounces on an inanimate object resembling its prey. It is closely related to anthropomorphism, the attribution of human characteristics to nonhuman events, and often overlaps it.

The observation that religion anthropomorphizes and animates dates back at least to the ancient Greeks, and assertions that religion *is* anthropomorphism and animism date back at least to Benedict de Spinoza and David Hume. Moreover, no less a biologist than Darwin asserted that the minds of humans and of other higher mammals are much the same and indeed that both humans and other animals are prone to animism. His dog, he noted, once behaved as though a distant umbrella moved by the wind "indicated the presence of some strange living agent" (1871). Nonetheless, the claims that religion *is* anthropomorphism and animism, and that these appear in nonhuman animals as well as in humans, have met skepticism.

Two apparent reasons for this skepticism are our attachment to religious views and our desire to place ourselves apart from other animals. Other reasons are that the pervasiveness of anthropomorphism and animism has not been appreciated and that why they occur has not been well explained. Writers have tried to explain both, either as comforting or as stemming from models of the world that are peculiarly available, namely those of ourselves. However, neither explanation is adequate because both anthropomorphism and animism may be frightening as well as comforting and because neither stems from a model that is necessarily more available than are competing models.

Instead, a single cognitive strategy better explains both anthropomorphism and animism. That is, when we do not know whether some ambiguous phenomenon—a bump in the night, a movement seen from the corner of our eye—reflects the presence of an agent, we tend to assume that it does, because the potential reward for being right outweighs the potential penalty for being wrong. It is better to mistake a boulder for a bear, a stick for a snake, or a thread for a spider, than the reverse.

Because perceptual data typically are ambiguous, perceptions necessarily are interpretations—that is, guesses. Because agents, both human and nonhuman, frequently conceal their presence (as by camouflage), we often miss them. In consequence, a better-safe-than-sorry strategy of guessing at the presence of agents is deeply engrained in us by evolution. Indeed we are perpetually on a hair trigger for detecting them: we scan for them unconsciously and have lenient standards for judging that we have found them.

What we call anthropomorphism and animism, then, are cases of seeing things or events as human, as humanlike, or as alive, which we later have come to see as nonhuman or inanimate: "I thought someone was in the house, but it was just the wind slamming a door," or "I thought it was an animal on the road but it was just a piece of truck tire." Animism and anthropomorphism thus are not themselves motivated (e.g., by any desire for comfort) but simply are mistakes. They are inevitable byproducts of our search for what is most important to us, namely other agents.

This account gives us no reason to think that animism, or even anthropomorphism, is peculiarly human. Instead, because all perceiving animals need to detect other agents in a perceptually ambiguous world, all employ the same strategy of guessing that agents are present, and all are subject to the same occasional failures.

Evidence of such failures is plentiful. Both humans and nonhuman animals are sensitive to a variety of potential signs of life, such as self-initiated motion and sudden noises, and frequently respond to them as though an agent is present. Predators that have discovered a class of prey animals in an area, for example, become especially sensitive to anything resembling the prey. Robins that have found caterpillars, coyotes that have found grasshoppers, and trout that have found midges respectively attack twigs, sticks, and lures resembling them.

Sensitivities to some important signs of life are built into us. Animals ranging from insects to fishes, reptiles, birds, and mammals are innately sensitive, for example, to visual patterns resembling eyes, which they typically avoid. Many prey animals exploit this sensitivity by displaying an outsize pair of "eyes," like those on the wings of some butterflies, to frighten potential attackers. Other signs may be learned. Animals evidently even attribute aspects of humanity to nonhuman things, as when birds or deer avoid gardens with scarecrows, or when caribou are driven into traps as they avoid humanlike rock piles (inuksuit) built by Inuit hunters.

Human Thought, Language, and Religion

Animism in the first sense mentioned, the attribution of the life of organisms to a vital principle distinct from the sum of their physical processes, also is widespread and probably universal in humans. Recent cognitive science and other research indicate that humans intuitively assume that the vital principle is more or less immaterial and is separable from the body that it inhabits. That is, a notion of something like spirit beings—disembodied but otherwise humanlike entities—apparently springs up unbidden in each of us.

This notion may be a consequence of the prominence in unconscious thought of "theory of mind," an intuitive understanding of how other human minds work. A central aspect of this understanding is our belief-desire psychology, which assumes that other people act so as to satisfy their desires, according to whatever beliefs they hold. Many cognitive scientists consider theory of mind both innate and fundamental to human social interaction. As a central component of human mental process, it apparently is applied readily to all agents and to any situation that may involve them. For the purpose of understanding the origins of beliefs in gods, spirits, and other frequently immaterial beings important in religion, it is revealing that representations of agency as immaterial are normal in theory of mind.

Combined with the proclivity to anthropomorphize and animate the world generally, this intuitive conception of agency as disembodied seems to provide the basis for our religious conceptions, in which much or all of the world, both cultural and natural, is influenced by humanlike but usually invisible agents. Thus a naturalistic account of religion, founded on evolved human capacities and tendencies, appears at hand. Completing its naturalism, however, requires showing some relation between religion as such—which after all surely is a human phenomenon, almost by definition—and its analogues in other animals, because evolutionary processes typically display gradual rather than radical change.

The closest analogues may be the occasional behavior of wild chimpanzees confronted either with the onset of thunderstorms or with waterfalls, the similar behavior of gorillas confronted with thunderstorms, and even the behavior of monkeys confronted with startling events. The chimpanzee behavior in question, namely the "rain dance" described by Jane Goodall and others, is exhibited by males. It closely resembles the threat behavior that they direct at predators and at rival chimpanzees. It consists of bristling body hair; bending, shaking, breaking, dragging, and throwing tree branches; stamping and slapping the ground; and hurling large rocks. Goodall writes that this display is as though the storm or waterfall was caused by some unseen force, and she suggests that if chimpanzees could speak, they might produce an animistic nature worship.

Goodall's suggestion about speech appropriately points to language as necessary to religion, as it is to much else that is distinctively human. Language and the broader symbolic capacities associated with it open, of course, a world of possibilities. With it one can, for example, communicate about unseen agents, about things and events of another time or place, and about abstractions such as the future and death—in short, about some of the common concerns of religion. In speaking of these, one objectifies, elaborates, and systematizes them, and so lays the foundations of religion proper.

Thus it may not be so much that religion distinguishes humans from animals as that language makes it possible to develop and systematize the seeds of religion, just as it makes possible so many other human developments. What language systematizes, moreover, apparently is a class of interpretations of the world that is not confined to humans. These interpretations, which we see as distinctive (i.e., as anthropomorphic or animistic) only after we have replaced them with alternatives, constitute attributions of more life and organization to the world than it has.

Such attributions are products of an evolved strategy of guessing at the most important possibility—the presence of agents—in an uncertain world. As such, they necessarily occur in a wide range of animals, including but not limited to apes and monkeys. To the extent that these animals behave as though inanimate natural phenomena embody, or are caused by, agents, we may see in them the beginnings of religious thought and action.

Further Resources

Burkert, W. (1996). *Creation of the sacred: Tracks of biology in early religions.* Cambridge, MA: Harvard University Press.

Darwin, C. (1871). *The descent of man, and selection in relation to sex.* London: Murray.

Goodall, J. (2005). Primate spirituality. In B. Taylor (Ed.), *Encyclopedia of religion and nature* (pp. 1303–06). London: Thoemmes Continuum.

Guthrie, S. E. (1993). *Faces in the clouds: A new theory of religion.* New York: Oxford University Press.

———. (2002). Animal animism. In I. Pyysiäinen & V. Anttonen (Eds.), *Current approaches in the cognitive science of religion* (pp. 38–67). London: Continuum.

———. (In press). Gambling on gods: Religion as anthropomorphism and animism. In D. M. Wulff (Ed.), *Handbook of the psychology of religion.* New York: Oxford University Press.

Stewart Elliott Guthrie

■ Culture, Religion, and Belief Systems
Rituals of Humans and Animals

Ethologists and anthropologists have long recognized that human and animal rituals have remarkably similar forms and functions. Consider the following rituals: In the Pentecost Islands of the South Pacific, adolescent males take part in an annual ritual called *naghol* aimed at ensuring a bountiful yam crop for the coming year. The ritual requires that adolescent and young adult males construct tall pole platforms, some reaching 25 meters in height. Accompanied by the singing and dancing of all the villagers, the males ascend these platforms and, with only a vine rope tethered to their ankles, plunge into the air. Males who successfully complete their first "dive" are considered to have made the transition from childhood to adulthood. They are congratulated by the villagers and, as adults, are able to take a wife.

Australian golden bowerbirds also construct tall pole platforms as part of their annual rituals. Bowerbird males gather brightly colored feathers, shells, and ornaments to create the most beautiful and eye-catching "maypoles" possible. As in the Pentecost Islands, bowerbird poles are used as staging areas for the male's ritual performance. This includes singing as well as an energetic dance performed before a single female. Successful bowerbirds conclude their courtship ritual with a new mate.

Rituals such as these are not confined to bowerbirds or an unusual human group; they occur across widely divergent animal species and in nearly all human cultures. Just as you find many Americans singing on a Sunday morning in church, humpback whales have collective singing rituals as well. In Western societies when we meet with friends we stick out our arms, clasp them, and pump our joined hands up and down a few times; chimpanzees have their own ways of ceremoniously greeting each other, which also includes hand-clasping. Why do humans and animals alike engage in these behaviors? And why do these rituals have similar forms and functions? In what ways do human and animal rituals differ? Fortunately, ethologists and anthropologists have begun to answer these questions.

Ritual is a universal feature of human behavior. Rituals differ from culture to culture, but the defining features that distinguish them from "ordinary" behaviors are surprisingly consistent across all human societies. Rituals tend to be formal, stereotyped, repetitive, and established by someone other than the performer. They are therefore easily distinguished from other behaviors. Rituals help pattern and predict social interactions. For example, when two people meet they have expectations about how the social interaction will proceed. In Western societies, meetings commence with a handshake and a simultaneous "How are you?" or some similar formality. While none of us invented the handshake, we all recognize it as a greeting ritual.

Religious rituals are particularly easy to detect as they tend to be more elaborate than other rituals. They also generally include music, chanting, or dance, which further distinguishes them from other behaviors. While religious rituals frequently appear to be shrouded in mystery, their formality and elaborateness make it clear to participants and observers alike that they are rituals. Nobody mistakes Sunday morning church for the Sunday afternoon football game.

These same underlying features of ritual also enable us to recognize ritual in nonhuman species, as well. Wolf spiders, salamanders, and Sandhill cranes all perform intricate dances as part of their courtship rituals. Parrots and Pacific humpback whales engage in improvisational, synchronized singing during mating and group rituals, and wild dogs, wolves, and chimpanzees all perform highly ritualized greeting ceremonies, including muzzle-to-muzzle contact and choral vocalizations whenever the members of a group meet. From bees, to birds, to baboons, ritual in both human and animal groups conveys significant information between individuals. Ritual permits and promotes social interaction by creating "frameworks of expectancy" that lay the foundation for the prediction of behavior by others. But to fully appreciate the similarities between human and animal rituals, and why they are similar, we first need to understand ritual's less complicated parent: signals.

Signals as Cooperative Communication

Animals and humans use many different kinds of signals to communicate with other members of their groups. The scent marking of dogs, the alarm calls of monkeys, and human facial expressions are all social signals that convey information to others. The information communicated may be about the condition, state, or intent of the sender, or it may be information about the surrounding environment. Some signals convey information about the sender's condition because they are a direct result of the physical and physiological characteristics of the sender. The croak pitch of male frogs is an example of such a signal. Croak pitch is a direct function of body size, with larger males producing deeper croaks. This direct relationship between body size and sound pitch makes it possible for both females and competitor males to estimate the size of unseen males based on their croaking. Such "indexical" signals convey reliable information about a signal sender because they are directly linked to attributes that cannot be concealed or manipulated.

Many signals used in human and animal communication are not indexical, but still provide reliable information. They have evolved over time because they benefit both the sender and the receiver. Numerous conventional signals, such as the pecking response of herring-gull chicks to red dots, are the result of genetically programmed fixed-action patterns. Such signals automatically elicit or "release" evolved preprogrammed behaviors in signal receivers. In the case of the herring chicks, pecking at the red dots on the mother's bill provides the chick with food. Such mutually beneficial signaling systems occur within many species, but they may also facilitate communication between species as well. Grouper fish exhibit innate responses to the "dance" performed by sucker fish. Even when reared in isolation, groupers exposed to the sucker fish dance lie down on the sand, spread their fins, and allow the sucker fish to clean the algae from their scales. Such fixed-action pattern signaling systems have evolved because the benefits they provide for both the sender and receiver outweigh the costs involved in signaling.

Although it was once thought that most animal signals result from these innate "fixed-action patterns," ethologists have since found that many animal signals incorporate both genetic and learned components. The alarm calls of vervet monkeys (*Cercopithecus aethiops*) provide an example of such a signal. Vervets inhabit woodland areas in eastern

Africa and use alarm calls to alert other members of the social troop to the presence of predators. Vervets emit a bark in the presence of a jaguar, a cough in the presence of an eagle, and a chutter in the presence of a snake. Young vervets have an innate tendency to respond to calls and to make different calls in response to different stimuli. However, young monkeys are not born with preprogrammed knowledge of jaguar, eagle, and snake calls. They must learn the specific call to emit for each particular predator. While vervets are "preprogrammed" to learn these calls, young monkeys must hear the different calls used within the appropriate context in order to learn the correct call for each particular predator.

This innate capacity to learn species-specific signals during a particular developmental period is seen in many other species. For example, the courtship songs of many birds exhibit this combination of genetic programming and developmental learning. While male birds are "primed" to learn their species' song, males must be exposed to that song during a specific developmental window in order for learning to occur. Research indicates that both human musical abilities and language learning also combine genetic predispositions and developmental windows for optimal learning.

Signals as Deception

Indexical signals and many conventional signals employed by animals and humans promote communication by providing reliable information that benefits both the sender and the receiver. Some signaling contexts, however, involve senders and receivers who have conflicting interests. The most obvious examples of such conflict are the male contests that occur in many species over females and food. Mating is also a potential source of conflict between males and females. Such conflicts can be expected to escalate as stakes increase and shared individual and genetic interests decrease. Under such circumstances, there is great incentive for signalers to use deception in order to influence receiver responses. The use of deceptive signals by humans to "bluff," cheat, and lie are well-known to poker players and soap opera viewers alike. Deceptive signals are also used by numerous animal species to achieve similar ends. Camouflage and mimicry are widespread throughout the animal kingdom. Many species have evolved color patterns and special structures to deceive potential predators and prey. Angler fish use a specially evolved mouth appendage to "lure" unsuspecting prey. Viceroy butterflies fool potential predators through their mimicry of the unappetizing Monarch. Deceptive signals also include behaviors. Cowbird chicks display open-mouth "begging" signals to induce their unsuspecting "adopted" parents to feed them first and most. Females of the predatory firefly genus *Photuris* mimic the mating flashes of the related genus *Photinus* in order to lure *Photinus* males close enough to attack and consume them. This exploitative use of another species' fixed-action pattern signals, referred to as "code breaking," is a relatively common form of signal deception. Equally common in humans and animals alike are "bluffing" behaviors and deceptive signals used by both males and females in order to manipulate potential and current mates.

Honest Signals

Signal receivers clearly have an incentive to prevent such manipulation. Receivers should seek out signals that provide honest information. In many species this has resulted in the evolution of "quality signals" that provide receivers with reliable information about the general condition of the sender. In birds the intensity of plumage color is negatively correlated with parasite load—the brighter the plumage, the healthier the bird. Females seek out males with the most brilliant plumage. As a result, the color brilliance of males has evolved to be a quality signal for females. In humans, a similar quality signal is provided by facial symmetry, which is positively correlated with health. Many studies have shown that males and females find symmetrical faces more attractive. Quality signals may be auditory as well as visual. In various passerine bird species, male song repertoire size is an important quality signal for females, because males with large song repertoires are less likely to be infected by malarial parasites and more likely to bring larger caches of food for their offspring.

While quality signals benefit the receiver, they frequently incur costs for the sender. The large, brilliantly colored tails of male peacocks, like flashy red sports cars, are effective signals, but they require the expenditure of resources. Male songbird acquisition of large song repertoires likewise requires high expenditures of time, energy, and learning effort.

In addition to their production and maintenance costs, these quality signals may also incur higher vulnerability costs. Male peacocks with the longest, brightest tails not only expend more energy on the development and maintenance of such tails, they also attract more predators and have greater difficulty escaping predation as a result of their heavy, cumbersome plumage. Songbirds with the largest, showiest repertoires not only alert potential mates, they alert potential predators. Biologist Amotz Zahavi has proposed that such high cost signals have evolved and are adaptive for signalers precisely *because* they "handicap" the sender. Because only those peacocks, bachelors, and songbirds with sufficient resources are able to produce and maintain the longest, showiest tails, most expensive sports cars, and the largest and most captivating song repertoires and still evade predation, it would be impossible for less fit competitors to "fake" these signals.

Ritual as a Signal

In terms of time, energy, and resources, rituals are often the costliest of signals. Ethologist W. J. Smith (1979) has described ritual as "behavior that is formally organized into repeatable patterns" (p. 51). This definition includes four basic features of ritual: (1) formality; (2) pattern; (3) sequence; and (4) repetition. These four features make up the structure of ritual in species as diverse as horned-toads, hens, and humans. These four structural components of ritual are optimally effective in engaging and focusing attention, heightening discrimination, enhancing multidimensional generalization, and improving associative learning, thus ensuring that the message sent by the sender is accurately perceived, assessed, and remembered by the receiver. The formality of ritual captures the attention of the audience and focuses it on the signal elements most likely to evoke receiver response. Ordinary traits and behaviors may be exaggerated in order to make them "extraordinary." The "eyes" of a peacock's long, iridescent tail prominently displayed during his ritual dance, the changing body colors of male squid as they gently jet water over a potential mate, and the extraordinary garments worn by human brides and bishops all represent formal elements of ritual that engage and focus the attention of ritual participants. Could anyone really turn their attention away from the Pentecost Island adolescents plunging into the air from their 25 meter poles?

By exaggerating and elaborating the ordinary, the formality of ritual alerts brain areas such as the reticular formation, the basal ganglia, and the amygdala. These neural structures are first-line responders to novel, unusual, and threatening stimuli. They prime emotions and prepare the body to react. Once attention is focused, the sequence, pattern, and repetition of ritual provide an opportunity for the message receiver to absorb, assess, and associate the information transmitted. The sequenced repetition of ritual creates focus time and processing time. Both are critical for memory and learning. At the cellular level, repetition induces changes in long term potentiation, while sequencing allows the brain to recycle CREB, a protein crucial to long-term memory function.

Ritual has other impacts on neuroendocrine function, as well. Changes in levels of neurotransmitters, neuromodulators, and hormones of both the sender and the receiver occur during ritual, resulting in changes in the physiological, immunological, and behavioral responses of ritual participants. Biologist Russell Fernald's studies of cichlid fish (*Haplochromis burtoni*) from Lake Tanganyika in Africa dramatically illustrate ritual's effects on physiology. He found that agonistic displays between cichlid males induce major changes in the hormones, external appearance, brain neuron sizes, and even the gene expression of winners and losers. Fernald observed aggressive and brilliantly colored black, yellow, blue, and red males almost instantly morph into much less aggressive drab brown "satellite" fish when ousted from their territories by rivals. If the "satellite" later acquired a new territory, his color, hormones, hypothalamic neuron sizes, and gene expression again changed. Similar neuroendocrine changes have been recorded in songbird responses to ritual, as well. The ritualized vocalizations of male songbirds impact female sexual receptivity by inducing hormonal changes in the female. They also impact the brain neurons and song-related genes of the signaler. In both wolves and nonhuman primates, ritualized dominance and submission behaviors alter participants' cortisol, dopamine, and testosterone levels. Across animal species, the ability of ritual to alter individual neurophysiology and behavior is critical to its adaptive value.

The Relationship of Human and Animal Signaling Systems

Human signaling systems share many features with those of nonhuman species. As with our chimpanzee cousins, we employ facial expressions, body gestures, and call systems to communicate important information about ourselves and the environment around us. Humans everywhere use similar facial expressions to convey and identify basic emotional states, and laughter, body language, and shouts of alarm are universally understood. These pan-human signals have deep phylogenetic roots shared with our closest primate kin. In contrast, the "distinguishing marks" of human religious ritual—chanting, music, and dance—are notably rare in other primates. While chimpanzees have been observed to engage in occasional "drumming" of tree trunks and sporadic "rain dances," the regular, ritualized use of song and dance is conspicuously absent in gorilla, bonobo, and chimpanzee societies.

Yet, song and dance are both found in many other animal species. Wolves and wild dogs engage in choral howling, humpback whales sing synchronized group songs, and a multitude of bird species chorus, sing, and dance. Understanding when and why these types of ritualized behaviors occur in nonhuman species offers insights into our own performance of these rituals.

Numerous bird species exhibit costly rituals, including both song and dance, as part of courtship rituals. Many of the species that engage in such rituals also pair-bond, with a single male and female forming a long-term relationship. Pair-bonding ensures adequate care for the highly dependent baby birds. Not only must young hatchlings be fed and protected, as they mature into fledglings, they must also be taught to fly. Males play an important role in caring for the offspring in these pair-bonded species. It is, therefore, very important for the female to ensure that the mate she chooses as her long-term partner is both fit and reliable. Courtship rituals that require males to sing, dance, and even provide "gifts" offer an opportunity for females to assess both the condition and intent of her male suitors. From the male's perspective, these rituals represent an opportunity to signal his fitness, but they are also critical in inducing female hormone changes that prepare her for mating.

Humans share several important features with pair-bonded birds. We too give birth to relatively undeveloped and highly dependent young that must be taught important life skills before maturation. Humans are also one of a very few primate species that establish long-term bonds between males and females. Because of the long-term investment involved in establishing a pair-bond, it is critical that males and females acquire reliable information about their potential mate. Singing, dancing, and gift-giving are all components of both bird and human mating rituals that serve as reliable indicators of fitness. Female satin bowerbirds evaluate the fitness of potential mates on the basis of their energetic dances. Anthropologists and biologists working in Jamaica have shown that among humans, too, better dancers are rated as more attractive and are preferred as mates. There is increasing evidence that singing and dancing have important neurophysiological effects on humans, as well.

Songs and Sentiments

Music is a fundamental component of human ritual, whether in industrial societies or in traditional cultures. In some societies, such as the Igbo of Africa, music and ritual are inseparable and are defined by the single word "nkwa" (Becker, 2001). Although music has been separated from ritual and secularized in contemporary Western societies, it remains the single most frequent element of religious worship across U.S. congregations. Western secularization of music has not diminished its importance, but, instead, has extended its impacts outside of ritual contexts. What are these impacts?

Music evokes and elicits emotions in listeners. Neuropsychologist Robert Levenson (1994) describes emotions as important evolved adaptations that "alter attention, shift certain behaviors upward in response hierarchies, and activate relevant associative networks in memory" (p. 123). Music has immunological effects on listeners. Music can boost immunological function and induces measurable physiological changes in listeners. Heart rate, pulse, blood pressure, and skin conductance are all affected by music. Traditional cultures have long recognized the ability of music-based ritual to alter emotional, physiological, and immunological systems. Throughout the world, such ritual is intimately associated with healing. In contemporary Western societies, music therapy is a growing field, and empirical studies show a significant, positive association between regular participation in weekly religious ritual and improved health and greater longevity. During ritual, the emotional and physiological effects of music are simultaneously experienced by all ritual participants. These shared neurophysiological states are fundamental to empathy and may contribute to the increased levels of cooperation associated with both music and ritual participation.

Birdsong and the music of human ritual appear to have similar effects on ritual participants. There is, however, an important difference. While most bird courtship displays involve only a single male and female, human rituals across cultures include many individuals. The singing, music, and dancing that accompany marriage rituals in most cultures are also found in many secular and all religious rituals across societies. These rituals, like the synchronized group songs of humpback whales, function to define group boundaries, integrate new members into the group, and strengthen group bonds.

The evolved signals, songs, and sentiments of ritual are shared by humans and animals alike. Across all species, these elements of ritual communicate important information between individuals. They also initiate emotional and physiological responses that impact individual health and behavior, and pattern social interaction. As our understanding of ritual continues to expand, the close relationship we share with all other animals becomes increasingly apparent.

See also

Music, Dance, and Theater—*Music and Animals*
Music, Dance, and Theater—*Music as a Shared Trait among Humans and Animals*

Further Resources

Alcock, J. (2005). *Animal behavior: An evolutionary approach* (8th ed.). Sunderland, MA: Sinauer Associates, Inc.

Alcorta, C., & Sosis, R. (2005). Ritual, emotion, and sacred symbols: The evolution of religion as an adaptive complex. *Human Nature, 16*(4), 323–59.

Becker, J. (2001). Anthropological perspectives on music and emotion. In P. Juslin & J. Sloboda (Eds.), *Music and emotion* (pp. 135–60). Oxford: Oxford University Press.

d'Aquili, E., Laughlin, Jr., C. D., & McManus, J. (1979). *Spectrum of ritual*. New York: Columbia University Press.

Darwin, C. (1965). *The expression of the emotions in man and animals*. Chicago: University of Chicago Press.

Fernald, R. (2002). Social regulation of the brain: Status, sex and size. In D. Pfaff, A. Arnold, A. Etgen, S. Fahrback, & R. Rubin (Eds.), *Hormones, brain and behavior* (pp. 435–44). New York: Academic Press.

Hauser, M. D., & Konishi, M. (Eds.). (1999). *The design of animal communication*. Cambridge, MA: The MIT Press.

Juslin, P. N., & Sloboda, J. A. (Eds.). (2001). *Music and emotion*. Oxford: Oxford University Press.

Levenson, R. W. (1994) Human emotions: A functional view. In P. Ekman & R. J. Davidson (Eds.), *The nature of emotion* (pp. 123–26). New York: Oxford University Press.

Smith, W. J. (1979). Ritual and the ethology of communicating. In E. G. d'Aquili, C. D. Laughlin, Jr., & J. McManus (Eds.), *The spectrum of ritual* (pp. 51–79). New York: Columbia University Press.

Sosis, R. (2004). The adaptive value of religious ritual. *American Scientist, 92*, 166–72.

Zahavi, A., & Zahavi, A. (1997). *The handicap principle: A missing piece of Darwin's puzzle*. Oxford: Oxford University Press.

Candace S. Alcorta and Richard Sosis

■ Culture, Religion, and Belief Systems
Shamanism

Shamans are religious specialists in a large number of cultures worldwide. Among their various functions, they may be called upon to mediate between humans and animals when interspecies relationships have broken down or to aid hunters by gaining knowledge of the whereabouts of distant prey-animals. They may also seek the aid of animals on behalf of their communities.

The word *shaman* comes from the language of the Evenks of central Siberia, among whom it refers to someone who learns in initiation rites and training to work with powerful other-than-human beings, sometimes called spirits, many living in otherworlds (e.g., the underworld or the upper world). Some such beings cause illness but may be persuaded to aid shamans in healing. Alternatively, they may seek out a shaman in order to offer aid in healing. Other beings control the presence and movements of animals, especially those hunted or herded by the Evenks. If offended, such beings may prevent hunters from finding or killing sufficient prey to feed their families or clans. Shamans are then called upon to "journey" while in trance or altered states of consciousness, projecting parts of themselves that may be called "souls" in order to reach locations inaccessible physically. Shamans may also seek knowledge of lost animals or distant events and pass on to their successors wisdom necessary for the survival and well-being of their communities. The shaman's relationship with powerful otherworld beings, especially during trance, is sometimes called possession, but if shamans become possessed they also claim to gain control. It may be better to see this as a parallel with the Evenki relationship with their reindeer: neither reindeer herding nor shamanizing are entirely one-sided matters of control; they are at least as much about symbiotic relationships between beings (herders and reindeer, or shamans and otherworld persons) who collaborate or negotiate with each other. In order to "journey" shamans may rely on the help of powerful helper-animals: to fly to the upper world, shamans may call on bird-helpers to carry them; to descend to the underworld they may call on digging animals. The wider context of Evenki shamans' work for their communities provides an example of animism (defined as a worldview and lifeway in which the world is treated as a community of persons, all of whom deserve respect, but most of whom are other-than-human, e.g., bear persons).

The Tungus term *shaman* was first recorded by Europeans in the late seventeenth century, although we can recognize shamans in some of the polemics about "devilish rites" and "conjuring tricks" in earlier travelers' and missionaries' tales. Once the word had been introduced to European languages, it was swiftly applied to religious leaders and activities far beyond Siberia. Many Native American healers, leaders, and teachers have been labeled shamans, especially when they make use of trance or altered states of mind. As in other places, indigenous terms or direct translations such as "doctor" can be preferred. A form of shamanism has become immensely popular among Europeans and Americans since the 1960s, most often when it is practiced as a form of personal growth and a quest for self-knowledge and holistic therapy. Care must be taken in using a term that is so clearly open both to applications in which it may be unwarranted and to restriction from use when comparison suggests it is helpful.

Indigenous shamans often but not always live within societies that subsist by either hunting and gathering, herding and hunting, or gardening and hunting. Thus they engage with animals in hunting and herding contexts and in support of others who are also interested in the well being of herds and the whereabouts of prey-animals. Many Amazonian peoples consider all living beings to be either predators or prey. They also believe all beings see themselves as humans who are predated upon by more powerful beings, "spirits" or "deities" perhaps, and who prey on less powerful beings (i.e., animals). For example, jaguars see themselves as people living in homes, eating cooked food, and hunting animals. However, what they see as "animals" may see themselves as humans. Simultaneously, humans see jaguars as predatory beings, more akin to "spirits" than to animals such as the peccaries, which are archetypal prey animals. Peccaries, however, may in fact be humans who have died and are returning to feed their new fleshy forms to hungry relatives. Shamans are people who have developed skills in perceiving the real nature of beings whose true form is disguised to ordinary people. Shamans' ability to perceive reality allows them to identify whether a jaguar is intent on predation among the shaman's neighbors or whether the peccary is a specific relative come to feed a specific family. In other words, shamans are experts both as repositories of cultural knowledge (the nature of reality) and as sources of practical information (the availability of prey).

In addition to seeking knowledge of the whereabouts of prey animals, shamans in hunting contexts may engage with beings who are believed to control or "own" animal populations. If, for example, a hunter offends such a being by acting inappropriately while hunting (perhaps joking about an animal and thus insulting it), all other animals of that species may become impossible to find or kill. Among the Arctic Inuit a powerful underwater female person, sometimes named Sedna, controls sea animals and objects to any mixture of sea and land animal prey or the hunting equipment appropriate to each. Failure by any member of the community to honor the distinction requires shamans to engage in dangerous conversations to placate her and gain the release of more prey for hunters. Among the Batak of the Philippines shamans mediate between humans and the Masters of Bees (and those of other animals and of plants), making sure that there are sufficient resources for hunters and gatherers.

Further Resources

Harvey, G. (Ed.). (2003). *Shamanism: A reader*. London: Routledge.

Lewis, I. M. (1989). *Ecstatic religion: A study of shamanism and spirit possession*. London: Routledge.

Vitebsky, P. (1995). *The shaman*. London: Macmillan.

Wallis, R. J. (2003). *Shamans/neo-shamans: ecstasy, alternative archaeologies and contemporary pagans*. London: Routledge.

Graham Harvey

■ Culture, Religion, and Belief Systems
Shapeshifting

Shapeshifting, the ability to shift one's form from human to animal or vice versa, is perhaps the most ancient of all human-animal relationships. Tens of thousands of years ago, Paleolithic hunters painted human-animal hybrids on the walls of deep caves. Such creatures as the famous figure from Les Trois Frères cave in France nicknamed "the dancing sorceror," which combines attributes of humans, deer, wolves, horses, and bears, likely depict shamans or other spiritually powerful humans shifted into animal form. Such images may also relate to widespread indigenous conceptions of a "time before time" when humans and animals were one people, easily able to shift from human to animal form (or vice versa). As Calvin Luther Martin (1992) put it in his book *In the Spirit of the Earth: Rethinking History and Time,* "once we were shape-shifters."

Shapeshifting: From Werewolves to Japanese *Kitsune* to Harry Potter

At any rate, shapeshifters are still very much with us today, whether depicted in the ever-popular werewolf or vampire films; in Inuit carvings of Sedna, the human girl turned mother of the sea mammals; or in contemporary literature from Franz Kafka's "The Metamorphosis" to K. A. Applegate's Animorphs and J. K. Rowling's Harry Potter series. This entry discusses some of the many shapeshifters found in human cultures, considers different modes of shapeshifting (physical as well as mental and spiritual), and concludes by exploring the functions of shapeshifting and stories about shapeshifting in humanity's past, present, and future.

There are nearly as many types of shapeshifters as there are human cultures to conceive of them. Werewolves (lycanthropes, named after the Greek word for "wolf," *lykos*) and vampires are perhaps the best known shapeshifters in Euro-American cultures, but virtually every species on the planet is honored by shapeshifting lore among one human culture or another. Many of these stories of transformation focus on the animals that particular peoples most fear, admire, or see as the most spiritually powerful. In China and Japan we find fox-human shapechangers, or *kitsune*; in India there are snake-humans, or *nagas;* in Central and South America, werejaguars, or *nahuales, runa-uturungu, yaguarete-aba;* in Africa, werelions; and in Scotland and Ireland, seal-people, or *selkies.*

Sometimes the changes possible for the shapeshifter are believed to be physical. For instance, in cultures around the world, witches transform into cats, owls, or other creatures of the night to work their magic. Sometimes the changes are conceived of as a sort of mental or spiritual shifting, of opening oneself to seeing the world through other than human animal eyes. Children seem instinctively to do this when they put on a lion mask, get down on all fours, and roar; Scandinavian warriors known as "berserkers" used to do this when they put on a "bear-sark," or bearskin shirt, and ran blindly into battle; and all humans do it when they listen to, watch, or read a story, film, or poem either depicting shapeshifting or simply written from the point of view of a nonhuman animal, such as Richard Adams's epic novel *Watership Down* or the stories in Ursula K. Le Guin's *Buffalo Gals and Other Animal Presences.*

Shapeshifting bridges or helps to bridge the gap between humans and other animals, both allowing humans to understand animals as well as they seem to understand us and allowing us to draw on their spiritual powers. Perhaps it is our awareness of our own lack of spiritual power as humans that explains why we so often make our gods shapeshifters. In ancient Egypt, many of the gods, such as the jackal-headed Anubis or the cat goddess Bastet, appeared in theriomorphic (animal shape) rather than human form. And in ancient Greece, the god Zeus was notorious for changing into various human and animal forms to pursue females who caught his fancy.

Indigenous Cultures

Similarly, in many indigenous cultures, such trickster figures as the Native American Raven or Coyote frequently change their own skins for that of another animal or human who will better enable them to feed their enormous appetites, sexual and otherwise. Even though they sometimes (mis)behave in ludicrous ways, tricksters are powerful spiritual beings because their ability to shapeshift between the human and animal worlds allows them to draw on the abilities of all species. In a story told by the Karok people of California, when Coyote sees that humans are suffering because they do not have fire, he sympathizes with them because of his human side, but it is his status as an animal that allows him to rally the other species, who ultimately succeed in helping him bring fire to humans. Buddhist Jataka stories retain these traditions as well.

Basil Johnston (2003) writes that for the Ojibway of the Great Lakes region of North America, "creation was conducted in a certain order: plants, insects, birds, animals, and human beings. In the order of necessity, humans were the last and the least; they would not last long without the other forms of beings." According to indigenous worldviews such as that of the Ojibway, because other animals have been on the earth longer, they have more wisdom and spiritual power than humans do. One of the main ways humans can access this greater knowledge is by shifting into animal form, often during a vision quest, dream, or ceremonial dance. By imitating an animal's cries and movements, by attempting to become it, even for a moment, a human can learn its language, sing its sacred songs, and share in its power. If you wanted to become a better hunter, you might seek the aid of a fearsome predator such as a wolf or eagle; if you wanted rain for your garden, you might ask for help from a frog or snake. The animal who shares its form and wisdom with you is known as your spirit guardian, familiar, totem, or, as among the Ojibway, a *manitou*. To represent and honor the animal guardian and what you learned from it, you might wear the animal's skin, feathers, or claws; each time you do, you symbolically shift into that animal's form and share again in the power of its teachings.

Seeking the Animal in the Human Spirit

Although shapeshifting continues to play an important role in indigenous cultures around the world, in other cultures shapeshifting has been relegated to the margins of society. Unlike indigenous peoples, who often maintain strong senses of kinship, reciprocity, and equality with animal peoples, contemporary Euro-Americans tend to view humans as superior to the other beings with whom we share the planet—an "anthropocentric," or human-centered, worldview. This sort of worldview places a nonpermeable membrane between human and animal, nature and culture, between what is seen as "wild" and what is seen as "civilized." Human hunters who run with the wolves to learn their ways are now demonized as werewolves out to devour civilized folk. Small wonder, then, that we now consign "stories of incarnation, of crossing the boundaries," "[s]tories explaining a different reality . . . that strange process of disappearing into the mysteries of the earth to be reborn back into human shape, knowing . . . the meaning of kinship," as Calvin Luther Martin (1999) describes them in his book *The Way of the Human Being,* to the realm of "children's literature" and fairy tales or to the marginalized adult literary genres of horror, fantasy, and science fiction.

An illustration of a medicine man wearing a wolf skin by George Catlin implies the close interrelationship between animals and the magical world for many Native Americans. ©The Art Archive.

Nonetheless, positive portrayals of shapeshifting are becoming more frequent. To give just a couple of examples of such marginalized but powerful portrayals, Lois McMaster Bujold's 2005 fantasy novel *The Hallowed Hunt* reclaims "the suppressed animal practices and wisdom songs of the forest tribes." Its protagonists, a werewolf and a wereleopard, are quite different from the slavering

Werewolf, *by Lucas Cranach der Ältere, 1512. Courtesy of the Dover Pictorial Archives.*

Hollywood stereotypes. The role-playing game *Werewolf: The Apocalypse* also reinvents traditional lore, recasting werewolves as noble warriors of Gaia fighting to save the world from the taint of the Wyrm, the technological, spiritually bankrupt cancer eating away at the heart of the wild.

These literary examples have been reinforced in recent years by more positive theoretical models of shapeshifting such as the concepts of "becoming animal" and "spiritual therianthropy." Gilles Deleuze and Félix Guattari (1987) based their idea of "becoming animal" on the beliefs of certain indigenous South American tribes. As Randy Malamud (2003) and Christopher Cox (2005) explain, in Deleuze and Guattari's conception, humans, as with all other beings, are not fixed essences but are constantly changing, in a state of becoming or shapeshifting. By becoming aware of this state of continual shapeshifting—"becoming animal"—that we share with other animal species, humans can understand the common ground that all living beings share and thus more easily open ourselves to the possibility of experiencing positive, reciprocal relationships with other animals.

Spiritual therianthropy is another way for humans to reconnect with other animals. The concept arose during the 1990s in various online discussion groups in the United States. Spiritual therianthropes, as with all other shapeshifters, serve as a bridge between human and animal realms. But rather than physically shifting between human and animal form, spiritual therianthropes instead shift mentally into the form of a particular animal with whom they feel a deep connection. They prepare for these transformations by studying their animal in detail, watching its behaviors, reading all they can about it in the scientific and mythic literature, perhaps even by imitating its movements and vocalizations (sometimes called "dancing your animal"). Thus, spiritual therianthropes actually *become* that animal in their minds, acting in and experiencing the world just as it would. Because of the close connection they feel to their chosen animal, spiritual therianthropes care deeply about it, and do all they can to protect the animal and educate others about its importance.

Modern humans have strayed so far from our original connection to the natural world that many of us no longer recognize either the evolutionary and spiritual kinship we share with other animals or our mutual dependence on a healthy environment. Because of this, our own survival and the survival of the other animals with whom we share the world may depend on shapeshifting in story or spirit helping us to "become animal" once again.

See also

Culture, Religion, and Belief Systems—*Totemism*
Culture, Religion, and Belief Systems—*Totems and Spirit Guides*

Further Resources

Adams, R. (1972). *Watership Down*. New York: Macmillan.

Aftandilian, D. (2005). Spiritual therianthropy. In M. W. Copeland (Ed.), *Experiencing animal presence: Totemism, shapeshifting. NILAS Newsletter, 3*(1–2), 39–43.

Bujold, L. M. (2005). *The hallowed hunt*. New York: HarperCollins Eos.

Copeland, M. W. (2005). Once we were shapeshifters. In M. W. Copeland (Ed.), *Experiencing animal presence: Totemism, shapeshifting. NILAS Newsletter, 3*(1–2).

Cox, C. (2005). Of humans, animals, and monsters. In N. Thompson (Ed.), *Becoming animal: Contemporary art in the animal kingdom* (pp. 18–25). North Adams, MA: MASS MoCA Publications.

Deleuze, G., & Guattari, F. (1987). *A thousand plateaus* (B. Massumi, Trans.). Minneapolis: University of Minnesota Press.

Diaz, N. G. (1988). *The radical self: Metamorphosis to animal form in modern Latin American narrative*. Columbia: University of Missouri Press.

Hall, J. (2003). *Half human, half animal: Tales of werewolves and related creatures*. Bloomington, IN: 1st Books.

Johnston, B. (2003). *Honour earth mother*. Lincoln: University of Nebraska Press.

Kafka, F. (2002). The metamorphosis (E. Muir & W. Muir, Trans.). In H. Kiesel (Ed.), *Kafka's the metamorphosis and other writings* (pp. 1–47). New York: Continuum.

Le Guin, U. K. (1987). *Buffalo gals and other animal presences*. Santa Barbara, CA: Capra Press.

Lopez, B. (1977/1990). *Giving birth to thunder, sleeping with his daughter: Coyote builds North America*. New York: Avon Books.

Malamud, R. (2003). *Poetic animals and animal souls*. New York: Macmillan Palgrave.

Martin, C. L. (1992). *In the spirit of the earth: Rethinking history and time*. Baltimore and London: Johns Hopkins University Press.

———. (1999). *The way of the human being*. New Haven and London: Yale University Press.

Otten, C. F. 1986. *A lycanthropy reader: Werewolves in western culture*. Syracuse, NY: Syracuse University Press.

Steiger, B. (1999). *The werewolf book: The encyclopedia of shape-shifting beings*. Detroit and London: Visible Ink Press.

White Wolf. (1994). *Werewolf: The apocalypse* (2nd ed.). Stone Mountain, GA: White Wolf.

Dave Aftandilian and Marion W. Copeland

■ Culture, Religion, and Belief Systems
Snakes as Cultural and Religious Symbols

Snakes have been used as cultural symbols since early civilization. They have been used to represent all aspects of human nature—both the good and the bad. Especially interesting is their relationship to a varying assortment of deities. Many cultures, both ancient and modern, have viewed the snake with ambivalence. The primary reason for these conflicting feelings toward snakes probably stems from our own fear and fascination with the bizarre. The tubular bodies of snakes and the amazing ways that they accomplish all of life's important functions without limbs is frightening and fascinating. Anyone who has seen a snake swallowing a meal without the benefit of hands knows how utterly bizarre it appears.

Snakes in Classical Greek, Judeo-Christian, and Native Peoples Symbolism

Among the earliest snake symbols is the staff of medicine, known as the staff of Asclepius, and the staff of Hermes, known as the Caduceus. The ancient Greeks used these symbols to represent two different organizations. Curiously, they are often confused with one another. The staff of Asclepius is characterized by one snake entwined around a gnarled tree branch or staff. The Caduceus is described as two serpents entwined in a double helix around a staff.

The staff of Asclepius represents the healing arts and is probably derived from a medical practice still preformed today to remove the parasitic Guinea worm (*Dracunculus medinensis*) from its human host. In a late stage of the parasites life cycle, the adult worm crawls beneath the surface of the skin. Physicians remove the worm by opening the skin and slowly winding the worm on a stick. This procedure requires great skill; hence the worm, or "fiery serpent," wrapped on a stick came to represent skilled doctors.

Caduceus is the name given to the magical staff of Hermes, the Greek messenger of the gods and protector of merchants and thieves. Over the years the Caduceus has come to represent merchants and commerce. The Caduceus became associated with medicine in the seventh century, when Hermes became linked with alchemy. During that time, alchemists prepared drugs and other medicines along with other chemicals and metallurgical materials. There are also occult associations with the Caduceus and therefore legitimate medical organizations do not identify with this symbol. The Ouroborus is also an ancient symbol. In Greek, Ouroborus literally means "tail-devourer." The symbol is a simple snake or dragon consuming its tail to form a ring or circle. Egyptian, Chinese, Hindi, and Norse mythology have similar symbols. Ouroborus represents eternity and the cyclic nature of life and the universe. The most recognizable Ouroborus in western culture is the wedding ring, which symbolizes the eternal commitment between a couple in marriage. Ouroborus also symbolizes death and rebirth.

An ambivalent view of snakes is also present in Judeo-Christian writings. In Genesis, chapter 3, the serpent symbolizes evil, guile, and deceit. The serpent tricked Eve into eating the forbidden fruit. In Revelations, chapter 12, Satan is metaphorically referred to as the serpent that deceived the entire world. However, in Numbers, chapter 21, and John, chapter 3, the serpent is a symbol of life. In the Old Testament God punished the Israelites by setting upon them fiery serpents. Later, God instructs Moses how to heal the Israelites by gazing upon a fiery serpent entwined on a staff. Most likely this serpent is a venomous snake. In the New Testament, the serpent symbolizes eternal life.

Native peoples of North America view rattlesnakes as symbols or messengers to weather deities, such as their association of the rattlesnake with thunder. It was believed that giant rattlesnakes in the heavens shake their tails to make the sound of thunder. The relationship between rattlesnakes and rain might also be a result of the association between rain and thunder. Rain brings life, especially to tribes in the arid southwestern United States. Snakes shedding their skin appear to be reborn. So rain and snakes may be linked as symbols of rebirth. Natives from the ancient civilization of Teotihuacan, Mexico, identified the men of the warrior class with the rattlesnake because of its fierce and defensive disposition.

British colonists in America identified with the rattlesnake for several reasons. First, they recognized that the rattlesnake never wounds its enemy unless given ample warning. Second, once engaged in attack, the rattlesnake never surrenders. Third, the concealed fangs make the rattlesnake appear defense-less, and even when exposed the fangs appear small and frail, yet when put into service can inflict decisive and fatal wounds. Fourth, each segment of the rattle is independent, yet firmly united to the other rattle segments. Fifth, the lack of eyelids makes the rattlesnake appear ever vigilant. Each of these traits relate to the colonies relationship with Great Britain and their struggle for liberty and independence. Perhaps most important, the rattlesnake was uniquely American and highlighted the differences between British and the new American cultures. But again, ambivalence is apparent. Many colonists considered the snake, especially rattlesnakes, a symbol of the wilderness that must be conquered and subdued. Snakes were routinely killed in this effort to conquer nature.

In conclusion, snakes have been used as cultural symbols since early civilization. They have been used to represent all aspects of human nature—both the good and the bad. Especially interesting is their relationship to a varying assortment of deities.

A statue of an unnamed goddess holding serpents, from Palace of Knossos, Crete, ca. 1500 BCE. ©The Art Archive/Heraklion Museum/Dagli Orti.

Further Resources

Klauber, L. M. (1972). *Rattlesnakes: Their habits, life histories, and influence on mankind* (2nd ed.). Berkley: University of California Press.

Morris, R., & Morris, D. (1965). *Men and snakes.* New York: McGraw-Hill.

Nissenson, M., & Jonas, S. (1995). *Snake charm.* New York: Harry N. Abrams, Inc.

Olson, L. C. (1991). *Emblems of American community in the revolutionary era: A study in rhetorical iconology.* Washington, DC: Smithsonian Institute Press.

Aaron J. Place and Charles I. Abramson

■ Culture, Religion, and Belief Systems
Taiwan and Companion Animals

Taiwan is a small island nation (area 36,000 km^2), which is bounded to the east by the Pacific Ocean, to the south by the South China Sea, to the west by the Taiwan Strait, and to the north by the East China Sea. During the early 1980s, many endangered species of exotic animals such as tigers, lions, monkeys, and orangutan were popular as pets and were kept in many homes across the island. The exotic primates such as orangutans were actively popularized by the Hollywood movie industry as adorable pets, which prompted people to purchase them. At that time, endangered species were not legally protected in Taiwan. Possessing an endangered animal as a pet became a status symbol, and it was easy to find infant monkeys, gibbons, and orangutans that were kept like puppies in local pet stores. When Taiwan enacted the Wildlife Conservation Act in 1989 because of international pressure, the trade in endangered species started to slow down. However, the exotic pet animal trade still goes on uninterrupted, and traders now import large volumes of unprotected species, including a variety of aquarium fish, amphibians, reptiles, birds, and mammals. Numerous pet shops can still be seen all across Taiwan that sell small and easier-to-handle exotic pets that are often unprotected by local and international laws.

Taiwan has evolved from an agricultural backwater status to a global economic giant over the last few decades, and the economic boom naturally changed the attitude of people from a dog-eating to a dog-loving society. Although eating dog-meat is illegal in Taiwan, some restaurants in rural areas do sell dog-meat discreetly. In recent years, people in the rabies-free Taiwan have started to own more and more domestic animals such as the dog and cat out of love and affection. Pet stores and veterinary clinics are located at several convenient locations in all major cities and towns. Besides, all major retailers and grocery stores carry pet supplies. However, the sudden surge in possessing dogs and cats also created a disaster for the island. During the 1990s, there was an abrupt increase in the stray dog population in Taiwan, particularly in urban areas because people tend to abandon their pets. The population of free-roaming dogs was estimated at 1.3 million, or 36 percent of the island's total dog population during the early 1990s. Poor welfare and the shortcoming of both public and private efforts to deal with the problem attracted global media attention and aggressive lobbying by local and international animal rights groups. As a matter of fact, the stray dog problem is a crisis not only for Taiwan but also in many countries worldwide. The current population of domestic dogs (*Canis lupus familiaris*) in the world is estimated at over 500 million, of which a substantial (unknown) proportion are poorly supervised or free-roaming.

Taiwan passed the Animal Protection Law in 1998 to prevent maltreatment of domestic and wild animals. According to the law, pets must be registered with an individual identification microchip. The registration must include a record of the animal's birth, acquisition, transference, loss, and death. If government officials later find a pet loose on the street, they run a scanner over the animal to locate the chip, and the owner is fined.

In general, human beings tend to require the companionship of other species, so a pet is a natural part of many households around the world. A pet can provide the owner some basic functions such as providing companionship, serving as a being to care for and touch, keeping a person busy, and giving the owner some exercise and protection. People keep various species of animals as pets, and almost any animal can serve the first three functions, but a dog is the usual pet to satisfy the last function.

Unless owners are responsible for their pet's behavior, society can suffer in many ways, such as cat or dog bites, road traffic accidents caused by or involving pets, damage to property and wildlife caused by free-roaming pets, dug-up gardens, disturbed bins of garbage, and uncontrolled population growth of stray pets that may lead to increased zoonotic diseases (transmitted from animals to human). Some of the common zoonotic diseases are ringworm, leptospirosis, dog roundworm, hydatids, and toxoplasmosis (carried by cats and is a potential hazard to pregnant women). The solution to these problems is sometimes difficult, and therefore, pet owners must be educated to understand the extent of the problems.

A pack of stray dogs resting in a parking lot in Kaohsiung city (Taiwan). Courtesy of Minna J. Hsu.

The government of Taiwan must work with community outreach groups to access ordinary citizens to enforce workable law-enforcement and responsible pet-ownership strategies, and only then can the problems related to pet abuse and irresponsible release of pets on the street be minimized in Taiwan. People should understand that responsible pet ownership is something to be carefully judged, not rashly and haphazardly assumed.

The government must also carefully review the currently uncontrolled exotic animal shipments for pet trade of unprotected species from overseas because people in Taiwan usually spend US$6 million annually to set free some 200 million wild animals for religious reasons, animals that range from insects to monkeys! Taiwan's two major traditional religions, Taoism and Buddhism, stress the importance of good deeds during a person's life, and the religions decree that returning animals to nature is one of the ways to garner good karma and will earn an individual's salvation in the afterlife. Temples usually provide religious animal-freeing services to the faithful. Because there is a huge market for this trade, various animals such as fishes, frogs, turtles, snakes, birds, mammals, and even insects are bought from pet shops by religious followers and set free around the island's rivers, mountains, forests, lakes, reservoirs, and even golf courses.

Taiwan's Council of Agriculture, which is the law enforcement authority, admitted that there is an immediate need to hammer out policies to end the practice of freeing animals, which at present is unregulated. Biologists fear that religious freeing of animals is leading to alien invasions of nonnative species, which is catastrophic to the delicate ecological food chain. So Taiwan must act quickly to counter the unmonitored release of wildlife that originates from local pet shops.

See also

Bonding—*Companion Animals*

Further Resources

Joshua, J. O. (1975). Responsible pet ownership. In R. S. Anderson (Ed.), *Pet animals and society*. London: Baillière, Tindall & Baltimore.

Minna J. Hsu

■ Culture, Religion, and Belief Systems
Totemism

In the totemic view of the world, certain animals are closely associated with human social groups, influencing each individual's identity and one group's relations with other social groups. Totemism is commonly defined as a set of beliefs concerning the relationship of members of a clan to a totem. Totems most often are animals, but might be plants (as among the Tikopia people of Polynesia) or other natural phenomenon. Each clan is named for a totem, and clan members might see themselves as descended from the totemic animal. Sometimes there is a prohibition or taboo against eating the totem.

For a brief time in the late 1800s and early 1900s, anthropologists developed ever more elaborate models of totemism as a set of customs that promised a window into the origins of human culture. Further study of the customs of traditional peoples eventually showed that such grand models did not fit the facts very well—but not before the most famous book of all, Sigmund Freud's *Totem and Taboo* (1913) began to connect "primitive" cultural ideas with modern mental illness in many people's minds.

What is Totemism?

Totemism is a form of animism, a view of reality in which many elements of the world, including animals and plants, are animated by spirits or a spiritual power. Through these powers they have a deep influence on human affairs. Totemism was once widespread among traditional peoples on several continents (especially North American and Australia). Specifics varied, but most totemic peoples had in common a belief that spirit helpers, in the form of a specific animal or plant, were connected not just with individuals (as might be the case following a vision quest) but with social groups as well. Not uncommonly, people saw themselves as descended from their totemic animal. For the bear clan, the bear was not just an abstract symbol or crest. Nor was there some essence of "bearness" that was important, but the clan felt real kinship with every particular bear to the extent that they carefully kept to a range of rules regarding relationships with this totemic animal. This sometimes (but not always) included taboos against killing or eating bears.

The word totem is derived from *ototemen* of the Ojibway language of Native North American and has been in use in literature at least since 1791, when John Long wrote in his *Voyages and Travels of an Indian Interpreter and Trader*, of the Chippewa belief in a guardian spirit that appeared in the form of an animal, after which the Chippewa people refused to eat or kill any of that kind of animal.

Totemism and the Evolution of Religion

European scholars had long been intrigued by reports of "animal worship" among traditional peoples around the world. But as Robert Alun Jones suggests, the development of theories of social evolutionism offered a final piece of the puzzle, making it possible to postulate a primitive totemic stage of society, to discuss questions of the origin and nature of totemism—for a time one of the most important topics in anthropology—and to propose models for totemism's impact on the course of humanity (Jones, 2005, p. 4).

Totemism as a stage in human social evolution, including the evolution of religion, involves at least the following features, as nicely summarized by Sigmund Freud in 1913:

> What is a totem? It is as a rule and animal (whether edible and harmless or dangerous and feared) and more rarely a plant or a natural phenomenon (such as rain or water),

which stands in a peculiar relation to the whole clan. In the first place, the totem is the common ancestor of the clan; at the same time it is their guardian spirit and helper, which sends them oracles and, if dangerous to others, recognizes and spares its own children. Conversely, the clansmen are under a sacred obligation (subject to automatic sanctions) not to kill or destroy their totem and to avoid eating its flesh (or deriving benefit from it in other ways). The totemic character is inherent, not in some individual animal or entity, but in all the individuals of a given class. From time to time festivals are celebrated at which the clansmen represent or imitated the motions and attributes of their totem in ceremonial dances. (Freud, 1913, p. 5)

But there were numerous variations of these ideas. James George Frazer, most famous for his immensely influential work *The Golden Bough,* also wrote a massive four-volume treatise on totemism. His work derives in part from his good friend and fellow Cambridge scholar William Robertson Smith, who studied totemism during his travels in Arabia. Here each clan worshiped its totemic animal and was not allowed to kill or eat it. Totemism was also linked to exogamy (where one is required to marry someone from outside his or her totemic group). Smith argued further, in *The Religion of the Semites* (1890), that ancient Hebrew sacrifices were surviving evidence of an earlier totemic stage in their history (Pals, 1996, p. 31). Frazer's work derives from these ideas as well as from then-newly published research on the Australian Aborigines by Baldwin Spencer and F. J. Gillen, who found that on special occasions the totem was killed and eaten. Frazer saw here, in what he took to be a very early stage in the evolution of religion, the first rites of religious sacrifice as well as the concept of the dying god. "By killing the totem," Daniel Pals summarizes, "primitives protect against the decline of power in their animal god; by eating it, they take its divine energy into themselves" (Pals, p. 40). Throughout his career, Frazer developed other theories of totemism. He understood this custom, as he understood most customs, to be a solution to an intellectual problem. For example, it was widely believed among European scholars at the time that primitive peoples did not understand how conception worked. Frazer believed that totemism was their answer to how conception works; in an intricate argument, he proposed that people believed it was the totem that entered the women at the time she first recognized the mysterious movement within her (see Jones, 2005, p. 170).

Emile Durkheim believed totemism was even more important than others, such as Frazer, thought. For Durkheim, it illustrated the concepts of the sacred and the profane that were so important to his understanding of religion. Durkheim believed that totemism is not simple animal worship. People instead were worshipping a kind of anonymous and impersonal force within each totem. Durkheim's most famous contribution was his attempt to identify this powerful and impersonal force. He believed that the totem is a concrete and visible representation of the clan; it is a kind of flag or logo. To put it more bluntly, if the totem "is at once the symbol of the god and of the society, is that not because the god and the society are only one?" (quoted in Pals, 1996, p. 104). Indeed it is one of Durkheim's most important, controversial (and in retrospect most implausible) contributions to see that all of religion really refers back to the social group, and its main purpose is to hold society together.

Returning to Freud, in *Totem and Taboo* he took the discussions in a different direction, using totemism as a basis for one of his "pet" theories: that the thought of "primitives" or "savages" has much in common with the thought of neurotics. To take one example, he was impressed with the idea, widely accepted at the time, that wherever one finds totems, "we also find a law against persons of the same totem having sexual relations with one another." All peoples strongly object to incest, but among totemic peoples, the prohibited group, those of one's own totem, was an unusually large portion of the total population. He describes in some detail the prohibitions or taboos on social interactions among various kinds of kin, or even non-blood kin. They would not have so many rules, he reasoned, if they weren't afraid of them being broken. Freud believed that the "horror of incest" expressed by these "savages" is strikingly similar to the mental lives of neurotic patients. Unlike most people, the neurotic people exhibit, in Freud's view, a kind of psychic infantilism in which incestuous fixations play a major part in their mental lives. Similarly, Freud thought that these "savages" with all their rules about avoidance within a totemic group were being, well, a little neurotic.

But totemism of the kind lauded by Frazer, Tylor, Durkheim, Freud, and numerous others may never have actually existed. It was instead "cobbled together" as one critic has it, from various ideas. Totemism in this sense turns out not to be as common among traditional peoples studied by ethnographers as was thought around 1900. At the same time, anthropologists have concluded as well that modern traditional peoples like the Australian Aborigines do not in any case represent some kind of "un-evolved" hold-out of early human life. They, too, have long and varied histories.

Claude Lévi-Strauss's 1962 book *Le totémisme aujourd'hui* is often credited with putting an end to the totemism of Frazer and colleagues, which it does very effectively indeed. But as Levi-Strauss himself points out, the passion had long since faded. Indeed, the last full-scale study of totemism in this mode was that of Van Gennep in 1919. For an intellectual history of totemism, Levi-Strauss's little book is enjoyable and helpful, but the thorough recent monograph by Robert Alun Jones is now the best place to start. He argues that these grand Victorian theories were not so much disproven as rendered useless—they were in effect answers to questions people no longer ask.

Totemism and the Place of Animals in Human Cosmologies

Nevertheless, totemism, in the basic form defined at the beginning of this essay, was quite widespread, especially in pre-European Australia, to some extent in Polynesia, and among native North Americans. If it is not fossilized evidence of the early conditions of humanity, and it does not (except in a few rare cases) form a complex of cultural traits including clans, exogamy, and so on, it does reveal some very interesting and widespread understandings of how humans and animals fit within a view of the world.

Totemism has sometimes been referred to as nature worship, but this is misleading. That label implies that these people worshipped inanimate objects and animals rather than spirits or a god, but in fact, to the totemic peoples themselves, our distinction of animate and inanimate is a false dichotomy for all animals, and many nonliving things were animated by spirits. To the extent these were worshipped, the people were worshipping spirits like most people everywhere.

From the totemic perspective, animals are seen as animated beings, not just animated with life in the sense of biological function, but spiritual beings. This does not mean they were never killed. It would only be your social group's totem that you would treat with special care, often (but not always) including avoidance of killing or eating it. You would not object to members of another clan hunting your totem, nor would they object to you hunting theirs. Although for some people, there might be a dozen or more different totem groups, there would be any number of usable resources, including edible animals, that anyone could use.

There might be other elements to a person's connection with his or her social group's totem. It is not always the case (as argued in the grand evolutionary theories of a century ago) that people thought themselves descended from their totemic animal, but sometimes this was indeed the view, and it indicates that their distinction between humans and other animals (even between humans and trees) was not so sharp as we sometimes see it.

A Note on Totem Poles

When many of us think of totems and totemism, we naturally picture the magnificent totem poles carved for centuries by the native peoples of the Northwest Coast of North America. This is somewhat misleading, however, because totemic thought is not a part of the culture of these peoples, and the carvings on these poles are not totems. They are much more like personal or family crests. As Philip Drucker said, a person with an Eagle, Raven, Bear, or other crest had no particular regard for animals of this species as would have been the case were they totems. "It was not the biological species in general that was of importance in his clan or lineage tradition, but a single, specific supernatural being who had used the form of an eagle, raven, or bear" (1963, p. 190). Poles might be erected to commemorate a former chief, to serve as the actual burial of a chief, or as part of a house.

The choice of subject depicted was very important, often including animals important to the lineage or family group involved and usually representing events in the lineage tradition. The totems were the property of one group who had the right to display them. For example, among the northern Tlingit and Haida peoples, spirit helpers (typically in the form of an animal) were routinely inherited in family lines, a fact displayed in a range of contexts, including totem poles and carvings within houses. People could also seek new spirit helpers through vision quests. Sometimes they came unbidden. Robert Torrance relates the story of Isaac Tens of the Gitskan people of British Columbia, who said that part of his call to becoming a shaman involved being chased by a large owl who tried to carry him off (1994, p. 178). Thus, although the creators of totem poles did not embrace the cosmology properly called totemism, they nevertheless expected supernatural beings to engage them in the form of animals.

See also

Culture, Religion, and Belief Systems—*Shapeshifting*
Culture, Religion, and Belief Systems—*Totems and Spirit Guides*

Further Resources

Drucker, P. (1963). *Indians of the Northwest Coast*. Garden City, NY: Natural History Press.

Durkheim, E. (1995). *The elementary forms of religious life* (Karen E. Fields, Trans.). New York: Free Press. (Original work published 1912)

Frazer, J. G. (1887). *Totemism*. Edinburgh: Adam & Charles Black.

Freud, S. (1913). *Totem and taboo: Some points of agreement between the mental lives of savages and neurotics* (J. Strachey, Trans.). New York: Norton.

Jones, R. A. (2005). *The secret of the totem: Religion and society from McLennan to Freud*. New York: Columbia University Press.

Lévi-Strauss, C. (1963). *Totemism* (R. Needham, Trans.). Boston: Beacon Press. (Original work published 1962)

Pals, D. L. (1996). *Seven theories of religion*. New York: Oxford University Press.

Spencer, B., & Gillen, F. J. (1899/1968). *The native tribes of Central Australia*. New York: Dover.

Torrance, R. M. (1994). *The spiritual quest: Transcendence in myth, religion and science*. Berkeley: University of California Press.

Van Gennep, A. (1919). *L'État actuel du probléme totémique*. Paris.

Paul K. Wason

■ Culture, Religion, and Belief Systems
Totems and Spirit Guides

Traditional Practices

The word "totem" derives from the Algonquin language and is closely associated with the beliefs of a range of Native American peoples. However, the phrase is now used more generally to refer to traditions of reverence for natural features (particularly animals but occasionally plants and geographical features of forces of nature) in a manner that is found in a wide range of cultures around the world.

Totemic practices are a type of animism (belief that animals and other natural objects have spirits) that seems to have its roots in ancient cultures, particularly those that depended primarily on hunting for subsistence and survival. The dependence on uncontrolled wild animals seems to have fostered a belief in and respect for animals' souls—both individually and in the form of an overall spirit representative of the species as a whole. This collective spirit, or representations of it, is referred to as a totem. The belief in and importance of this spiritual aspect of an animal species is the defining feature of totemism, whereas a "spirit guide" is a particular animal soul that may act as a companion or teacher during a spiritual quest.

In many cultures, totemism is related to the belief that a certain animal species is the ancestor to, or brother of, the members of a tribe or clan. This led to groups and individuals having "totem" animals that become part of their identity and define their group or subgroup. Often these animals are not killed or exploited by the groups who used them as a totem, except under certain important ritual circumstances. This sign of respect may form the basis of many dietary taboos in modern religions, such as halal and kosher requirements.

The totem animal contributes to the bonding of a group and serves many functions related to holding a group together. For example, people are generally prohibited from marrying within their totem group, which aids in avoiding inbreeding. Totem markings help identify people and places, such as in the case of the totem poles of the Pacific Northwest that indicate a person's lineage and place of residence or burial.

The choice of totem for the group sprang from the bond between the human tribe and the animals they depended on, and as such, the bond was strong and reverential. As many cultures became more agriculturally based, this totemic tradition seemed to wane, perhaps because once animals could be controlled and reliably exploited, they came to have a more mundane role in human daily, and spiritual, life. There are, however, many cultures that continued to revere at least some highly important and high-status domesticated animals as symbolic of important qualities and forces (for example, the importance of horses to many early Celtic people, symbolic of leadership and fertility).

General Modern Practices

In many modern, Westernized cultures, animals continue to feature strongly in stories, visual arts, and music—as part of a general tendency for all human cultures to attach spiritual meaning to different animal species. Animals appear in tales with deeper morals, and important heroes are often shown with animal companions or taking on animal forms. In modern Western cultures, animals are used as important symbols and teaching figures, from Aesop's fables to Stuart Little.

It is now common for people to have strong feelings about animals such as whales, lions, or wolves, which they may never have seen in the wild. Charismatic animals, however, even in this abstract context, can have genuinely totemic roles as the mascots of sports teams, national symbols, and corporate logos, with the animals serving as symbols of ideals and rallying points for certain groups. For example, the bald eagle is emblematic of the United States. Its use identifies U.S. citizens and is used to represent virtues such as patriotism, valor, and teamwork—whereas the portly, flightless Kiwi is a beloved symbol of New Zealand for its uniqueness and dogged persistence. Human communities naturally seek out identifying symbols, and animals are still frequently used.

New Age Revival

There has also been a revival in the practice of forming a personal relationship with a particular animal species (totem) or individual animal spirit (spirit guide). Totems and guides are used informally by many people, and a number of modern rituals and systems of belief have been developed—often inspired by more ancient traditions. The traditions referred to are normally from Native American cultures but also sometimes from Celtic culture or African sources. Often, these popular revivals are somewhat shallow, confused, or exploitative of the deep belief systems from which they borrow.

There is often a lack of a coherent relation between the animals revered in new totemic systems and those animals with which followers have day-to-day experience. For example, widespread reverence is shown toward the wolf in many cultures or regions where wolves are no longer, or have never been, found; and in areas where wolves and coyotes are endemic, they are more likely to be seen as an inconvenience. The abstract virtues associated with different animals seem to reflect the strongest traditional systems and the most charismatic large animals found worldwide rather than serve to connect people to their natural environment and its animals.

Yet many people find new age totemism an inspiring part of their personal spirituality. The works of writers such as Brad Steiger and John Matthews draw upon a wide range of native beliefs as well as personal intuition and answer a need for many modern neo-pagans. This general tendency reverses a colonial trend to treat totemism as primitive and idolatrous and animals as either pests or resources. Totemism seems to be part of a general interest in the symbolic roles of animals from many cultures, including widespread knowledge of Chinese year animals, North American totems, Australian dreamtime animals, and specific spiritual animals from ancient and modern traditions, such as the New Zealand taniwha, and various "water monsters," such as the Loch Ness Monster (Scotland) and Ogo-pogo (Canada). It seems that there is a general human desire to reconnect with animals as representations of our deepest values and our spiritual connection to the natural world.

On a personal level, people may seek to form a spiritual relationship with animals so that they can better emulate the animals' behavior and share in their understanding of the natural world. Totems, guides, mascots, and familiars rarely fill their old roles of aiding us to acquire food, shelter, and other necessities of life. They are now more likely to be employed on a more abstract level and to help enrich our day-to-day experiences and give them deeper meaning. In more artificial environments, totems and guides have become more a part of spiritual development, teaching broader lessons such as patience, persistence, or teamwork that can be applied to our relationships, workplace, and general life experiences.

Manipulating Animal Meaning

Some believe that the most valid answer to accusations that modern totemism is shallow and exploitative is to encourage a reconnection of the abstract symbolic use of animal species to the actual animals from which it developed. That is, people would realize that the spiritual meaning of a species must be seen as emerging from the individual souls of the individual animals of that type. As such, there is some potential for the symbolic strength of animals to feed back into conservation of these species and their habitats. Some also believe that it would be useful to encourage people to investigate the meanings of animals endemic to their area rather than those of traditional importance to their ancestors who may have lived in a very different part of the world.

The symbolic importance of animals has shown some flexibility in this regard. For example, in Australia there is a movement to replace the Easter bunny (the rabbit being an imported pest) with the Easter Bilby (an endangered native animal with similar appearance and lifestyle). This change would help raise support both for rabbit eradication and bilby conservation, without giving children conflicting and distressing messages about how rabbits should be regarded and while respecting the bilby's role in the rapidly vanishing native ecosystem.

Another successful example can be found in a previously disregarded bird, the Chatham island robin. This species was once the most endangered bird in the world (with five individuals remaining and only one, elderly female), but they now number in the hundreds. Part of the dramatic turnaround involved the adoption of this species as an important symbol by the Chatham island population, which has seen the bird adopted as an emblem for beer, sports teams, and major companies. This tiny (15 cm high) bird is not a typical totem, but its very ability to come back from the brink of extinction became a powerful symbol for residents of this remote cluster of islands—establishing a real connection between human and animal residents.

Totemic practices originally evolved as a reflection of a deep bond necessary for physical survival. Animals had to be closely observed and understood by people subsisting on the land. Modern new age totemism may be a means to reestablish this bond for the good of these animals we profess to admire. Reverence for animals as individuals with souls, and as represented by a totem, might be one way to encourage more ethical use of domestic animals and conservation of the habitats of their wild brethren.

See also

Culture, Religion, and Belief Systems—*Shapeshifting*
Culture, Religion, and Belief Systems—*Totemism*

Further Resources

Matthews, J. (2002). *Celtic totem animals.* San Francisco: Red Wheel/Weiser.
Steiger, B. (1997). *Totems: The transformative power of your personal animal totem.* San Francisco: HarperSanFrancisco.

Emily Patterson-Kane

■ Culture, Religion, and Belief Systems
Witchcraft and Animals

During the Middle Ages and the Renaissance, a variety of animals were associated with witchcraft in Europe. The nature of these associations, however, depended greatly on whether they derived from official or popular sources. During the fourteenth and fifteenth centuries, ecclesiastical and judicial authorities in continental Europe began persecuting witches as heretics, accusing them of membership in organized Devil-worshipping cults that were believed to exist in direct opposition to Christianity. According to this official or "establishment" view of witchcraft, the Devil was widely reputed to appear to his followers in the shape of an animal—usually a monstrous dog, cat, goat, or ram—and witches were thought to be in the habit of riding or flying to orgiastic gatherings or "Sabbats" on the backs of demons disguised as animals (as well as on pitchforks and broomsticks).

This conception of witchcraft as an orchestrated, anti-Christian religious conspiracy is thought to have been largely a product of the papal Inquisition's practice of extracting confessions from suspects under torture. Assuming this is correct, organized diabolism of the kind envisaged—together with most of its animal trappings—was probably a fantasy engendered by the fevered imaginations of the prosecuting authorities. Presumably, it also served a valuable propaganda role. Official efforts to eliminate nonconforming religious beliefs during this period laid particular emphasis on supposedly zoophilic or zoocentric behavior: between the twelfth and the fourteenth centuries, nearly all the major heretical sects, including the Templars, the Waldensians, and the Cathars, were accused of worshipping and having carnal relations with the Devil in the form of an animal of some kind. Given the marked medieval horror of mixing human and animal categories and attributes, such accusations of demonic bestiality were guaranteed to excite popular condemnation.

Long before European witches became the objects of official persecution, a number of other curious animal affinities were also attributed to witches, including the practice of shape-shifting. At the level of popular or "folk" culture, it was common, especially in northern Europe, for people to believe that witches were capable of transforming themselves into animals in order to travel about incognito and engage in acts of *maleficium*—that is, the practice of harming people, their property, or livestock by supernatural means. As early as 1211 CE, Gervase of Tilbury attested from personal experience to the existence of witches, "prowling about at night in the form of cats" who, when wounded, "bear on their bodies in the numerical place the wounds inflicted upon the cat, and if a limb has been lopped off the animal, they have lost a corresponding member." Stories of this type are extremely widespread in medieval and post-medieval witchcraft folklore. Most relate instances of witches adopting the form of either cats or hares in order to engage in the theft of milk from their neighbors' cows. If the animal was pursued and wounded, the witch invariably suffered an equivalent parallel injury and was thus identified as the true perpetrator. The geographic distribution of these folktales of shape-shifting in England—particularly in the counties of Northumberland, Yorkshire, and Lincolnshire—suggests that the original concept may have been imported by Scandinavian colonists from Norway and Denmark where similar stories were widespread. Further south, in East Anglia, Kent, and the Home Counties, a subtly different animal-witch association developed with the concept of the witch's "familiar."

Briefly defined, the familiar or "imp" was a demonic companion whom the witch dispatched to carry out her evil designs in return for protection and nourishment. Although it cropped up from time to time in other parts of Europe, the concept of the familiar achieved its most elaborate and vivid expression in southern England in the late sixteenth and seventeenth centuries. According to the surviving records of the English witch trials, familiars only rarely appeared in human form. The overwhelming majority manifested themselves as commonplace animals, especially as cats, dogs, ferrets, mice, rats, toads, poultry, and wild birds. Familiars could be acquired by witches from a variety of sources. Many were represented as gifts of the Devil, given in return for a promise of allegiance. Others were obtained from fellow witches or were passed around and shared between groups of witches, like some sort of useful household implement. Frequently, they just appeared out of nowhere, like stray cats, offering their services and demanding to be fed. Familiars also acquired a variety of interesting names, many suggestive of pet names. Sometimes the witches themselves bestowed these names on the familiars, but in other cases, the Devil assigned a name, or the familiar chose its own name. Such attributes have led some historians to conclude that most so-called familiars were simply miscon-strued pets that were implicated by their affectionate associations with the kinds of solitary old women who were most often accused of witchcraft. Other evidence, however, suggests that this interpretation is overly simplistic.

In trial depositions and testimonies, familiars were usually represented as relatively autonomous beings whose function was to serve as the witches' magical agents or emissaries in the performance of acts of *maleficium*. In return for these services, witches provided their familiars with shelter, often in boxes or pots lined with wool, and food— occasionally milk, oats, bread, cheese, cake, or other scraps, but more usually blood sucked from reddish spots or swellings on the witch's own body. A reasonably typical example of this kind of evidence was provided by the son of an accused witch who was tried for witchcraft in the county of Essex in 1582:

> He saith that his mother, Ursula Kemp, alias Grey, hath four several spirits, the one called Tyffin, the other Tyttey, the third Pygine, and the fourth Jacke: and being asked of what colours they were, saith that Tyttey is like a little grey cat, Tyffin is like a white lambe, Pygine is black like a toad, and Jacke is black like a cat. And hee saith, hee hath seen his mother at times to give them beere to drinke, and of a white Lofe or Cake to eat, and saith that in the night time the said spirites will come to his mother, and sucke blood of her upon her armes and other places of her body.

Although familiars appeared to possess a degree of autonomy, the line separating them from the "animal-as-transformed-witch" sometimes became blurred, at least in the popular imagination. In some cases, witches were reported to suffer parallel injuries when their familiars were wounded, and sometimes it is clear that prosecution witnesses believed that the familiar was simply the witch herself transmogrified. In the notorious case of the Walkerne witch, Jane Wenham, in 1712, several witnesses not only testified to being visited and "tormented" by her cats, but also reported that one of these cats had the face of Jane Wenham.

This apparent confusion provides a clue to the true origins of the animal-witchcraft connection. Belief in the supernatural origins of sickness and misfortune was evidently prevalent during this period, and it is clear that many people believed in the existence of malevolent "spirits" that assisted or represented witches in their magic and that prefer-entially adopted the physical form of animals while doing so. The fact that these beings were often referred to as "imps" (literally "grafts") also indicates some ambiguity in the popular mind concerning whether such creatures existed independently of the witch or were alien offshoots of her own persona. According to anthropologists, similar ideas are almost universal among "shamanistic" societies throughout the world. For example, one account of shamanism among the Penobscot Indians of New England reiterates precisely the same kind of ambiguity:

> Every magician [shaman] had his helper which seems to have been an animal's body into which he could transfer his state of being at will. The helper was virtually a disguise, though we do not know whether the animal was believed to exist separately from the

shaman when not in the shaman's service or whether it was simply a material form assumed by the shaman when engaged in the practice of magic . . . Direct information from Penobscot informants says that the *baohi'gan* [familiar] could be sent to fight or to work for his master the shaman. It could be sent on any mission whatsoever according to the shaman's will. We are told, too, that the owner remained inert while his *baohi'gan* was away.

Such obvious parallels suggest that at least some of the animal associations in European witchcraft had their roots in archaic shamanism, particularly the apparent belief in the witch's (or shaman's) ability to shape-shift, or to perform magical acts by sending his or her spirit out of the body in the form of an animal. Establishing the existence of pre-Christian shamanic traditions in Europe is difficult because of the lack of clear documentary evidence, but there are abundant circumstantial clues. According to legend, for example, the sixth-century Welsh shamanic bard Taliesin was able to transform himself into a long list of different animals, including cats, dogs, foxes, goats, hares, snakes, squirrels, cranes, and eagles. Similarly, the hagiographies of many early Celtic saints are replete with accounts of their shamanic ability to communicate with animals and control them.

Although toad familiars are featured in several prominent trial records, there is some evidence to suggest that these animals may also have occupied a unique position as part of the witches' or shamans' pharmacopoeia. Toad's blood and other secretions, together with henbane, belladonna, thorn apple, mandrake, and other "witch herbs," were among the staple ingredients of the various brews, potions, love philtres, and hallucinogenic "flying ointments" supposedly concocted by witches and sorcerers during the Middle Ages and early modern period. The skin and glandular secretions of toads, genus *Bufo*, are known to contain a variety of toxins, such as bufotenine, bufagenins, and bufatoxins, many of which are pharmacologically active and which exert powerful effects on the heart, the blood vessels, and the peripheral vascular and nervous systems. These effects— comparable to those of the drug digitalis—are highly toxic and can induce heart failure in humans and other animals at high doses. Toad venom has a long history of use in Europe and Asia as a diuretic, expectorant, local anesthetic, and aphrodisiac and for the treatment of a wide variety of ailments including dropsy, heart failure, and nose bleeds. Bufotenine and related compounds are also known to have antibacterial and antifungal properties. Despite belief in the hallucinogenic properties of toad venom, only one species in the genus, the Sonoran Desert toad (*Bufo alvarius*), has been found to secrete a psychoactive bufotenine derivative (5-MeO-DMT) from its paratoid glands. None of the European *Bufo* species appear to produce this compound, but it is possible that medieval witches had found a way to create it chemically from toad-derived precursors.

In short, although the evidence is limited and often distorted by the extreme methods used by prosecutors to obtain information, it appears likely that medieval and early modern herbalists, cunning folk, and shamans were well aware of the toxic and medicinal properties of toad venom and knew how to obtain it from live toads that they kept for this purpose. Folk healers of this kind were a common focus for accusations of witchcraft at this time, and it was probably this that gave rise to the popular association between toads and witches.

Further Resources

Cohn, N. (1975). *Europe's inner demons*. New York: Basic Books.

Serpell, J. A. (2002). Guardian spirits or demonic pets: The concept of the witch's familiar in early modern England, 1530–1712. In A. N. H. Craeger & W. C. Jordan (Eds.), *The animal/human boundary* (pp. 157–90). Rochester, NY: Rochester University Press.

Speck, F. G. (1918). Penobscot shamanism. *Memoirs of the American Anthropological Association, 6,* 238–88.

Thomas, K. (1971). *Religion and the decline of magic.* London: Penguin Books.

James A. Serpell

■ Culture, Religion, and Belief Systems
The Zodiac and Animals

Humans are a storytelling species, and the animal kingdom has served as a source of characters for many of these stories. The term "zodiac" literally means "circle of little animals" (Greek: zodiaion—"little animal"; kyklos—"circle"). Perhaps as long ago as 5000 BCE, ancient cultures developed zodiacs to tell stories and to predict future events. This indicates that animals have been inextricably linked to human storytelling, fortune-telling, astrology, and astronomy for a very long time.

Relationship between Animals, the Zodiac, and the Stars

The metaphor is a powerful storytelling device whereby humans take two unrelated aspects of their environment and associate them in order to increase understanding. For example, a person may be described as being a "busy bee," a "wise old owl," or a "sly fox". These metaphors assume that animals have certain stereotyped characteristics and that we recognize these characteristics in the animal. As long as we buy the assumptions, the metaphor provides a vivid image to facilitate a story.

An important aspect of the metaphor is that the person must have an understanding of the associated objects. For instance, early astrologers in Egypt did not name constellations after bears, and Scandinavian astrologers did not name constellations after scarab beetles. Furthermore, the attributes associated with a rat in the Chinese culture may be very different from the attributes associated with a rat in the Greek culture. People relate to the metaphor better if the comparison is made between things that they have experienced. This may be why as cultures grew, intermixed, and evolved, the animals associated with the zodiacs and constellations changed.

One of the most intriguing aspects of the metaphor is that it occasionally switches from being a figurative device to an absolute truth or literal reality. The zodiacs are such a case. For instance, in the Chinese zodiac, the animal-human metaphor conveys predictive power about a person's nature. A person is said to have characteristics of a given animal because the person was born in the year of that animal. In this metaphor, animals, humans, and the universe become more than just figuratively connected. Rather, the zodiac becomes a fortune-telling tool for understanding deep-hidden, cosmic relationships between animals, the stars, and the fate of human lives.

The nature of the night sky reinforces the predictive power of the zodiac metaphor as follows. For species to survive, it is important for them to observe the surrounding environment and determine what is predictable and what is not. Somewhere in our past, we humans became aware of stars and their predictive power. Because of the earth's orbit around the sun, certain stars can only be seen in the night sky during certain seasons. The night sky allowed humans to recognize and predict seasonal patterns and changes.

By adding names to groups of stars, people could recognize large patches of sky, and the concept of constellations was born. Stars became a powerful tool for forecasting weather changes, anticipating the growing and harvesting seasons, and providing a means for navigating at night. The patterns of constellations could be effectively passed on through generations, especially if their recognition was linked to memorable stories.

But the ancients did not distinguish between "correlation" and "causation" where stars were concerned. Certain stars and certain seasons were not just coincidentally appearing together. Rather, the ancients believed that the stars were causing the seasonal changes. And if the stars could cause season changes, it was likely that they could also affect human destiny. Today, people still look to the stars for natural and supernatural answers. In particular, the zodiacal signs were, and for some people still are, believed to carry significant supernatural powers. The zodiac can be used to discover fundamental essential truths about humanity as encoded in animal characteristics and time. However, there is no scientifically sound basis for anything other than a figurative, metaphorical connection between the zodiacs, animal characteristics, and a person's or civilization's fate.

Foundations of the Zodiac

Both Western and Eastern zodiacs exist, and it is likely that other ancient civilizations such as those in Mesoamerica had zodiacs as well. It is not entirely clear precisely how or when the zodiacs originated and how the zodiacs and night constellations became linked. Suffice it to say that the zodiacal elaborations and their ties to animals and constellations began somewhere between 5000 BC and 2000 BC and remain a work in progress.

Both Chinese and Indo-European zodiacs have twelve constellations, signs, or symbols associated with them, and this may be a result of the relationship between lunar cycles and the year, there being about twelve lunar cycles per year. Alternatively, early commerce between Chinese and Indo-European cultures also likely occurred, and so many of the shared zodiacal stories and symbols, including the use of the number twelve, may represent a byproduct of this commerce.

The Greek Zodiac

The Western or Greek zodiac consists of the following twelve members: Aries (the ram), Taurus (the bull), Gemini (the twins), Cancer (the crab), Leo (the lion), Virgo (the maiden), Libra (the scales), Scorpius (the scorpion), Sagittarius (the centaur-archer), Aquarius (the water-bearer), Capricornis (the horned goat or sea goat), and Pisces (the fish). Some of the Greek zodiac members are not animals, but most are. All of the zodiac members stem from Greek myths, as described for the animal members of the zodiac in the following paragraphs.

Aries, the ram, comes from Greek mythology, but the name was used in Persian, Babylonian, and Egyptian mythology as well. The ram was supposed to bear Phrixus, son of Athamas, away from his queen stepmother, Ino. Upon safe arrival, Phrixus sacrificed the ram and hung its fleece up in the garden of Ares, whereupon it turned to gold. This became the Golden Fleece, which Jason and his Argonauts sought in later tales.

Taurus, the bull, represents the form adopted by Zeus when he attempted to seduce Europa. But the bull in general was a sacred animal in ancient times, and there is some evidence that the constellation Taurus may have predated even ancient Greek mythology.

Cancer, the crab, is a dim constellation, and the constellation was various invertebrates at different times and in different cultures. It was given the form of a lobster or water beetle by early European astronomers and was considered a crayfish by Persian astronomers, and in Egypt, it was depicted as a scarab. It was a vertebrate in Babylonia and at one time in Egypt, where the constellation appears to have represented a turtle.

Leo, the lion, represents the lion slain by Hercules as his first of twelve tasks. In myth, this Nemean lion's hide was supposed to be impenetrable, making it impossible to kill him with weapons, even for Hercules. Hercules only succeeds in killing the animal by strangling it, and then he has to use its own claws to remove the hide. Hercules wears the hide, making him nearly indestructible. It is noteworthy that this constellation is known as the lion in Persian, Syrian, Jewish, Babylonian, Turkish, and Greek astronomy. It is said that real lions were drawn to the Nile when the sun was in this constellation—hence, a possible non-mythological explanation for why this constellation was so named.

Scorpius represents the scorpion sent by Gaia, Hera, or Apollo (depending on the myth) to kill Orion, the hunter. In the night sky, Scorpius rises in the East as Orion sets in the west, suggesting that the hunter is still running away from the stinging beast. The tail of the scorpion is easy to identify in this constellation.

Sagittarius is a centaur, or man with a horse's body. There are two constellations in the sky representing the same mythical beast, Sagittarius and Centaurus. Centaurs were considered wild, uncivilized, and lawless creatures because of their animal natures. In ancient Babylonia, the centaur was anatomically given even more animal parts, having wings and a lion's head.

Capricornis, the horned goat, is also known as the sea goat because it is located in the part of the sky that represents water. It is among the oldest if not the oldest constellation known. It is often depicted as having a fish tail. Capricornis may have been the goat that suckled Zeus as a child. Another myth suggests that Capricornis was the god, Pan, who leapt into the Nile River to avoid an attack by the monster Typhon. The upper part of Pan, which remained above water and dry, stayed goat, whereas the submerged part of Pan became a fish. The Capricornis's broken horn became the cornucopia, the horn of plenty.

Pisces, the fish, is a long, meandering constellation, which may represent two fish. In one Greek myth, these fish represent Aphrodite and her son, Eros, who have transformed themselves into fish to avoid the monster Typhon. The constellation was not always recognized as fish, however, and earlier mythology places these stars with one of the twelve tasks of Hercules involving the capture of a three-headed dog at the gate of Hades.

The Chinese Zodiac

The Chinese zodiac consists of twelve animals, which in order are the rat, ox, tiger, rabbit, dragon, snake, horse, sheep (or goat), monkey, rooster, dog, and pig (or boar). Chinese astrology, which is far more complex than that typically depicted in the West, pairs these animals with elements, times of the month, and times of the day to describe various aspects of a personality. The Chinese zodiac has been used by other Asian cultures (although the cat replaces the rabbit in the Vietnamese zodiac) and even by early Bulgarian Pagans.

The mythological story of how the Chinese zodiac came to be is told in the following extracts:

The rat was given the task of inviting the animals to report to the Jade Emperor [ruler of Heaven] for a banquet to be selected for the zodiac signs. The cat was a good friend of the rat, but the rat tricked him into believing that the banquet was the next day. The cat slept through the banquet, thinking that it was the next day. When he found out, the cat vowed to be the rat's natural enemy for ages to come.

Another popular legend has it that the Jade Emperor decided to have a race to determine which animals would be in the zodiac. In order to participate, the animals had to cross a river. The cat and rat, being good friends, decided that the best and fastest way to cross the river was to hop on the back of the ox. The ox, being a naïve and good-natured animal, agreed to carry them across. However . . . the crafty rat decided that in order to win, it . . . [had to] push the cat into the river. Because of this, the cat has never forgiven the rat and . . . hated the water [ever since].

After the ox had crossed the river, the rat jumped ahead and reached the shore first, and it cleverly claimed first place in the race! Following closely behind was the strong ox, and it was named the 2nd animal in the zodiac.

After the ox, came the tiger, panting away while explaining to the emperor just how difficult it was to cross the river with the heavy currents pushing it downstream all the time. But with powerful strength, it made [it] to shore and was named the 3rd animal in the cycle.

Suddenly, from a distance came a thumping sound and out popped the rabbit. It explained how it crossed the river: by jumping from one stone to another in a nimble fashion. Halfway through, it almost lost the race, but the rabbit was lucky enough to grab hold of a floating log that later washed him to shore. For that, it became the 4th animal in the zodiac cycle.

Coming in 5th place was the gallant dragon, flying and belching fire into the air. Of course the Emperor was deeply curious as to why a strong and flying creature such as the dragon should fail to [finish] first. The mighty dragon explained that he had to stop and make rain to help all the people and creatures of the earth... Then on his way to the finish line, he saw a little helpless rabbit clinging to a log, so he . . . gave a puff of breath to the poor creature so that it could land on the shore. The emperor was very pleased with the actions of the dragon and he was added into the zodiac cycle.

As soon as [the Emperor] had done so, a galloping sound was heard and the horse appeared. Hidden on the horse's hoof was the . . . sneaky snake, whose sudden appearance gave the horse a fright, thus making it fall back and giving the snake 6th spot whilst the horse took the 7th.

Not long after that . . . the sheep, monkey and rooster came to the shore. These three creatures helped each other . . . . The rooster spotted a raft, and took the other two animals [along] with it. Together, the sheep and the monkey cleared the weeds, tugged and pulled and finally got the raft to the shore. Because of their combined efforts, the Emperor was very pleased and promptly named the sheep as the 8th creature, the monkey as the 9th, and the rooster the 10th.

The 11th animal was the dog. His explanation for being late, although he was supposed to be the best swimmer amongst the rest, was that he needed a good bath after a long spell, and the fresh water from the river was too big a temptation. For that, he almost didn't make it to [the] finish line.

Just as the emperor was about to call it a day, an oink and squeal was heard from a little pig. The term "lazy pig" is due here as the pig got hungry during the race, promptly stopped for a feast, [and] then fell asleep. After the nap, the pig continued the race and was named the 12th and last animal of the zodiac cycle. The cat finished too late (thirteenth) to win any place in the calendar, and vowed to be the enemy of the rat forevermore.

In . . . another variation, each animal was called before its peers and had to explain why it deserved a position at the top of the Zodiac. The Boar, at a loss, proceeded to claim that the meat on its bones "tasted good." This explanation was apparently considered unsatisfactory, because the Boar was placed at the very end of the Zodiac.

The Chinese zodiac does not correspond to the Greek constellation regions of the sky at all. Rather, the regions of the sky occupied by the Greek zodiacal constellations are divided into four segments in ancient Chinese astronomy. These regions are represented by four animals, the Azure Dragon of the East, the Black Tortoise of the North, the White Tiger of the West, and the Vermillion Bird of the South.

Constellations and Animals

The skies are populated by animal stories. There are presently eighty-eight named and recognized constellations. Each constellation occupies its own territory in the sky, and jointly, all eighty-eight constellations exhaustively span the sky, north to south, east to west. Exactly half of the constellations represent real or mythical animal creatures. Constellations representing real creatures include invertebrates (a fly, scorpion, and crab); fish (including a flying fish); reptiles (snakes, a chameleon, and a lizard); birds (including an eagle, crane, bird of paradise, peacock, crow, dove, swan, and toucan); and numerous mammals (giraffe, bears, wolf, fox, bull, lions, lynx, hare, horse, dolphin, dogs, and a ram). The mythical animal constellations include those in the Greek zodiac, mentioned previously, as well as a dragon, unicorn, phoenix, sea monster, and winged horse. Three more constellations are humans whose occupations are strongly associated with animals—Boötes the herdsman, Orion the hunter, and Ophiuchus the snake bearer. Clearly, the animal kingdom has played a central role in human stories that have been immortalized in the constellations. The rich details of many of these stories are lost to the ages, but many delightful remnants remain for the interested reader to discover (see Further Resources).

The Chinese zodiac is depicted on this elaborate bronze disc. ©Csaba Fikker/ Dreamstime.com.

Naming of Constellations

Two major groups of astronomers are responsible for naming, compiling, condensing or revising the constellations. The first group consisted of astronomers from ancient cultures, including the Greeks, Romans, Babylonians, Egyptians, Chinese, and Scandinavians. These were the ancient cultures that took serious steps to map the stars. An Egyptian-Greek astronomer named Ptolemy, living in the second century CE, was responsible for compiling forty-eight of the "modern" constellations from works of the Indo-European astronomers. Ptolemy may have revised these constellations from the work of Hipparchus, a Greek astronomer who lived several centuries earlier than he.

The second group of astronomers consisted of explorers in the 1500s to 1800s. These explorers were looking to fill in the gaps in the star map for easier navigation, and they documented "new" constellations in so doing. Johann Bayer published a book in 1603 called *Oranometra* that contained twelve new constellations and four modified ones. Of the twelve new constellations, ten were named after animals. *Usus Astronomicus Planisphaerii Stellati* was published in 1679 by Jakob Bartsch, who added two new constellations named after a unicorn and a giraffe. In 1690 Johannes Hevelius added ten new constellations, seven that were animal-related; three of these are no longer recognized. The final astronomer to add or change constellations' names was Nicolas Louis de Lacaille in 1763. He published a book called *Coelum Australe Stelliferum*, in which he outlined fourteen new constellations, split two constellations, and altered two more. He was the first and only astronomer not to use animal references. Most of his new constellation names were tools or boat parts.

The Anatomical Man, *from* Les Très Riches Heures du duc de Berry, *shows the relationships between areas of the body and astrological entities as imagined in the Middle Ages. ©The Art Archive/Victoria and Albert Museum London/Eileen Tweedy.*

Further Resources

Aveni, A. A. (1993). *Ancient astronomers*. Montreal: St. Remy Press.

Brundige, E. N. (1995). *Inventing the solar system: Early Greek scientists struggle to explain how the heavens move.* Retrieved from Tufts University, Greek Science, Web site: http://www.perseus.tufts.edu/GreekScience/Students/Ellen/EarlyGkAstronomy.html

Chinese astrology. http://en.wikipedia.org/wiki/Chinese_zodiac

Kaler, J. (n.d.). *The constellations*. Retrieved from University of Illinois, Urbana-Champaign, Astronomy Department Web site: http://www.astro.uiuc.edu/~kaler/sow/const.html

Legg Middle School Planetarium. (n.d.). *Constellation mythology*. Retrieved from http://www. coldwater.k12.mi.us/lms/planetarium/myth/index.html

Lim, A. (2003). *Legend of the Chinese zodiac*. Retrieved from http://www.thingsasian.com/goto_article/ article.2137.html

Ridpath, I., & Tirion, W. (1985). *Universe guide to stars and planets*. New York: Universe Press.

Snodgrass, M. E. (1997). *Signs of the zodiac: A reference guide to historical, mythological and cultural associations*. Westport, CT: Greenwood Press.

Geoffrey E. Gerstner and Ben Dirlikov

Disease.
See Health

Domestication
Dog Breeds

Where Did That Dog Get Its Breeding?

The time and place, or places, of the dog's divergence from the wolf remain the subject of intense debate, with estimates, based on genetic analysis, of around 15,000 years ago in East Asia from a single population of wolves; 27,000 years ago from one or more small populations; and 40,000 to 135,000 years ago in as many as four different locations. Archaeologists generally support the most recent date because it coincides with the available archaeological record. The older dates are based on assumed rates of genetic mutation in the dog.

Certainly, since hitching its evolutionary fate to humans, the dog has proved the most useful and malleable of creatures, traveling with its people to every corner of the globe and adapting to all their social transformations. Dogs have joined in the move from hunting and gathering to agricultural, industrial, urban, and sprawling suburban societies.

During that journey, through what Charles Darwin called "methodical," or "conscious" and "unconscious" selection, humans have suppressed, emphasized, and even remixed in often subtle ways various wolfish traits to create the roughly 400 breeds of dog extant today, out of perhaps as many as 1,000 that have existed worldwide. These include dogs that are smaller, larger, and more varied in shape and appearance, more specialized in behavior, and more adaptable to human society than their wolf forebears. There are midget dogs, dwarf dogs, and giant dogs; hairless dogs and dogs with dense corded coats, not to mention all styles between; dogs that look like perpetual puppies and dogs that look like wolves; brachycephalic dogs whose extremely shortened noses give them flatter, more human-looking faces; and dogs with Cyrano de Bergerac noses.

The Origins of "Scientific Breeding"

The breed estimates are broad not only because an unknown number have vanished but also because dog fanciers and scientists disagree among themselves over what constitutes a breed. The modern definition dates from the growth of "scientific breeding" in the eighteenth and nineteenth centuries, when people in England, Europe, and North America began self-consciously to work at improving their dogs and livestock. They came to define a breed according to a written standard, describing its ideal physical and behavioral characteristics, and a registry, or studbook, recording the genealogy of each dog. A certificate of pedigree and conformity to the standard were required for membership in the breed. Although the studbook is usually closed soon after the breed is established, it can be, and on rare occasions is, reopened by breeders to incorporate fresh bloodlines.

This definition of breeds is wrapped up in nineteenth century notions of blood purity, racism, and eugenics—the belief that quality is in the "blood" and can be passed on or diluted through breeding; that well-bred people or animals are superior in all regards to their rustic, low-born kin; and that individuals can improve on nature by breeding the "best" to the "best"—in animals and people. Wealthy sportsmen formed kennel clubs for sponsoring dog shows where their dogs competed for top honors, as the "best in breed" or "best in show," and other events. In turn, the successful breeder derived status from his champion.

Breeds were most commonly formed—and still are—through consolidation of several existing, similar types of dog; through deliberate crossings of different types or breeds; and through splitting an established breed on the basis of coat texture, ear shape, or some other physical characteristic. To fix the traits they desire, breeders rely on intensive inbreeding of a small number of founders and the use of "favored sires," who are bred repeatedly. The resulting dog is accorded the finest abilities and physical characteristics of the type it represents. Thus, the purebred sheepdog or bloodhound is deemed superior to its rustic progenitors, although prior to their consolidation into a breed, that original stock might have been praised for its sagacity and intelligence.

By this modern definition, a distinctive group of dogs with unique traits cannot be considered a breed, even if they breed "true" and are reproductively isolated, because there is neither a standard nor registry to guarantee the purity of their bloodlines. They can only be a "type." The Alaskan husky, star of sled dog racing, is a "type" of "purpose-bred" dog because although it generally fits a Northern phenotype with tough feet, a double coat, and a capacity to pull hard and run at a fast aerobic pace, there is no registry or written standard. Breeders regularly bring other types of dogs into the Alaskan husky gene pool to improve performance.

Similarly, if a breed is created from a broader population—for example, the Canaan dog from the so-called pariahs common in Palestine, the basenji from the hunting dogs of Central Africa, or the Catahoula leopard dog from the leopard curs of the Gulf of Mexico coastal states and the broader gene pool of curs—the remaining indigenous dogs are not considered members of the breed. Or various kennel clubs will disagree on the breed standard or some other issue and refuse to cross-register dogs from another club.

Thus, a more expansive definition essentially says that a "breed" is a cultural and biological entity, a reproducing group of domesticated animals created, defined, and maintained by humans.

Over the years, individuals and kennel clubs have attempted to group dog breeds broadly by their functions: hunting by sight and high-speed pursuit, hauling, herding, retrieving, hunting by scent, guarding, fighting, rousting game from a den, pointing, and serving as companions. Northern dogs or huskies or Eskimo dogs are often placed in a separate category. As society has changed, dogs, with an assist from their human companions, have turned their talents from those traditional tasks to many different purposes—serving the disabled, competing in adventure races, running agility courses, chasing Frisbees, and detecting explosives and other objects.

Genetics and Dog Breeding

Although their tasks might have changed, the underlying behaviors have not. Each dog breed represents "one of many individual behavioral variations" found in the ancestral wolf, said J. P. Scott and John L. Fuller in their landmark 1965 book, *Genetics and the Social Behavior of the Dog*. Thus, terriers are often more aggressive than wolves, fighting with minimal provocation from an early age, whereas hounds retain a wolfish talent for hunting sociably in packs, and border collies and certain other herding dogs and pointers show an extreme prey "stalking" behavior, including intent staring or "showing eye."

The researchers also demonstrated that breeding does not guarantee a dog's behavior and physical abilities; it simply increases the odds that a Labrador retriever will like to fetch objects and swim, for example. Indeed, a central paradox of the dog world lies in this observation: there is more variation in behavior and temperament among dogs of the same breed than there is between breeds of dog.

On a fundamental level, there is a link between morphology, the form and structure, and behavior. A basset hound with its short, bowed legs cannot run as fast as a long-legged coonhound, for example, while the long-legged hound cannot roust a fox from its den the way a little terrier can. Factors such as the shape of a dog's head, whether it has floppy ears or no tail, and the size of its dewlaps, the number of folds or wrinkles in its skin, and the length of its hair affect its ability to communicate. Moreover, while dogs have developed the wolf's little-used bark to a high, if sometimes grating, form of communication, some breeds—basenjis, for example—bark little, if at all, and others do not howl or in other ways lack a full range of vocalization. Like physical and behavioral characteristics, approximately 400 genetic diseases, including cytoskeletal abnormalities, also are associated with specific breeds. Dogs share more than 350 of those conditions with humans.

Geneticists have recently shown that a number of physical characteristics are inherited together—because, it is assumed, the genes affecting them are closely positioned on the dog genome. Thus, breeding two Australian shepherds with distinctive merle coats increases the odds of producing puppies with merle coats—and deafness. Other genes that move together seem to work through "trade-off mechanisms" to affect the cytoskeleton. For example, University of Utah geneticist Karl Gordon Lark has shown that one genetic locus controls muzzle length and leg thickness in such a way that as the snout becomes elongated, the limb bones become less thick and more suited for running. This mechanism seems to account for the stout limbs, short muzzle, and powerful jaws of the mastiff as well as the long nose and long, thin legs of the hound.

An analysis of the dog genome, published in the journal *Nature*, on December 8, 2005, revealed that after passing through the foundation genetic bottleneck, early dogs bred rather freely. Because in sexual reproduction, the chromosomes from the sire and the dam are recombined to form the single chromosome passed on from each and because alignments over time become misaligned due to mutation and genetic drift, haplotypes—large segments of DNA that can be as large as a chromosome and are identical on both chromosomes in a pair—can be broken up, becoming shorter and more scattered over the generations of random breeding.

Most breeds also pass through a genetic bottleneck at their formation, and so each breed has a distinctive pattern of large homogeneous haplotype blocks and shorter heterogeneous ones, which selective breeding has sustained. Although each breed shows a distinctive combination of haplotypes, it also shares its haplotypes with other breeds. The proportion of homogeneity to heterogeneity within each breed is nearly the same across breeds—about 62 percent to 38 percent, respectively.

Through inbreeding from a small gene pool and the overuse of "favored sires" to create their particular breed, humans unknowingly selected a small group of overlapping chromosomes carrying the genes for the traits they wanted—and some diseases they did not want.

Over the years, a number of theories have been put forth to account for the great diversity of breeds. A few have been discredited; to date, none have been proved. Darwin and others proposed evolution of "Southern" breeds from the jackal and "Northern" breeds from the wolf. Since confirmation of the wolf as the ancestor of all dogs, some experts have speculated that a large portion of the dog's genetic diversity derived from the different wolf populations that gave rise to it and regular, if erratic, interbreeding. Another group of experts holds that if domestication involved only one population of wolves, then something inherent in the process itself must be at work (a view that does not preclude multiple origins, because presumably, the molecular mechanisms involved in domestication would, at least, be similar).

A popular hypothesis posits natural and artificial selection for tameness, or a similar characteristic produced over the course of generations, changes in the timing and phasing of the dog's physical development, called "heterochrony," and in the levels of hormones and neurotransmitters affecting aggression and fright responses.

Delayed physical development relative to sexual maturation (called "paedomorphosis") is believed to produce dogs with more domed heads; shorter, broader muzzles; and overall reduced size and slighter build than the wolf forebear. When maturation is stopped early enough, the resulting animal resembles a "neotenic," or perpetually juvenile dog, and in the past 150 years, as the dog has moved from city to suburbs, breeders have created dogs that look, when grown, like puppies. Accelerated physical development relative to sexual maturation (hypermorphosis), on the other hand, produces dogs larger than the progenitor wolf.

Another theory suggests that domestication was a dynamic process involving wolves and people and that the early dog probably remained nearly indistinguishable from the wolf until people began to move into villages, taking their dogs with them and limiting their access to wolves for breeding. The smaller gene pool forced inbreeding that along with changing environmental conditions somehow "destabilized" the genome, thereby freeing the diversity inherent in it. The dog merely represents variations, some of them freakish, others more common, such as dwarfism and giantism, which people have maintained through artificial selection.

Some of the mutations might involve just a few dominant genes, which become quickly established once they occur: for example, a red or yellow coat, a curved tail and floppy ears, and some forms of dwarfism. Other mutations are recessive and require close breeding to fix. Hairlessness, on the other hand, is a lethal dominant gene, so that breeding two hairless dogs is an exercise in death. The breed is maintained only by breeding a hairless dog to a hair-bearing carrier. Other mutations are multifactoral, involving a number of genes in a complex dance, influenced by environmental and nutritional as well as biological factors.

John Fondon III and Harold Garner proposed in the December 28, 2004, issue of the *Proceedings of the National Academy of Sciences* that the length of short repetitive sequences of DNA in certain genes, called tandem repeats and formerly considered "junk DNA," were responsible for many of the morphological features of purebred dogs, such as snout length and inclination. In their view, breeders quickly capture and through the use of favored sires and inbreeding spread through their breed a naturally occurring mutation that underlies a morphological change.

In the December 2005 issue of *Genome Research*, Wei Wang and Ewen F. Kirkness of the Institute for Genomic Research proposed that slightly longer repetitive sequences of DNA, called SINE elements, were a primary source of genetic and phenotypic variation in dogs. Research has already shown that when inserted into genes, SINE elements can cause diseases, such as narcolepsy in Doberman pinschers and centronuclear myopathy (a congenital muscle disease) in Labrador retrievers.

These theories have aroused cautious interest among other canine geneticists, but no one knows which is correct, if any.

The History of Dog Breeds

The early history of dogs is no more clear. Most experts assume that from the beginning, dogs have bred according to their rules, whenever possible, because that is the nature of dogs, and that one of the early basic types of dogs, if not the only one, was much like today's pariah or dingo, prick-eared, short-coated, lupine in movement and demeanor, and with a coat ranging across the full spectrum of colors. But also from an early date, even without multiple origins, small groups of dogs that became reproductively isolated while wandering with their people could develop distinguishing features and characteristics through mutation and inbreeding.

It is difficult to determine how great these differences were. At a basic level, there was probably an early division between large and small dogs, but other splits may also have occurred. In a 2004 paper for *Science*, Elaine Ostrander and her colleagues, while documenting how on the basis of an analysis of nuclear DNA it is possible to identify a dog's breed with 99 percent accuracy, divided the eighty-five breeds they used into four major clusters—"ancient breeds"; mastiff types; herding, or working, dogs; and hunting dogs, including terriers. Most of the hunting dog breeds came into existence during the past 200 to 300 years. The "ancient breeds" apparently had not interbred with other dogs much since their appearance. That group, which may have gone back at least 2,000 years, included the akita and shiba inu of Japan; the shar pei, shih tzu, Pekingese, and chow chow of China; the Tibetan terrier and lhasa apso of Tibet; the Siberian husky and samoyed of Siberia; the Alaskan malamute; the saluki and afghan hounds of the Middle East; and the central African basenji.

Some anthropologists believe that, by the time they began domesticating other animals with their dogs' help, ancient agriculturalists some 10,000 years ago may have begun to sort their dogs according to their appearance and abilities. The evidence for that occurring wherever people were creating more structured societies, however, is slim. On the other hand, hunting and gathering people were clearly moving into the Americas with their dogs, who began to differentiate almost immediately, if they were not already divided into distinct types.

By the dawn of civilization, the situation was clearer. There were in Egypt sight hounds, mastiffs, companion dogs, small hounds resembling the basenji, and basset-like hounds about 5,000 years ago. Around the same time, according to Frederick E. Zeuner in his 1963 study *A History of Domesticated Animals*, hounds and sheepdogs appeared in Europe and England, and the ancient Greeks sorted their big dogs by color. They used white dogs to guard the flock, in the belief they wouldn't scare the sheep, and black dogs to protect the home, with the knowledge that they were more likely to scare intruders. The distinction aside, white and black dogs frequently came from the same litter.

The Romans recognized companion, war, hunting, draft, and guard dogs, as well as scent and sight hounds—the basic divisions that exist today among Anglo-Euro-American dogs. But dogs were often sorted according to their predilections, size, and color, as much as breeding. That rule seems to have held around the world. Various societies had their own dog or dogs, who served their purposes and bred more or less to type. In what is now Mexico, around 250 BCE, the Colima appear to have isolated a hairless dog, which the Aztec later named *xoloitzcunitli*. At an unknown date, on islands in Puget Sound, the Clallam Indians kept a dog solely for its wool-like black and white hair, which they spun into clothes and blankets, until the arrival through trade of the Hudson's Bay Company's woolen goods.

Pekingese were the constant companions of Chinese emperors for hundreds of years. They gained fame in the West when a pair was presented to the Duchess of Richmond; they were later to become the foundation of the breed in England. Courtesy of Shutterstock.

The Crusades exposed Europeans to sight hounds and other dogs from the Middle East, and returning Crusaders brought some back to Europe. Through the Middle Ages and into the Renaissance, a few monasteries and noblemen maintained their own lines of hunting hounds, war dogs, and companion animals, some of which became famous, like the Saint Hubert bloodhound from the eponymous Benedictine monastery in Belgium and the Talbot hound from England—or infamous. By the fifteenth century, the Spanish were keeping large livestock-guarding dogs, smaller herding dogs, fighting mastiffs, and greyhounds and were concerned about maintaining the blood purity of their lines. They deployed their war dogs to devastating effect in the conquest of the New World.

In 1920 Harvard zoologist Glover M. Allen estimated that there were three main types of Native American dog—the Eskimo dog; smaller wolf-like dogs; and much smaller dogs in the Southwest, Southern Mexico, and Caribbean. He divided those into seventeen distinctive types, while saying that it was no longer possible to find purebred Native American dogs because of interbreeding with those from Europe.

By the time Johannes Caius attempted to classify British dogs in 1576, breeds were proliferating, often through crossbreeding. Caius's breeds had to keep their "type" when reproducing. Those that did included the sight and scent hounds; the poacher's Tumbler and "Theevishe dogge," which would today be called curs; spaniels; fowling dogs; comforters; French dogs; setters; finders, or archaic poodles; and rustic dogs of the "courser sort"—the sheepdogs and mastiffs.

Some of those dogs could trace their existence to the nobility they served, and others were kept by the peasantry for poaching and protection. By the late eighteenth and nineteenth centuries, gun dogs were being developed, and the drive to improve breeds of livestock and dogs gathered momentum. In 1859 the Kennel Club of England was founded and along with it the sanctity of the breed standard and registry.

The "fancy," or the sport of purebred dogs, remained largely an upper-class enterprise, in no small measure because of the costs involved. The common free-breeding or purpose-bred dog continued to represent more than 95 percent of the dogs in the county. One of the most common was the big yellow dog, a cur; another was the generic little dog—a feist in the South.

Since World War II, purebred dogs have gone from representing less than 5 percent of all dogs in America to more than 50 percent, and their numbers are on the rise around the world. The Labrador retriever has become the most common "big" dog.

The overall increase is largely a result of the status that having a purebred dog is believed to impart—a phenomenon that helped inspire Thorstein Veblen to coin the phrase "conspicuous consumption" in his 1899 book *The Theory of the Leisure Class*—and of the (erroneous) belief that a purebred dog will behave according to the standards of its breed.

This proliferation of purebred dogs has brought an epidemic of behavioral and health problems, many of them congenital. Finding the genes associated with those diseases, many of which are shared with humans, is a major goal of a number of dog geneticists. They also hope to use the dog genome to locate genes responsible for specific behaviors and physical characteristics of the various breeds and individual dogs, as well as for the transformation from wolf to dog. Correcting problems to create the most sound dog possible is something only breeders can do.

Further Resources

Caius, J. (1576/1969). *A treatise of Englishe dogges*. Amsterdam: Theatrum Orbis Terrarum and New York: Da Capo Press.

Derr, M. (1997). *Dog's best friend: Annals of the dog-human relationship*. New York: Holt.

———. (2004). *A dog's history of America: How our best friend explored, conquered, and settled a continent.* New York: North Point Press.

Lindblad-Toh, K., et al. (2005). Genome sequence, comparative analysis and haplotype structure of the domestic dog. *Nature, 438*(8), 803–19.

Morey, D. F. (2006). Burying the evidence: The social bond between dogs and people. *Journal of Archaeological Science, 33*(2), 158–75.

Parker, H. G., Kim, L., Sutter, N., Carlson, S., Lorentzen, T. D., Malek, T. B., Johnson, G. S., DeFrance, H. B., Ostrander, E. A., & Kruglyak, L. (2004). Genetic structure of the purebred dog. *Science, 304,* 1160–64.

Parker, H. G., & Ostrander, E. A. (2005). Canine genomics and genetics: Running with the pack. *PLOS Genetics, 1*(5), 507–13.

Ritvo, H. (1987). *The animal estate: The English and other creatures in the Victorian Age.* Cambridge, MA: Harvard University Press.

Scott, J. P., & Fuller, J. L. (1965). *Genetics and the social behavior of the dog.* Chicago: University of Chicago Press.

Wayne, R. K., & Vilà, C. (2001). Phylogeny and origin of the domestic dog. In A. Ruvinsky & J. Sampson (Eds.), *The genetics of the dog.* Cambridge, MA: CAB International.

Zeuner, F. E. (1963). *A history of domesticated animals.* New York: Harper & Row.

Mark Derr

■ Domestication
The Domestication Process: The Wild and the Tame

In the Western world today, animals are divided into three basic groups, the wild, the tame, and the domestic, but these divisions are fluid and more interchangeable than they seem at first. It is difficult to define what is a wild and what is a domestic animal. A wild animal is usually thought of as one that is fearful of humans and that runs away if it can. But this fear of humans is in itself a behavioral pattern that has been learned from experience of human predation over countless generations. A "wild" animal that has no contact with humans has no fear of them and can be quickly exterminated, as was the dodo on Mauritius. This large flightless bird evolved without any predators, so when Portuguese sailors landed on the island for the first time in about 1507, they only had to knock the dodos on the head to get much-needed fresh meat.

However, for perhaps the past 150,000 years, humans have become so supremely successful at killing other species that there are rather few "wild" animals left on Earth that do not attempt to escape from us as the master predator. On the other hand, it is remarkable how many species of wild animals can be tamed, and taming is not a modern phenomenon; it has probably always been a very important and essential part of human behavior and an adjunct to hunting. Young animals whose mothers were killed in the hunt would have been nurtured and reared by people, and it is not only in modern times that wild animals were captured and tamed as symbols of status, as shown by this anecdote recorded by the Greek writer Diodorus Siculus and written in the first century BCE (Oldfather, 1979, p. 2187). It is about the capture of a python in ancient Egypt for King Ptolemy's zoo in the middle of the third century BCE:

> Observing the princely generosity of the King in the matter of the rewards he gave, some hunters decided to hazard their lives and to capture one of the huge snakes and bring it

alive to Ptolemy at Alexandria. . . . They spied one of the snakes, 30 cubits long, as it loitered near the pools in which the water collects; here it maintained for most of the time its coiled body motionless. . . . and so, since the beast was long and slender and sluggish in nature, hoping that they could master it with nooses and ropes, they approached it the first time, having ready to hand everything which they might need . . . but the beast, the moment the rope touched its body whirled about and killed two of the men.

Nevertheless the hunters did not give up . . . They fashioned a circular thing woven of reeds closely set together, in general shape resembling a fisherman's creed and in size and capacity capable of holding the bulk of the beast . . . and so soon as it had started out to prey upon the other animals as was its custom, they stopped the opening of its old hole with large stones and earth and digging an underground cavity near its lair they set the woven net in it and placed the mouth of the net opposite the opening . . . And when it came near the opening which had been stopped up, the whole throng, acting together, raised a mighty din and so it was caught.

When they had brought the snake to Alexandria they presented it to the king . . . and by depriving the beast of its food they wore down its spirit and little by little tamed it, so that the domestication of it became a thing of wonder. (Book III, p. 36)

The Process of Domestication

In one sense, it can be said that a domestic animal is just one that has lost its fear of humans like that snake, but true domestication involves much more than this.

The process of domestication is subject to two profound, overriding and interlocking influences, the biological and the cultural (Clutton-Brock, 1999a). The biological process of domestication begins when a small number of animals are separated from the wild species and become so tame that they lose all fear of the humans around them and are said to be habituated. For domestication to follow from taming, the animals have to go through a series of morphological and behavioral changes, which in mammals broadly follow the same pattern in succeeding generations, irrespective of the species. In general, the characteristics of the juvenile animal are retained into the adult state, a process that is known as neotony. Thus, domestication of the wolf, the wild cat, the wild sheep, and the wild boar all led in the initial stage to reduction in size of the skull, skeleton, and brain. This was followed in succeeding generations by an increase in the proportion of fat to muscle in the body, to changes in the coat and in the carriage of the ears and tail, and to loss of the "wild" temperament.

When a small population of animals that have undergone the first stages of domestication is bred over many years in isolation from the wild population, it may form a founder group that is changed both in response to natural selection under the new regime of the human community and its environment and by artificial selection for economic, cultural, or aesthetic reasons.

Once a species of animal has become fully domesticated—as with the domestic dog, *Canis familiaris*—new breeds are produced by further reproductive isolation. The founders of the new breed contain only a small fraction of the total variation of the parent species, and it becomes a genetically unique population that continues to evolve under natural and artificial selection. At any point, the process can begin again, and further new breeds can be developed by crossbreeding. A breed can be defined in the following way: a group of animals that has been bred by humans to possess uniform characters that are heritable and that distinguish the group from other animals within the same domestic species.

There are many anomalies in the interface between the wild and the domestic. For example, domestic rats, mice, and rabbits can be adored animal companions and laboratory animals that are highly valued for medical research, but their wild counterparts are universally treated as vermin and killed on sight.

The Cultural Process of Domestication

The second fundamental side to the process of domestication is the equally important cultural process, which affects both the human domesticator and the animal domesticate. Domestication begins with ownership. In order to be domesticated, animals have to be incorporated into the social structure of a human community and become objects of ownership, inheritance, purchase, and exchange. The relationship between human and animal is transformed from one of mutual trust in which the environment and its resources are shared to total human control and domination.

The process of taming a wild animal, whether it is a wolf or a wild goat, can be seen as changing its culture. The term "culture" has many meanings, but here it can be defined as a way of life imposed over successive generations on a society of humans or animals by its elders. Where the society includes both humans and animals, the humans act as the elders.

The animal is removed from where, in the wild, it would learn from birth either to hunt or to flee on sight from any potential predator. The tamed animal is brought into a protected place where it has to learn a whole new set of social relationships as well as new feeding and reproductive strategies, and under domestication, this "culture" is passed down from generation to generation.

A domestic animal is a cultural artifact of human society, but it also has its own culture, which can develop, say in a cow, either as part of the society of nomadic pastoralists or as a unit in a factory farm. Domestic animals live in many of the same diverse cultures as humans, and their learned behavior has to be responsive to a great range of different ways of life. In fact, so closely do many domestic animals fit with human cultures that they seem to have lost all links with their wild progenitors. The more social or gregarious the progenitors are in their natural behavioral patterns, the more versatile the domesticates will be, with the dog being the earliest animal to be domesticated (around 14,000 years ago) and an extreme example of an animal whose culture has become humanized.

It is not fully understood why the wide-scale domestication of livestock animals, these being sheep, goats, cattle, pigs, and equids in the Old World and camelids in South America, occurred progressively from 8,000 years ago, but this was the basis of the so-called Neolithic revolution when the fundamental change in human societies occurred and groups of hunter-gatherers became farmers and stockbreeders. Archaeologists in the past have hypothesized that there was a natural progression first from generalized or broad-spectrum hunting in the Paleolithic, at the end of the last ice age, to specialized hunting and herd following, say of reindeer or llama. It was believed that this stage was then followed by control and management of the herds, and then there was a move to controlled breeding and finally to artificial selection for favored characteristics. However, the sequence would very rarely have been so smooth, for the social implications of ownership by a social group of hunter-gatherers are a bigger hurdle to domestication than they may seem. Many hunter-gatherer societies that could have domesticated animals never did so, and this was probably for cultural as much as for many other complicated reasons. Why, for example, was the big horn sheep never domesticated in North America?

Tim Ingold has argued that for hunter-gatherer societies, there is no conceptual distance between humanity and nature, and the boundary is easily crossed. The animals in the environment of the hunter act with the hunter in mind and present themselves to him. The hunter believes that if he is good to the animals, they will be good to him, and if he maltreats them, the animals will desert him. Animals to be hunted are not seen as "wild," but as individuals that allow themselves to be taken. The best-known survival of this belief is seen with the Ainu of Hokkaido (Japan), who still practice a bear sacrifice in which a bear cub is nurtured for months and then killed in an elaborate and ancient ritual.

In the pre-domestication world, humans and animals lived in mutual trust, but all was changed by the herding of animals and even more so by full domestication. Herdsmen do care for their animals, but it is quite different from the care of the hunter because equality is lost, and domination takes over from trust. By 8,000 years ago, domination of the natural world was already well under way, and by the period of the ancient Egyptians and the capture of the python described previously, agriculture and the breeding of livestock were the established foundations of all the ancient civilizations of the Old World. The transformation in attitudes toward the animal world from those of the hunter-gatherer to those of the farmer and stockbreeder was epitomized by Aristotle (384–322 BCE), who wrote about more than 500 kinds of animals, all of which, he believed, existed for the sake of men (Clutton-Brock, 1999b). This belief that the world exists for the benefit of humans has persisted until the present day and is imbued in the worldwide sport of hunting. But the wild places and their faunas are shrinking fast, and increasingly in the future, biologists will have to tackle the great problems of their conservation and management. Whether these faunas include African elephants, Asian lions, or giant tortoises, they are all becoming increasingly hedged in. In order to survive, the wild will have to merge with the tame, and as a result of morphological and behavioral changes brought about by human ownership and control, the "wildlife" may even become domesticated.

Sami herder Nils Peter lassos a reindeer in a corral near Sapmi in northern Norway.
©Bryan & Cherry Alexander Photography/Alamy.

Further Resources

Clutton-Brock, J. (1999a). *A natural history of domesticated mammals* (2nd ed.). New York: Cambridge University Press/The Natural History Museum.

———. (1999b). Aristotle, the scale of nature, and modern attitudes to animals. In A. Mack (Ed.), *Humans and other animals* (pp. 5–24). Columbus: Ohio State University Press.

Ingold, T. (1994). From trust to domination: An alternative history of human-animal relations. In A. Manning & J. Serpell (Eds.), *Animals and human society: Changing perspectives* (pp. 1–22). London & New York: Routledge.

Oldfather, C. H. (Trans.). (1979). *Diodorus Siculus.* Cambridge, MA: Harvard University Press and London: Heinemann.

Juliet Clutton-Brock

■ Domestication
The Wool Industry and Sheep

Sheep have been bred for their wool since ancient times. The Romans were selectively breeding sheep as early as 200 BCE, and fragments of woolen fabrics have been found in the tombs and ruins of Egypt, Peru, Babylon, and Britain. Wool was once obtained by plucking it from the sheep during molting seasons. Breeding for continuous fleece growth began after the invention of shears.

Different breeds of sheep produce wool in varying textures. Coarser fleece is used to make blankets and carpets, whereas finer fleece is used in clothing. Most wool used in clothing comes from Merino sheep raised in Australia. Other major wool-producing countries include the United States, Argentina, New Zealand, Russia, South Africa, Uruguay, Great Britain, China, and India.

With about 100 million sheep, Australia produces 30 percent of all wool used worldwide. Flocks usually consist of thousands of sheep, which outnumber the human population of Australia 5 to 1. Large flocks of sheep are more profitable, but they also make it more difficult to individually monitor sheep. Every year, hundreds of lambs die before the age of eight weeks from exposure or starvation, and mature sheep die every year from disease and lack of shelter.

Merino sheep are prized not just for their fine wool, but also for their wrinkly skin, which means more wool per animal. But this excess wool can cause problems for sheep. Unsheared animals may die of heat exhaustion during hot months, and the wrinkles also collect urine and moisture. Attracted to the moisture, flies lay eggs in the folds of skin, and the hatched maggots feed on the sheep's flesh, causing severe wounds and even death. To prevent "flystrike," many Australian ranchers perform an operation named after its inventor, John Mules, called "mulesing." Mulesing involves slicing strips of skin and flesh off the backs of lambs' legs and around their tails. This is done to cause smooth, scarred skin that won't harbor fly eggs, but the surgery is thought to be extremely painful, and the wounds often get flystrike before they heal.

Alternatives to mulesing that are being explored and are already being used by some farmers include genetic selection for less-susceptible breeds, increased monitoring and treatment of sheep, applying pesticides to sheep during seasons of high blowfly activity, applying a topical protein to the breech that causes the hair to fall out, and managing blowfly numbers by releasing sterile male blowflies and using baited fly traps.

In addition to mulesing, other common husbandry methods include hole-punching ears for identification, amputating tails so that they don't become caked with urine and feces, and castration. Male lambs are castrated when they are between two and eight weeks old, with a rubber ring used to cut off blood supply.

Sheep are sheared each spring, after lambing, just before some breeds would naturally shed their winter coats. Timing is critical: Shearing too late means loss of wool, but if it is done too early, sheep may die from exposure. Shearing has to be done carefully, or sheep may be cut or seriously injured.

Sheep raised for their wool are also slaughtered for their meat. In Australia, millions of sheep are shipped thousands of miles to the Middle East, where live sheep are preferred to packaged meat, which has to be refrigerated, and because many consumers prefer for their meat to come from sheep who have been slaughtered according to halal ritual. Critics of "live export" say that the ships are too crowded, and the voyages are too long, resulting in sheep deaths due to trampling, heat exhaustion, and starvation or dehydration resulting from limited access to feed and water troughs. Shipboard mortality ranges up to 10 percent, and 14,500 sheep reportedly died from heat stress while in transit to the Middle East in 2002.

In the Muslim nations of North Africa and the Middle East, ritual slaughter is exempt from humane slaughter regulations. Some sheep are slaughtered en masse in lots; others are taken home to be slaughtered by the purchasers.

In response to concerns about the distances sheep are sent to slaughterhouses in Europe, the European Parliament adopted a report in 2001 calling for journeys of a maximum of eight hours in livestock export, the first step toward creating a law.

Because sheep are usually raised in large flocks and live mainly in large, remote pastures, they are subject to predation by wildlife such as coyotes and wolves. In the United States, ranchers are permitted to kill coyotes that they believe are threatening their flocks, but other farmers protect their flocks through the use of guard dogs and even "guard" donkeys and llamas. In Australia, kangaroos compete with sheep for grazing land and are widely regarded as "pests," so the Australian government permits the killing of more than 6 million kangaroos a year.

Alternatives to Wool

The demand for sheep's wool has been declining since 1990, with Australia's former near-total dominance of the world market falling by about 35 percent in a decade.

Many people who are allergic to wool or who are concerned about sheep welfare use alternatives to wool clothes and blankets, including cotton, cotton flannel, polyester fleece, synthetic shearling, and other fibers. Tencel—breathable, durable, and biodegradable—is one of the newest wool substitutes. Polartec Wind Pro—made primarily from recycled plastic soda bottles—is a high-density fleece that has four times the wind resistance of wool and that also wicks away moisture.

Further Resources

Akin, C. (2004, March 24). The urgent need for a permanent ban on mulesing and live sheep exports in the Australian wool industry based on animal welfare concerns. People for the Ethical Treatment of Animals.

Encyclopedia.com. *Wool*. Retrieved on March 13, 2007, from http://www.encyclopedia.com/html/section/wool_HistoryofWoolProduction.asp

Fitzpatrick, J., Scott, M., & Nolan, A. (2005). Assessment of pain and welfare in sheep. *Small Ruminant Research, 60,* 153–66.

Gibson-Roberts, P. A. (2000). Scandinavian sheep. *Knitters Magazine,* pp. 19–20. See http://www.icelandicsheep.com/scand.html

Maat, A. J. (2001, September 3). Commission Report, Committee on Agriculture and Rural Development.

Read, J. (1998, December 18). The effect of ewe iodine supplementation on perinatal lamb mortality. *Meat New Zealand.*

Ruibal, S. (2001, November 23). Edge of Winter: Beauty, Danger; Layering Clothes Essential for Sudden Temperature Shifts. *USA Today,* p. 8C.

Townend, C. (1985). *Pulling the wool: A new look at the Australian wool industry.* Sydney: Hale & Iremonger Pty.

Alisa Mullins

■ Ecofeminism
Ecofeminism and Animals

Vegetarian ecofeminism is a way that some people expand their understanding of feminism, animal liberation, environmentalism, and democratic theories of social justice by recognizing the conceptual and experiential connections among the oppression of women, nonhuman animals, people of color, and the earth—and by creating theories and strategies for ending that oppression. Ecofeminists believe that understanding these connections advances the understanding of the pervasiveness of animal oppression by showing how the domination of nonhuman animals reinforces other forms of social domination. Vegetarian ecofeminists argue that no democratic movement for social and ecological justice will be complete unless it addresses the position of nonhuman animals.

Women and Animals, Feminism and Ecofeminism

Is there a connection between women and animals? And if so, is that connection inherent in our bodies, or is it inherent in our culture and our social consciousness? In feminist history, there have been two very different responses to these questions.

One response has argued that women's association with animals and with bodily functions (menstruation, pregnancy, childbirth, lactation) has been used to support arguments that women are more "animal-like," less "rational," and thus inferior to men. Such arguments have been used to justify social relations that keep women tied to the private realms of housekeeping, reproduction, and child rearing and out of the public sphere of education, economics, and politics. This branch of feminism (liberal feminism) has sought to sever women's presumed association with animals, to accept the opposition of mental versus bodily functions—with reason defined in contrast to emotions—and to emphasize women's rationality, so that women are seen as similar to men and thus worthy of human rights.

Another response to the woman-animal association has also accepted the opposition of minds versus bodies, reason versus emotion, and men versus women, but has sought to reverse the dominant valuation and celebrate the strengths of the physical side of that opposition, using the woman-animal-nature association as an opportunity to speak out in defense of women, animals, and the earth. This branch of feminism (radical feminism) developed more fully in the 1970s, and through works such as Andrée Collard and Joyce Contrucci's *Rape of the Wild* (1989) and Susan Griffin's *Woman and Nature* (1978), it developed analyses that explored women's connection with not just animals, but all of the natural world.

In the nineteenth century, however, most feminists did not ask such questions about whether or in what ways women and animals might be connected—they simply wrote, spoke out, and acted to promote the welfare of other animal species. Prominent feminists such as Charlotte Perkins Gilman, Margaret Fuller, Mary Wollstonecraft, Harriet Beecher Stowe, Sarah and Angelina Grimké, Frances Willard, Frances Power Cobbe, Anna Kingford, Victoria Woodhull, Elizabeth Blackwell, and many others were active in advocating either vegetarianism or animal welfare reform. In the twentieth century, feminists such as the British author Virginia Woolf wrote essays condemning the practice of using birds to adorn women's hats, and American author Alice Walker has written essays empathizing with horses ("Am I Blue?") and chickens ("Why did the Balinese chicken cross the road?") and aligns herself with those who feel that "eating meat is cannibalism."

In the second half of the twentieth century, the woman-animal connection was more carefully scrutinized by feminists such as Carol Adams, Josephine Donovan, Susan Griffin, Marjorie Spiegel, and others who noted the ways that the English language—and therefore, our conceptual perspective—similarly positioned not just women and animals but also children, slaves, and nature itself. Joan Dunayer's essay in *Animals and Women* explores insulting slurs that associate women with animals—terms such as catty, dumb bunny, cow, bitch, old bat, sow, and chick. For example, calling a woman a "dog" implies that it is a woman's duty to be attractive and that she has failed in her duty. But the term is also a direct insult to dogs, who are deprived of individuality and are merged into one mass term of "ugly." "Dumb bunny" implies females are mindless and hyper-reproductive; it also rests on the speciesist assumption that rabbits are stupid. Calling a woman a "cow" implies she is fat and dull, but like the preceding terms, it also insults the cow by assuming that a cow's primary purpose is pregnancy and lactation, reproductive capacities that are seen as opposed to intelligence. Many of these pejoratives reveal the association between women and domesticated animals, particularly the females of the species, who are exploited for their reproductive capacities.

Other insulting terms and treatment reveal the conceptual associations between domesticated animals and enslaved humans. When obedient to their white masters, Negro slaves were compared to household pets and described as "loyal dogs," but when seeking their own freedom, they became "beasts." In *The Dreaded Comparison* (1989), Marjorie Spiegel discusses the feminist concept of "tokenism" in terms of the plantation system and the ways that white slave owners often kept house slaves who received better food, clothing, and housing and better treatment than the majority of other enslaved Africans. Tokenism is a strategy that perpetuates an oppressive system by allowing the oppressor to show kindness to a small, select group within an oppressed class, while continuing to exercise domination over and violence on the class as a whole. The house slaves were still slaves subject to sexual violation, whippings, and sale based on the whims of the Master—but the illusion that they were better treated allowed the slave owners to use them as tokens, examples of the slave owners' "kindness" to their slaves.

Tokenism is also at work in the pet industry, where humans have domesticated a few select species for companionship while the majority of animals continue to suffer in zoos, scientific laboratories, fur ranches, and factory farms. Even the privileged pets are not free: their wings are clipped, they live in cages and on leashes, their meals and their sexual behaviors are controlled, and even their lives and deaths may be decided for them.

Like domesticated pets, women in systems of male dominance receive token privileges through chivalry, but the most important decisions of their lives often remain under the control of men. Door-opening rituals, dinner-table courtesy, and occasional offers to carry heavy grocery sacks do not offset the more significant inequalities in wage-earning potential (women still make an average of $0.75 for every dollar men earn), retirement and elder care (women over age sixty-five account for 70 percent of elders in poverty), control over one's own body (through reproductive choice), and freedom from the pervasive threat of rape and sexual assault.

Through these analyses, feminists of the 1970s and 1980s noted parallels between the animal rights concept of *speciesism* (the unjust dominance of humans over other animal species) with *sexism* (the unjust dominance of men over women) and *racism* (the unjust dominance of one race over all others). These analyses tended to focus on the ways that women, people of color, and nonhuman animals were seen as both associated with and "closer to nature" than to Euro-American men and to culture. In the 1980s and 1990s, ecofeminists brought together these socially constructed connections among women, animals, people of color, and nature, suggesting that the strategic importance of these connections was not in the alleged character of the oppressed, but in the conceptual structure itself.

That conceptual structure relied on three steps: first, a construction of selfhood that defined the human self as fundamentally separate from, rather than connected to, all others ("alienation"); second, a logic that claimed uniqueness and superiority for whatever qualities were associated with this separate self ("hierarchy"); and third, reasoning that asserted that whatever was superior was justified in using and controlling that which was inferior ("domination") to meet the needs, goals, and purposes of the superior being. Using this conceptual structure as justification, European nations had sent explorers out to discover and colonize other lands, people, and animals—a practice that regularly included the rape of colonized women along with the enslavement and export of indigenous humans and animals, who were used for their productive and reproductive labor with little regard for their interests and needs.

Rather than defining humans as unique, superior to, and separate from animals, ecofeminists recognize the biological fact that humans are animals and thus part of both culture and nature. Based on this connected sense of self and the connections to other animals, ecofeminists argue that they can challenge the dominant ways of thinking that justify oppression and restore right relations with other humans as well as other animals and nature as well.

Experiential Data: The Woman-Animal Connection

Vegetarian ecofeminists have uncovered the connections between women and animals in food production, preparation, and consumption; in cleaning products, makeup, and the beauty industry; and in the areas of hunting, animal experimentation, domestic violence, reproductive choice, and the practice of keeping domesticated animals.

Food Production, Preparation, and Consumption

In the practices of most industrialized animal production ("factory farming"), female experience is horrific. Dairy cows are regularly separated from their newborn calves so that the milk can go to humans, while their infants are often chained in tightly fitting crates for four months and fed an iron-deficient diet until they are slaughtered for "veal." Female pigs are kept in a continual cycle of pregnancy, birth, and artificial insemination, their piglets taken away from them before they have even had the chance to suckle. Female birds ("battery hens") are kept twelve to a cage the size of typing paper, never allowed to spread their wings, their beaks sheared off with a hot iron, all so that they will produce eggs that they will never be allowed to warm to life. Vegetarian ecofeminists have recognized these practices as a form of *compulsory motherhood* that uses the female's body for reproduction without regard for her own interests and needs, and without regard for her offspring either.

Once the living animals have been turned into dead meat, their bodies are available for preparation and cooking—a job socially assigned to women. In *The Sexual Politics of Meat* (2000), Carol Adams explores the ways that in male-dominated Western industrialized cultures, meat is seen as the centerpiece of the meal, just as men are made to be the center of culture: every other dish (the appetizer, soup, or salad) leads up to the meat, augments the meat (the vegetable), or follows the meat (dessert). The English refer to meat as "the essence of a thing," while vegetables connote passivity (i.e., "to vegetate"). Cultural stereotypes imply that men need meat—whether for the extra protein, for hard physical labor, or for the more elusive qualities of virility or masculinity. One outcome of these gendered associations is that around the world, men often eat the first and best foods, and women and children may eat last and least. Animal-based agriculture and first-world dietary practices, especially in the context of these cultural assumptions about women, show that food production is an ecofeminist concern.

Cleaning Products and the Beauty Industry

In cleaning products as well as makeup research and production, animals are used to test the safety and effectiveness of products, and cultural norms about women's role as homemakers and the importance of women's impeccable appearance pressure women into being consumers of the products of animal suffering. Writing in *Ecofeminism: Women, Animals, Nature* (1993), Lori Gruen describes numerous experiments requiring animals to drink, inhale, or be injected with the toxic chemicals that are used in cleaning products. These products are often environmental toxins as well, poisoning the air, water, and wildlife long after their uses in the home.

After testing on animals, the beauty industry markets these makeup products to women through advertising, implying that women's natural physical appearance is inadequate, unlovely, and unacceptable without some kind of augmentation—whether through makeup, clothing, or plastic surgery—that men are the proper judges of women's appearance, and that women's acceptability and worth rest heavily on their physical appearance. The manipulation of women's self-image and the physical mutilation of animals come together again in the fur industry, which also plays on class differences for profit. Wearing fur is promoted as making women beautiful and glamorous and giving them "high-class status" as well. But behind the manipulated image of women's glamour in fur lies the torture of women's minds and animals' bodies: in the wild, fur-bearing animals are hunted using steel-jaw leg-hold traps, where trapped animals may chew off a leg in order to break free or may lie for days without food or water until the trapper returns to kill them. Fur "ranches" confine the animals in small wire cages their entire lives, until they are killed in the least expensive way possible—usually through electrocution. Gruen's work reveals the connections among cleaning products, cosmetics, and furs that make women complicit in their own oppression as well as in the suffering and death of other animals.

Hunting

In *Rape of the Wild*, Andrée Collard and Joyce Contrucci examine the way that Western cultures' "origin stories" trace humans' most significant cultural developments back to men's hunting behaviors, rather than to women's gathering, food preparation, shelter-building, childbearing, child care, or care for the sick. These origin stories have been used to portray weapons, violence, and the domination of nature, animals, and women as not only legitimate but also inevitable and pivotal in the achievements of civilization. Collard and Contrucci turn these stories inside out by exposing them as the fantasy of controlling that which is uncontrollable: the inevitability of having to live with a body subject to the natural laws of birth, old age, sickness, and death. Hunting reveals a cultural mentality that associates the human (primarily male) hunter only with animal predator species, a strategic association so embedded in Western male-dominant cultures that it has become hard to recognize in its various manifestations of militarism, the build-up of nuclear weapons, and the economic structures and behaviors that invoke hunting and predation as metaphors of "good business." Ecofeminists contribute to the mainstream animal-rights analysis of hunting by unmasking its concurrent association of women, animals, and nature as "prey" and by exploring the ways that hunting and predation have sometimes been used to legitimate the violent behaviors of rape, sexual assault, and male dominance.

Animal Experimentation

Animal rights theorists have long opposed the inhumane and needlessly repetitive practices of animal experimentation. Ecofeminists contribute to these critiques by noting how scientific researchers are trained to shut off their emotional connections with animals in order to become "objective" and perform experiments on animals. This privileging of reason and denial of feeling is a key feature of male-dominant ideologies, which identify maleness with reason, intellect, the mind, and superiority and which associate femaleness with emotion, the body, animals, and inferiority. Describing this psychological separation of self, both Marti Kheel and Chaia Heller have examined the heroic ethics of Western scientific research and its claims that experimentation on animals is "required" to advance knowledge that will save human lives. Kheel's work in *Ecofeminism: Women, Animals, Nature* has focused on the ways that masculinist ethics rely on crisis situations (i.e., "lifeboat" ethics) to enact their rule-based thinking, without ever asking how these epic situations arose (i.e., "how did we all end up in a lifeboat with only one chance for a survivor?"). Kheel calls these heroic situations "truncated narratives" because they tell only part of the ethical story. Instead, ecofeminist ethics seeks to restore and "re-story" the whole narrative of animal lives, so that humans can make better choices for interspecies justice and ecological sustainability, choices that keep us out of the epic situations that are used to justify such heroic choices.

Domestic Violence

Carol Adams's activism and research with battered women uncovered a shocking wealth of co-occurrences of violence against women, children, and companion animals. In these incidences, the companion animal was abused (tortured, maimed, mutilated, or murdered) as a surrogate for the battered woman or sexually abused child, as an "example" of what would happen to the woman or child if she did not obey the abuser and as a means of manipulating her behavior through fear, coercion, and the threat of violence. In child sexual abuse, children are coerced through threats and the actual abuse of their pets and are forced to decide between their own victimization or the pet's death. Adams's research shows that these abused children often become abusers themselves: serial murderers and others who rape and mutilate often acted out these behaviors in their early teens, torturing and killing cats and dogs. The link is so significant that the American Psychiatric Association now recognizes childhood cruelty to animals as a Conduct Disorder indicating potential of future violent behaviors.

An ecofeminist perspective enables recognizing the connections between violence against women, violence against children, and violence against companion animals as a pervasive hostility to the body. The problem is not just that women and children are equated with their bodies, objectified, and abused; the problem is also that animals are equated with their bodies as well, as Adams explains. This hostility to the body, called somatophobia, is at the root of many forms of oppression, including homophobia, racism, and ageism.

The conceptual connections among these oppressive systems make it clear that ending the violence against one oppressed group will not uproot the thinking that condones oppression. To be effective, activist strategies must make these connections evident and challenge the thinking that authorizes domination and violence against all these oppressed groups. One project of the vegetarian ecofeminist organization Feminists for Animal Rights has involved efforts to link animal shelters and women's shelters as a strategy for freeing women, children, and animals from oppression and for educating activists in the feminist and animal right movements.

Reproductive Choice

Complementing Marti Kheel's work on truncated narratives, Carol Adams has argued that the heart of the abortion rights–animal rights connection involves telling the whole truth of women's lives and animals' lives. Just as pro-life arguments focus on the fetus without regard for the context of fetal life—the womb, the woman, and the circumstances under which she became pregnant and would have to bear and raise a child—"meat"-eating cultures ignore the contexts of animals' lives, in experimental laboratories, zoos, rodeos, "pet" stores and breeders' shelters; in leg-hold traps and on fur "ranches"; and in the cages and stalls of industrialized animal production. Yet the moral dilemmas of abortion rights and animal rights are different, Adams maintains: women and animals are already subjects of a life, living in specific social and natural (or unnatural) environments. A fetus is not yet a subject of a life in the same way. Until women's personhood is legally beyond dispute, it makes little sense to grant legal personhood to the fetal life contained in a woman's body. Instead, the logic of pro-life arguments would make more sense if applied to fully sentient beings of all animal species and if used to defend females' rights to their own bodies, including freedom from experimentation and pregnancy without consent. As long as women and animals are thought of as objects for others' use, both sexually violable on the one hand and consumable on the other, both abortion rights and animal defense will be necessary.

A Final Note: Can Men Be Ecofeminists?

To date, at least two men have made important contributions to ecofeminist thought. Deane Curtin (1991) has developed a theory of contextual moral vegetarianism that recognizes that the reasons for moral vegetarianism may differ by locale and by gender, as well as by class. The context in which moral vegetarianism is completely compelling as an expression of an ecological ethic of care, Curtin argues, is for economically well-off persons in technologically advanced countries—Euro-American men in particular. Because of the many associations between women and animals, vegetarianism is not a gender-neutral issue. "For men in a patriarchal society," Curtin concludes, "moral vegetarianism can mark the decision to stand in solidarity with women."

Brian Luke has also used his position as a man to challenge the social construction of masculinity as inherently rational and carnivorous and to address the importance of interspecies sympathy. Noting that mainstream animal rights thought tends to emphasize rational arguments as a primary persuasive strategy and to background or omit appeals to sympathy and emotion, in *Animals and Women,* Luke shows how institutionalized animal exploitation does not so much result from a lack of human sympathies for animals as it continues in opposition to and despite these sympathies. In order for us to tolerate the institutions of factory farming and animal experimentation, we must have social mechanisms in place to forestall our sympathies for exploited animals, along with mechanisms for overriding (i.e., preventing us from acting on) any sympathies that might remain. "To cut off our feelings and support animal exploitation *is* rational, given societal expectations and sanctions," Luke argues, "but to assert our feelings and oppose animal exploitation is also rational, given the pain involved in losing our natural bonds with animals. So our task is not to pass judgment on others' rationality, but to speak honestly of the loneliness and isolation of anthropocentric [human-centered] society, and of the damage done to every person expected to hurt animals."

Ecofeminists of all genders argue for recognizing and respecting the interdependence—humans with nonhuman animals, reason with emotion, culture with nature—by developing relational structures that promote compassion, equality, and sustainability for all species on earth.

See also

Ethics and Animal Protection

Further Resources

Adams, C. J. (1990/2000). *The sexual politics of meat: A feminist-vegetarian critical theory.* New York: Continuum.

———. (1994). *Neither man nor beast: Feminism and the defense of animals.* New York: Continuum.

Adams, C. J., & Donovan, J. (Eds.). (1995). *Animals & women: Feminist theoretical explorations.* Durham, NC: Duke University Press.

Collard, A., with Contrucci, J. (1989). *Rape of the wild: Man's violence against animals and the earth.* Bloomington: Indiana University Press.

Curtin, D. (1991). Toward an ecological ethic of care. *Hypatia, 6*(1), 60–74.

Gaard, G. (Ed.). (1993). *Ecofeminism: Women, animals, nature.* Philadelphia: Temple University Press.

———. (2003). Vegetarian ecofeminism: A review essay. *Frontiers, 23*(3), 117–146.

Griffin, S. (1978). *Woman and nature: The roaring inside her.* New York: Harper & Row.

Spiegel, M. (1989). *The dreaded comparison: Human and animal slavery.* New York: Mirror Books.

Walker, A. (1988). *Living by the word.* San Diego: Harcourt Brace Jovanovich.

Greta Gaard

■ Ecotourism
Ecotourism

Although ecotourism continues to increase in popularity and indeed is touted by tourism organizations as a growth industry, tourism focused on wildlife is not new. In the days when Africa was divided into European colonies, for instance, safaris were very popular with wealthy European travelers. Opportunities to hunt and kill "big game" such as lions, leopards, elephants, rhinos, and buffalo are still available in Africa, but there has been a decided shift toward "nonconsumptive" enjoyment of these animals, paralleling the rise of environmentalism and animal activism in the West. Whether wildlife tourism, even of the nonconsumptive variety, is more benefit than cost to wildlife remains open to debate. Ecotourism is therefore a hot topic for those concerned with human-wildlife relations.

In 1990, as the idea of ecotourism (also sometimes known as nature-based tourism or sustainable tourism) was growing in prominence, the World Wildlife Fund published one of the first serious analyses of the phenomenon. In this report, Elizabeth Boo defined ecotourism simply as "traveling to relatively undisturbed or uncontaminated natural areas with the specific goal of studying, admiring, and enjoying the scenery and its wild plants and animals" (p. xiv). Wildlife tourism is seen as a subset of the broader phenomenon of ecotourism.

Since Boo's 1990 report, there has been much debate about what gets to count as ecotourism. Much of the discussion involves where to draw the line on various continua, usually placing ecotourism at one end and traditional "mass" tourism at the other. Each continuum examines a different criterion. Examples include dependency on natural systems, particularly wilderness; duration of stay in natural areas; level of infrastructure support needed; group size; physical challenge and accessibility; amount of economic leakage out of local communities; local involvement in planning, development, and implementation; and impacts on the natural environment and wildlife. Other continua have been developed to describe the quality of the experience, including intensity of interactions with nature, including wildlife; levels of interaction with local cultures; emphasis on learning; and opportunities for tourists to become actively involved in conservation initiatives during the expedition as well as upon their return home.

These various criteria demonstrate that there is no single, universally accepted definition of ecotourism. Some commentators distinguish between "shallow" and "deep" ecotourism or "passive" and "active" ecotourism. Others distinguish between ecotourism focused on "consumptive" activities such as hunting and fishing, "low consumptive" activities such as visiting zoos and aquaria or swim-with-the-dolphin ventures, and "non-consumptive" activities such as traveling through natural areas to view and photograph wildlife. Still others argue that any form of consumptive activity that leads to the death, harm, or incarceration of wildlife should not be considered ecotourism and that attempts to do so are an example of the marketing phenomenon called "greenwashing."

A wide range of animal species on every continent have been targeted for ecotourism. They include animals that most would expect to be of interest to tourists, such as the charismatic megafauna (e.g., great apes, big cats, elephants, whales, and son on) and birds (e.g., penguins, puffins, and parrots), but also those that some may find more surprising, such as sharks, bats, and reptiles (e.g., crocodiles, Komodo dragons, and garter snakes). Paul Reynolds and Dick Braithwaite (2000) argue that, ideally, wildlife targeted for tourism should be easily seen in the daytime, predictable in activity or location, approachable, and tolerant of human intrusion. They also suggest that animals that are thought to be rare have a certain cachet for tourists, whereas animals that are locally superabundant ensure that tourists have a good chance of actually seeing the animals they desire.

Regardless of what species is targeted, the dominant rationale for ecotourism, even among conservationists, is economic. Akin to the "use it or lose it" argument, it is thought that if wildlife and wildlands become economically valuable as tourist attractions, they will be more likely to be protected. Supporters note that in some countries, wildlife tourism is now the leading foreign exchange earner and thus contributes to continued support for parks and protected areas. Karen Higgenbottom (2004) reports, for example, that money earned through gorilla tourism has contributed to anti-poaching initiatives and habitat conservation, turtle-watching programs have funded patrolling of nesting beaches and predator control, and penguin-watching at Phillip Island Reserve has financially benefited the reserve. Economic analyses have also demonstrated that a lion or elephant in Amboseli, for example, is worth far more alive as a continuing target of ecotourism than shot dead and stuffed.

Detractors, however, argue that it is often multinational corporations who control air travel and accommodations and not the local communities who profit: this is known as economic leakage. Further, some worry that the economics of ecotourism has inherent traps, casting wildlife and wildlands as commodities to be exploited for the viewing pleasure of tourists and, ultimately, for profit. This commodification is seen as dangerous because, if animals are imagined solely as sources of economic gain, other activities and uses (e.g., sport hunting, cattle ranching, logging, mining) may be seen as perfectly logical and acceptable alternatives to ecotourism initiatives that become less profitable. In this rationale, then, the intrinsic value of animals is disrespected.

The other dominant rationale for ecotourism is educational. The fascination that many people have for wildlife is seen as an entrée for teaching tourists about the lives of these animals, about the threats facing them, and about ways to contribute to their conservation. It is thought that direct experiences with wildlife will lead to increased knowledge about wildlife, which will lead to feelings of caring for and commitment to wildlife, and this in turn will lead to action on behalf of wildlife. Encouraging tourists to become involved in protecting a target "flagship species" is also seen as potentially beneficial to other, less popular or less conspicuous species.

As yet, there has been little research conducted on the educational rationale, and what research has been done has demonstrated that, thus far, the rhetoric outstrips the reality. Environmental education researchers have pointed out that the linear relationship proposed in this rationale (experience → knowledge → caring → commitment → action) is simplistic and does not heed the growing body of literature investigating the "gap" between environmental knowledge and responsible behavior. In our own experience with orangutan tourism, for example, we have too often witnessed how the knowledge of the dangers of disease transmission did not prevent some tourists from holding formerly captive, now free-ranging and friendly, orangutans in their arms, potentially infecting them. Knowledge was not able to trump desire in determining action. Attention, then, needs to be paid to the expectations and desires that tourists bring with them and to how these contribute to or interfere with educational goals. Attention also needs to be paid to the "hidden curriculum" of ecotourism—that is, not only to what tourists may be explicitly taught on expeditions (e.g., orangutans are very much like humans and are susceptible to human diseases), but also to underlying messages conveyed through actual practices (e.g., orangutans want to be held, are harmless, and are here for your enjoyment).

Research assessing the economic and educational rationales for ecotourism is mixed at best. There is much more evidence of negative than positive impacts of ecotourism, both on humans and on other animals. Negative impacts on humans include theft (e.g., orangutans stealing cameras or other equipment), injuries (e.g., bites), and death (e.g., a dingo targeted for tourism killed a boy, and villagers were killed by lions, buffalos, and elephants who moved outside of parks), as well as the displacement of human communities for the creation of parks and protected areas. Negative impacts on animals can be divided into direct and indirect impacts. Examples of direct impacts include death or injury from being hit by vehicles; disease (e.g., gorillas being infected with influenza and orangutans with polio); stress (e.g., cheetahs failing at hunting because of tourist presence and penguins staying out at sea because of the presence of tourists, thereby delaying changeover in sharing the care of young); behavior change (e.g., courting behavior of birds and feeding behavior of fin whales, vervet monkeys, macaques, dolphins, and Komodo dragons); discontinued use of habitat (e.g., disruption of nesting sites of birds) or other changes in habitat use (e.g., habituated gorillas spending more time outside parks in agricultural lands); increased predation (e.g., the eggs or young of disturbed crocodiles, dolphins, puffins, razorbills, and penguins are more vulnerable to predators); and habituation to human presence. Indirect impacts include damage to vegetation or coral, soil erosion, fires, introduction of nonnative species, damage to habitat for building infrastructure of tourism facilities, and pollution.

Of course, one cannot assume all changes associated with ecotourism are necessarily detrimental to the animals. Much depends on the particular species or population of animal. David Newsome, Ross Dowling, and Susan Moore (2005) suggest that in order to determine potential or actual impacts, researchers should know the following about the animals targeted for ecotourism: their social behavior; their natural movements and use of various areas; their critical habitat requirements; their life history parameters (e.g., reproductive fitness, survival rate); and their responses to tourism pressures and activity. Even vocal proponents of wildlife tourism note the importance of well-managed ventures that pay close attention to the unique context of targeted animals. For example, some species of bats are more easily disturbed than others, and some bats live in caves where the presence of too many humans would quickly increase the cave temperature and harm the bats. Targeting species like these would need to be carefully considered before proceeding, and if a decision were made to do so, the number of visitors and their behavior would need to be closely regulated.

For some species, assessing potential or actual impacts can be very challenging. Thus far, little hard data exists on the impacts of whale-watching on whales, for example, although a number of researchers are concerned. In the absence of proof but with doubts remaining, there has been some movement toward adopting the "precautionary principle" as an approach to ecotourism. This principle, enshrined in a number of international declarations and agreements and applied to a variety of sectors including environmental law and health, asserts that if there is a reasonable probability that certain human actions will be detrimental to the environment or human health, it is inappropriate to wait for conclusive scientific evidence before responding. Applying the precautionary principle to ecotourism would mean that if particular ventures seem likely to harm the species targeted, the ventures should be carefully regulated or prevented from operating at all. Although those interested in the establishment of guidelines or regulations for various ecotourism sectors discuss the potential of the precautionary principle, they also recognize that applying it in practice is very challenging.

Ecotourism, then, is a complex phenomenon. It continues to elicit strong reactions, both for and against. Some advocates see it as almost a panacea for many wildlife conservation woes. Others argue that the benefits of ecotourism outweigh the costs and that the problems encountered thus far can be managed. Still others, such as Brian Wheeler (2005), are dismissive of the entire idea; he has famously dubbed the phenomenon "egotourism," suggesting that ecotourism, in the end, favors human interests over the interests of other animals or the environment in general and is thus counterproductive to the conservation agenda. Regardless of the position one takes on ecotourism that targets wild animals, it is undeniable that it continues to grow in popularity, and research from a wide range of disciplines is still much needed to help us understand its benefits and its costs.

Further Resources

Boo, E. (1990). *Ecotourism: The potential and pitfalls.* Washington, DC: World Wildlife Fund.
Buckley, R. (2004). *Environmental impacts of ecotourism.* Cambridge, MA: CAB International.

Epler Wood, M. (2002). *Ecotourism: Principles, practices and policies for sustainability*. Paris, France: UNEP and Burlington, VT: International Ecotourism Society. Available at http://www.ecotourism.org/

Higginbottom, K. (2004). *Wildlife tourism: Impacts, management and planning*. Gold Coast, Australia: Common Ground.

Newsome, D., Dowling, R., & Moore, S. (2005). *Wildlife tourism*. Toronto: University of Toronto.

Reynolds, P., & Braithwaite, D. (2000). Towards a conceptual framework for wildlife tourism. *Tourism Management, 22*, 31–42.

Russell, C. L. (1995). The social construction of orangutans: An ecotourist experience. *Society and Animals, 3*(2): 151–70.

Russell, C. L., & Hodson, D. (2002). Whalewatching as critical science education? *Canadian Journal of Science, Mathematics and Technology Education, 2*(4), 485–504.

Shackley, M. (1996). *Wildlife tourism*. London: Thomson.

Weaver, D. (2003). *The encyclopedia of ecotourism*. Cambridge, MA: CAB International.

Wheeler, B. (2005). "Ecotourism/egotourism and development." In C. M. Hall & S. Boyd (Eds.), *Nature-based tourism in peripheral areas: Development or disaster?* Toronto: Channel View.

Constance L. Russell and Anne E. Russon

■ Ecotourism
The Whale and the Cherry Blossom Festival: A Personal Encounter

This was Yoshiko and Masumi's first trip from Japan to Mexico's remote San Ignacio Lagoon to encounter what scientists call "the Friendly Whale Syndrome." At last our plane descended toward the dirt landing strip bordering the lagoon's turquoise waters and gentle pink salt mountains. Excitedly, this mother and daughter scanned below for the heart-shaped blows of the many mother-calf pairs of gray whales.

The Mexican government has shown considerable foresight in setting aside San Ignacio lagoon as El Vizcaino Biosphere, the first whale sanctuary in the world. This will help ensure the future health of this last population of healthy gray whales. And thanks to Mitsubishi—who decided *not* to build a salt factory on San Ignacio—this birthing lagoon is still pristine. Yoshiko and Masumi, along with their longtime friend Maureen and her teenage daughter Alexa, decided to have an international reunion here in the quiet tents and small boats of this faraway camp. Eighteen of us had signed up with SummerTree Expeditions to have what we called a "pod" reunion this spring. Some of us have been visiting San Ignacio for many years to study this remarkable yearly ritual: wild animals seeking physical contact with the very species that has twice brought gray whales to the brink of extinction.

The next morning, we floated in the skiff with our boatman Renulfo. In the 1970s, his father, Francisco Mayoral, was the first man in the lagoon to touch these forty-five-foot whales, thus attracting scientists who still do not know why the "Friendlies" approach our boats of their own volition. Are they curious? The seventy-year worldwide ban on gray whale hunting has allowed their population to return to what we hope are healthy numbers. Could this hunting ban also be the reason the whales here trust us? What would be possible if this moratorium, which Japan,

Norway, and Russia defy by continuing to hunt whales, spanned another century? Would we have friendly whales all along the West Coast?

Rocking in the lagoon, we scanned the horizon for the familiar blows of mother-calf pairs. All night, we had listened to the whales breathing under a dark dome filled with more stars than many of us have seen in a lifetime. First, the bass blasts of the bull whales; then the long mezzo-soprano sigh of the mother whale; and almost in synch, the treble, brief blow of the baby.

"Whales at one o'clock!" Yoshiko happily called out as a mother-calf pair surfaced half a lagoon away from us.

Motoring parallel to the whales, Renulfo kept his steady eye on the mother-calf pair. Were they Friendlies? Would they approach us and seek our outstretched hands—here, where once Yankee whalers had brutally harpooned calves to attract the loyal, and more lucrative, mothers? In the nineteenth-century whaling days in San Ignacio Lagoon, the mothers were so fiercely protective of their calves that they earned the nickname "Devil Fish." Over a hundred years later, we sit in a small eighteen-foot skiff, awaiting a visit from a forty-five-ton mother whale and her 1,500-plus-pound newborn.

Renulfo throttled back the motor, allowing only enough noise to let the whales know we were nearby, and we drifted apart, but alongside them, waiting hopefully. Not everyone here in these pristine lagoons will have a physical encounter with a wild whale. It will be the whale's choice.

First Contact is a phrase usually meant for encountering extraterrestrials. But for those of us who keep our eyes and hopes focused on this majestic and wondrous blue planet, First Contact means the first time we humbly accept that we are not the only intelligent species, that we have much to learn from other animals, that we are not alone in this sea-encircled world. We stretch out our hands and our imaginations and hope to meet the Other in the animal's own element.

Over the years, we have noted that gray whales often approach boats in which people are singing. After all, acoustics are any whale's main sense. It is how they navigate, find food, find each other, and communicate. For the gray whale, add to acoustics a marvelous sense of earth's electromagnetic fields. A sense of always knowing exactly where they are.

"Let's sing a Japanese song," someone said, in honor of Yoshiko and Masumi's "First Contact."

Yoshiko, a lively woman in her mid-fifties, smiled at her daughter, who is a flight attendant for Japanese Airways, and they both nodded and together suggested, "Sakura!"

"This is our Japanese song of the cherry blossom festival that comes every spring," Yoshiko explained to us as she taught us the words and haunting melody of the song. "It's also a lullaby a mother might sing to her child."

A perfect song for a birthing lagoon, for mother-calf pairs. As we leaned over the boat, splashing with our hands to call the whales nearer, I realized that on our Pod trip, ten out of twenty of us were mother-child pairs. Our eldest was eighty-four and our youngest one year old. Masumi had been present when Maureen gave birth to Alexa—a water birth.

So here we were at another kind of water birth of newborn calves.

"Sakura, Sakuraaaaaaaa, Yayoi no sora wa," we all sang out, our voices harmonizing. "Cherry blossoms, Cherry Blossoms / as far as I can see."

Suddenly, the mother-calf pair veered right toward us, as if hearing our song. "Keep singing!" Yoshiko encouraged.

Though we did not get all the Japanese lyrics at first, the melody was a universal language. Even the whales seemed to understand this was a lullaby between a mother

Yoshiko sings to call the gray whales for first contact. Courtesy of Robin Kobaly, SummerTree Institute, a nonprofit environmental organization.

and her newborn. With a mighty whoosh from her double blowhole, a rainbow prism of light and sweet mist, the mother surfaced, her huge snout mottled with white barnacles.

"Cherry blossoms!" I sang out. Those barnacles looked just like the luminous blooming flowers I see everywhere in Seattle—first signs of spring in my hometown.

Yoshiko nodded, smiling, "Sakura!"

We sang even stronger now as the mother whale turned on her side to study us with her great eye, as big as a softball and wide open. Looking up at these humans, what did she see? Why did she trust us, when humans still hunt grays up in the Arctic? One of the mother whales we saw this spring had a harpoon scar slashed along her side. And yet these whales are friendly, approaching us to present their precious newborns.

"Coming up!" Renulfo cried out, and we all scrambled to one side of the boat, singing at the top of our voices.

"Sakuraaaaa, hana zakari!"

And then the mother whale did something I had not before witnessed in the five years I had been visiting San Igancio. She turned belly-up under our small boat, floating us all very gently on her belly. Her blowhole upside down and closed, she held her breath as a huge pectoral fin lifted her baby up to our outstretched hands. Eye-to-eye with a newborn calf, we hushed, gazing into that dark, deep eye. Were we the first humans this calf had ever seen? Would we be the last? In this gray whale's long life— some scientists say that grays can live up to 150 years old—will this whale always remember us?

Mother-calf pair of spyhops approaches the boat in Baja, Mexico, spring 2005. Courtesy of Jose Angel Sanchez Pacheco, SummerTree Institute.

We will always remember her trust and tenderness, as she turned to let us scratch her tiny whiskers and smooth, new skin; it is so sensitive, like touching silk and cool melon.

"Keep singing," Yoshiko called out, as she stroked the baby's snout. "She likes it."

The mother and calf liked our lullaby so much that for the rest of our expedition, everyone wanted to be in same boat as Yoshiko and Masumi; theirs was the most popular boat of all for whale encounters. Everyone learned to sing "Sakura."

I could still hear the mesmerizing melody several weeks later when I read the astonishing news that during the Japanese cherry blossom festival that year, a single gray whale had swum into Tokyo Bay. It was as if the gray whale had shown up to attend the spring celebration of their annual "Golden Week"—Sakura, the cherry blossom festival.

"TV footage showed holidaymakers on nearby wharves cheering wildly as the whale came into sight and blew water high into the air," reported the *Mail & Guardian*.

This Japanese cherry blossom festival from March to May coincides with the birthing season and visit of the friendly whales in Baja, Mexico. Immediately, I e-mailed the news report to Maureen, who forwarded it to Yoshiko in Japan. I felt so happy to send this good news along.

The last e-mail I had sent to Yoshiko was about the Japanese announcing their plans to seek "broader and more comprehensive" research whaling in the Antarctic at the June meeting of the International Whaling Commission in Ulsan, Korea. Already, Japan saddens the world by killing almost one thousand minke, sperm, sei, and rare Bryde's whales in the northwestern Pacific under the label "research whaling." Many of the whales end up in Japanese meat markets. Under Japan's expanded plan, their whalers will double their yearly catch of minke whales and begin hunting endangered humpback and fin whales.

"They do not tell us about Japanese whaling in our newspapers," Yoshiko wrote back. "They will not publish anything critical about our whaling industry. But we are telling everyone we know about the friendly whales in Baja. We are speaking for the whales here in our country. And many Japanese are listening—especially the young people."

In late spring of 2005, Japan did not succeed in influencing many smaller nations to overturn the worldwide hunting moratorium on gray whales and so officially return to commercial whaling in the twenty-first century. But every year Japan, Norway, and other whaling nations lobby relentlessly to return to commercial hunting. If they succeed, what will become of this trust that has developed over the last seventy years in San Ignacio and all along the West Coast migration path? Will our grandchildren ever again be able to encounter friendly whales? Or will that become another legend of interspecies trust—a long-lost time when the bond between humans and animals was strong, respectful, and far-sighted?

Yoshiko and Masumi encounter their first mother gray whale, Baja, Mexico. Courtesy of Brenda Peterson.

When that lone gray whale showed up for the spring cherry blossom festival with the Japanese people, was it perhaps a sign? A plea? A possibility?

The once-plentiful Western Pacific gray whale, or Korean population of gray whales, is now almost extinct, down to 200 whales. There have been only twelve sightings of gray whales around Japan since the 1960s. What if Japan, like its multinational corporation of Mitsubishi, made another decision—a decision *not to hunt*? A decision for the future generations of people and whales together. Who knows what might happen if instead of seeing harpoons in gray whale mothers, we recognized the symbols of cherry blossoms—rebirth. Might more whales find their way back to the islands of Japan? Might there one day be friendly whales swimming in Japanese seas?

Less than a week after that solitary gray whale in Tokyo Harbor surprised Japanese celebrating their Sakura Cherry Blossom Festival, the same whale was found dead. Floating silently, the barnacle-mottled body was entangled in a fishing net near the town of Tomiyama. How did the whale die? Perhaps she was drowned by the fishing net or had been struck by the crush of harbor vessels before drifting into the net. It is a mystery.

But here is the biggest mystery of all: How can a nation that rushes to welcome and cheer a rare, visiting gray whale in Tokyo Harbor also at the same time begin serving "whale burgers" to their schoolchildren? How can a nation that so prides itself on culture not also recognize the culture of cetaceans? Scientists have now documented that whales and dolphins have unique cultures: humpbacks pass down songs from their elders to their young; dolphins use sponges as tools; gray whales, the elder of all whales, migrate 10,000 miles round-trip from Baja to Alaska, following the complicated electromagnetic grids under our seas. And most poignant of all, the friendly whales in Baja are increasing in numbers every year, this trusting behavior passed down generations.

It is humans who are not adapting to a changing and more threatened whale, a compromised and often fathomless ocean. Humans are the animals who must evolve if we are to survive within the healthy web of all creatures. Humans must find the humility in the "humus," or common matter, of our names.

"All things are connected," said the great Chief Seattle, whose birthplace is across the Salish Sea from my home, "like the blood that unites one family. All things are connected."

Is there a connection between a gray whale mother and calf who chose Yoshiko and Masumi singing a Japanese folk song as their favorite boat of the season—and the fact that a single gray whale visited Tokyo during their cherry blossom festival?

I reflect upon all of this as I take my daily walks, strolling every spring past the beautiful Japanese ornamental trees in my neighborhood, where the wind steals away the last of the luminous flowers. A blizzard of blossoms. A lone whale visits Japan.

Brenda Peterson

■ Ecotourism
Whales, Dolphins, and Ecotourism

The relationship between humans and cetaceans (dolphins and whales) is perhaps nowhere as variegated as within current trends in ecotourism, running a difficult gamut from benign, potentially educational observation to disruptive, even harmful intrusion.

Expanding quickly into arctic, temperate, and tropical waters, this type of ecotourism encompasses excursions that involve whale or dolphin watching or swimming with wild cetaceans. Although some viewing operations are shore-based, many involve voyaging into coastal bays and inlets, up river tributaries, or out upon the open seas. Tourists may choose from guided or unguided tours on all kinds of vessels, from tiny kayaks to huge viewing ships, on excursions ranging in length from a couple of hours to several weeks on a live-aboard ship.

According to a recent report by author and researcher Erich Hoyt, some eighty-seven countries now offer whale-watching activities, attracting over 9 million people globally each year, in a billion-dollar industry. Although some view whale watching as just another novel or interesting outdoor activity, many others share a deep, or even spiritual, fascination with dolphins and whales. Wild ecotourism encounters also hold appeal as an ethical alternative to those who object to the confinement of cetaceans in marine parks.

Our relationship with cetaceans, particularly the great whales, has a dark past. The human onslaught against whales began in earnest in the 1600s, and with the advent of the industrial revolution, modern mechanized whaling practices decimated populations of great whale species until they were threatened with extinction. Only in 1986 was an international moratorium on commercial whaling instated, but by the time whales were offered this protection, it was almost too late, with no more than 5 to 10 percent of the original populations remaining. The present moratorium is only a temporary halt. Japan,

Norway, and Iceland (among others) have all resumed killing whales. Other nations are pressing for the full-scale resumption of commercial whaling, and both careless and deliberate fishing practices continue to claim the lives of many thousands of dolphins every year.

With this in mind, the flourishing of ecotourism demonstrates a *positive* trend in the human-cetacean relationship, where whales and dolphins are valued more alive than dead. While offering opportunities for people to observe and connect with wildlife, whale and dolphin watching also provides considerable economic benefits for many coastal communities. The advent of whale watching is particularly important in areas where whaling is still practiced. Author Jim Nollman observes that the growth of cetacean ecotourism in Japan is influencing the Japanese to appreciate *living* whales and dolphins, increasing local resistance to their killing more effectively than many years of foreign protest. Many conservationists take it as a hopeful sign that, like Japan, the whaling nations of Iceland and Norway now support increasing numbers of whale watchers. The beneficial aspects of cetacean ecotourism are further revealed in how many whale-watching epicenters, such as New England, were once whaling towns where the idea of *intentionally* killing a whale has now become unthinkable.

However, increasingly overenthusiastic or exploitative ecotourism has created adverse impacts on cetaceans of a different kind. Crowding tour vessels can disrupt dolphins' and whales' vital natural behaviors, including migrating, mating, resting, foraging, navigating, and nursing young. In areas where whale watching has become a competitive industry, as the numbers of recreational and tour boats increase, dolphins and whales are being harassed by aggressive boaters who crowd, chase, corral, and even hit cetaceans as they compete to get close. Areas in both Atlantic New England and Vancouver/Puget Sound in the Pacific Northwest are experiencing such expanding problems with overcrowding. More than seventy-five boat operators take some 472,000 people whale watching on the Pacific Northwest Coast each year, which experts think may simply leave too many people on the water. Another facet of the disturbances caused by human presence frequently not taken into account is the noise pollution produced by boat motors. Magnified up to four times underwater, the whine and roar of nautical traffic is known to interfere with cetacean navigation as they forage and communicate. Such acoustic harassment may be enough to drive cetacean species away from their traditional habitats.

Some overcrowded areas are making efforts to limit or regulate the numbers of whale-watch boats and manage human impingement by enforcing protective regulations. In the United States, legal whale-watching guidelines are designed to reduce potential disturbance by vessels, prohibiting any approach closer than 110 yards. Fines for breaching these parameters or otherwise harassing cetaceans run as high as $25,000.

As with whale watching, the effects of "swim-with-the-dolphins" tours have also become controversial, especially in heavily visited areas. The results of human presence on specific wild dolphin populations are in some cases measurably disruptive. Many people, ingrained with unrealistic images of dolphins as cuddly, smiling clowns, as they are frequently portrayed in marine parks, fail to realize that they are, in reality, powerful and sometimes unpredictable *predators*. Commonly aggressive gestures, such as attempting to grab hold of a dolphin's dorsal fin for a ride, not only are inappropriate and intrusive, often causing wild dolphins to leave the area, but also may also provoke an aggressive or defensive response, resulting in human injury or even death. Because of their enormous size and strength, whales always have the potential to be dangerous to swimmers or small craft. Similarly, smaller cetaceans can also be dangerous when people are ignorant of their behavior signals.

Many swimmers are simply unaware that dolphins do not necessarily share in the human pleasure and excitement of joining their company. Researchers working with the Ocean Mammal Institute ran a project to study the effects of swimmers, kayaks, and motorboats on pods of spinner dolphins off the Big Island of Hawaii. From cliff-tops 150 feet over the water, researchers were able to follow the movements of human activity and the dolphins' resulting behaviors. After a busy night of deep diving and feeding, spinner dolphins seek shallow sandy bays in which to rest, but by midmorning, assorted boats, tourist kayaks, and dolphin-swim workshop participants all converge in eager pursuit of the dolphins. Though the next several hours were when the dolphins most needed quietude, it was also the height of demanding and ceaseless human presence, often causing the dolphins to take evasive action.

Off Kaikoura, New Zealand, scientists studying dolphins' reactions to in-water tourists and the presence of tour boats found that when given the choice, bottlenose dolphins approached swimmers only 34 percent of the time. The rest of the time, the dolphins merely continued with whatever activities they had been involved in prior to a swimmers' entrance into the water. Unfortunately, instances of both tour boat operators and swimmers acting in an intrusive and aggressive manner were not rare. People were observed leaping off boats directly on top of dolphins, lunging at them in the water, and driving vessels in high-speed circles in an attempt to get them to jump. The study concluded that not all dolphins are interested in engaging human activity, that they prefer to choose whether or not to interact among swimmers, and that the majority of the time, dolphins appear to prefer going about their normal lives to engaging humans.

On the other side of the globe, Dr. Kathleen Dudzinski, who studies wild spotted dolphins in the Caribbean, writes, "I could not help but witness the aggressive tendencies of the swimmers toward dolphins. Mostly people thought they could catch a dolphin and, once caught, the dolphins would just love to be touched all over." Understanding that the popular desire to encounter cetaceans at close range is likely only to grow, Dudzinski feels the best way to deal with the expanding human demand is to make every effort to ensure people are properly educated on how to act *responsibly* in the presence of wild cetaceans. Experts such as Dudzinski remind people that they are *guests* in the dolphins' world and that it is a *privilege,* rather than a right, to observe these wild animals in proximity.

Humans may have adverse impacts in other ways when seeking cetacean encounters. On the western coast of Australia, a remote bay has become one of the country's leading tourist attractions, where wild bottlenose dolphins have been greeting people for some forty years in the sandy shallows of Shark Bay. These dolphins draw over 100,000 visitors each year, as daily crowds wade eagerly into the water hoping to touch and feed the dolphins a fish. The feeding of wild dolphins is illegal in many parts of the world, and dolphin researcher Dr. Rachel Smolker is not alone in voicing concern over the level of dependency developed by some of the Shark Bay dolphins on human handouts. This dependence has been implicated in the unusually high infant-mortality rate of dolphin mothers seeking handouts. One infant was savaged by a large shark very close to shore while its mother was distracted, begging for fish from tourists. Additionally, a number of dolphins that once frequented the Shark Bay beach appear to have succumbed to infections and died as a result of the pollution being flushed into the bay—human effluent from the dolphins' many visiting admirers.

In the United States, swimming with cetaceans is illegal and officially considered harassment. Yet removing any legal option for humans and cetaceans to meet eye-to-eye at sea may serve to increase a deadly disassociation in the relationship between humans and wild creatures in their natural world. Researcher and author Erich Hoyt maintains that whale watching need not be a harmful exercise if conducted responsibly, with the cetaceans' needs set before human preference.

Responsible ecotourism requires that people approach cetaceans in a careful, informed, and respectful manner, preferably in small numbers. To reduce the disruption of human presence, experts advise selecting a smaller tour vessel with an experienced naturalist or biologist on board to better interpret cetacean behavior. Experienced, considerate, and reputable tour operators who are familiar with an area and its inhabitants will err on the side of caution when approaching cetaceans, and a good skipper is knowledgeable in maneuvering around cetaceans safely with due care and attention. An accidental encounter with the propeller blades of a boat can leave a dolphin or whale injured or dead. Vessels under sail create far less noise pollution and are much less likely to disrupt or injure cetaceans. Over-trafficked areas should be avoided.

A number of nonprofit organizations offer educational opportunities for the public to participate in scientific cetacean research. In addition to viewing wild dolphins and whales up close, people may learn much about cetacean biology and behavior, enhancing the experience considerably.

For those encountering cetaceans without the benefit of an experienced and responsible guide, taking the time to absorb some factual, practical knowledge regarding cetacean signals and behavior is highly advisable. As a general rule, small, quiet groups are less disruptive, and employing a thoughtful, informed, and respectful approach improves the likelihood of prolonging a close encounter. Dr. Toni Frohoff, a scientist who has studied human-dolphin interactions closely for over a decade, strongly advises people to *never* feed or chase wild dolphins; to avoid abrupt changes in boat speed and direction when cetaceans are nearby; and to refrain from separating or scattering dolphins or whales. She further recommends against interrupting cetaceans that are resting, feeding, mating, or fighting and stresses that one should never grasp, grab, or attempt to impede a dolphin's movements in any way. Experts maintain that the tendency to have too many people in too many boats on the water, all trying to get too close, should be avoided, and guidelines that maximize cetacean comfort and minimize human disturbance should be adopted and enforced. Because strategic land-based watching sites have the lowest impact, their growing popularity is encouraged.

In a world where whales and dolphins are still very much at risk from human hunting practices, ecotourism that is educational and respectful, rather than exploitative, appears to encourage an increasing empathy toward cetaceans and their marine environment. Though the potential for harm should not be ignored, when managed and conducted properly, such wild encounters represent a hopeful turn, enriching the ancient relationship between dolphins, whales, and humankind.

Further Resources

Carwardine, M., & Hoyt, E. (1998). *Whales, dolphins and porpoises.* San Francisco: Time Life Books.

Corrigan, P., & Roger, P. (1999). *The whale watchers' guide.* Minnesota: Northword Press.

Frohoff, T., & Peterson, B. (2003). *Between species: Celebrating the dolphin-human bond.* San Francisco: Sierra Club Books.

Leach, N. (1999). *Whale watching*: Discovery Travel Adventure Insight Guides. New York: Discovery Communications.

Morton, A. (2002). *Listening to whales: What the Orcas have taught us*. New York: Ballantine Books.

Simmonds, M. (2005). *Whales and dolphins of the world*. Cambridge, MA: MIT Press.

Smolker, R. (2001). *To touch a wild dolphin: A journey of discovery with the sea's most intelligent creatures*. New York: Doubleday.

Leah Lemieux

■ Education
Alternatives to Animals in Life Science Education

Within biological science, medical, and veterinary medical education, animals often play a central role in laboratory practical classes. They are used live in experiments to illustrate physiological and pharmacological principles and for acquisition of a range of clinical and surgical skills. They are killed for their tissue and organs and for students to perform dissections in anatomy classes. Tens of millions of animals—perhaps more—are used each year across the world.

In this conventional use of animals for education and training, the relationship between the animal and the student is one of harmful animal use. Animals suffer harm in various forms during capture, breeding, and incarceration, and they suffer pain, injury, and distress in experiments. These may be conducted without anesthetic and with lasting negative impact on the individual animal (if he or she survives). Killing is also a significant form of harm because the most significant freedom that each individual animal has—his or her life—is denied.

Such use is not what most students are expecting when they choose to study life (through biology) or to train to heal people or animals (through medicine). Harmful animal use is counterintuitive for the life sciences, and it may create a learning environment for all students that is not conducive to effective acquisition of knowledge, skills, and responsible attitudes. It can facilitate the process of desensitization, and it may teach that the instrumental use of animals is acceptable and that ethical concerns are unimportant or irrelevant. Harmful animal use may be against the ethical or religious beliefs of some students, and although conscientious objection to such practice may bring success, it may also bring academic or psychological penalty.

The replacement of harmful animal use has been gaining momentum across the world, and progressive, humane alternatives have now fully replaced animal experiments and dissections in many university departments.

Technological innovation, particularly the use of multimedia software to support the learning process, has played a major role in this revolution. The availability of new tools supports good curricular design: a considered approach to meeting teaching objectives effectively. The limitations of harmful animal use and the advantages of new approaches are illustrated by the many published academic studies comparing conventional methods with alternatives: in almost all cases, the alternatives are shown to be equal or superior in terms of student and trainee performance.

The economic advantages of using alternatives play a role as important as the pedagogical. Student pressure and the broader social and cultural changes concerning animals and their ethical treatment contribute further to the ongoing transformation. Alternative models include the following examples.

Mannequins and Simulators

By allowing repeated practice, these alternatives help students gain confidence and competence. Lifelike mannequins can support effective training of clinical and surgical skills such as drawing blood, intubation and critical care, and the perfusion of waste organs in advanced simulators allows for realistic surgery practice.

Multimedia Software and Virtual Reality

Visualization and understanding of anatomical structure and function can be enhanced through the high-resolution graphics, video clips, and other tools available in multimedia software. Virtual labs can encourage an understanding of the interplay between complex and related phenomena and can support the development of problem-solving skills. In true virtual reality (VR), specific clinical and surgical procedures, such as endoscopy, can be practiced in a highly immersive environment, and even the sense of touch—haptics—can be simulated. Just as an airline pilot is expected to train using flight simulators in order to be fully versed with all likely scenarios, so must all students and professionals who will be working with patients have achieved the required mastery.

Ethically Sourced Animal Cadavers and Tissue

All future veterinarians will require hands-on experience of animals and animal tissue. The use of ethically sourced cadavers and tissue is an alternative to the killing of animals for dissection and surgery practice. The term "ethically sourced" refers only to cadavers or tissue obtained from animals who have died naturally or in accidents or who have been euthanized because of terminal disease or serious injury. Body donation programs can provide cadavers in an ethical way.

Clinical Work with Animal Patients

Student access to clinical learning opportunities could be significantly increased to replace animal experiments and to better prepare students for the professions. A progressive approach to learning veterinary surgery might involve mastering basic skills using non-animal alternatives, then using ethically sourced cadavers for experience with real tissue, and finally performing supervised work with animal patients. Shelter sterilization programs are an important potential resource, with castration and spays observed, assisted, and then performed by students. The clinic can also teach students many other skills that the lab cannot, such as postoperative care and recovery of patients.

Student Self-Experimentation

For further experience of the whole, living body, the consenting student is an excellent experimental animal, particularly for physiology classes. The intense involvement and self-reference of such practicals makes them highly memorable and supports effective learning.

In Vitro Labs

The rapid development and uptake of in vitro technology in research and testing needs to be supported by student familiarity with the techniques. Animal tissue and cells used for in vitro practicals can be sourced ethically, and within some cell biology practicals, the use of animal tissue can be replaced directly with plant material.

Field Studies

Students may study animals in a laboratory setting as a model for nature or may face invasive or otherwise harmful interaction with wild animals. However, biology is not just experimentation, nor does its study require harm. Studying animals within their natural environment can be a particularly rewarding alternative.

In a fully humane education, the negative relationship with animals can be transformed. The important cultural values and skills of compassion, empathy, respect for life, critical thinking, and ethical decision-making are valued and developed alongside the acquisition of more standard knowledge and skills.

The International Network for Humane Education (InterNICHE) works with teachers to introduce alternatives and with students to support freedom of conscience. Resources developed by InterNICHE to catalyze change include the information-rich multi-language book *From Guinea Pig to Computer Mouse* (2nd ed.), which presents case studies, background information on curricular design and assessment, and details of over 500 alternatives; the Alternatives Loan System, an international library of over 100 mannequins, simulators, and software items available to teachers and students; the Humane Education Award, an annual grant program of 20,000 euros for the development and implementation of alternatives internationally; the multi-language Web site http://www.interniche.org, with a wide range of news, information, and resources available for free download; the major InterNICHE Conference; and outreach visits and training across the world.

Alternatives to harmful animal use are possible for all practical classes within the life science disciplines. In many departments, the word "alternative" may not even be used because these are now the standard teaching approaches—in some cases, examples of best practice—sometimes backed by laws and regulations stating that alternatives should be used wherever possible. The multiple positive impact of alternatives means that replacement can be to the benefit of students, teachers, animals, the life sciences, and society itself. Continuing effort is required to replace the remaining harmful animal use internationally, but the growing success of implementation illustrates how science education and ethics can indeed be fully compatible.

Further Resources

Jukes, N., & Chiuia, M. (Eds.). 2003. *From guinea pig to computer mouse: Alternative methods for a progressive, humane education* (2nd ed). Leicester, UK: InterNICHE.

Nick Jukes

■ Education
A History of Human-Animal Studies

Human-animal studies (HAS) is a rapidly growing field that is devoted to the study of human-animal relationships. Because it is an interdisciplinary field, scholars from many different fields contribute to it. As African American studies and women's studies paralleled the rise, respectively, of the civil rights movement and feminism, HAS arises at a time when another progressive social movement, the animal protection movement, has become part of the political agenda.

HAS examines and critically evaluates the complex and multidimensional relationships between humans and animals. Although some fields largely are defined by preferred method of investigation, HAS is defined by its subject matter. Through exploring human-animal relationships, we acquire a greater understanding of the ways in which animals figure in our lives and we in theirs.

The term "animal" in "human-animal relationships" refers to nonhuman animals, although we humans are also members of the animal kingdom. In fact, one area of study in HAS attempts to understand the historical origins of this peculiar way of categorizing the two groups and the impact of that "category mistake" on human-animal relationships. For example, historians of religion describe how, in the early Judeo-Christian tradition, a strong distinction was made between humans and other animals. Considered the only beings made in the likeness of God and provided with souls, humans were given stewardship over all the other animals. Lisa Kemmerer, a philosopher interested in language, has suggested a new word to deal with this "lexical gap"—there is no one word in the English language for any animal who is not a human. Anymal (pronounced ene-mel) is a contraction of "any" and "animal" and is a shortened version of the concept "any animal that does not happen to be the species that I am."

In addition to over a million known species, there are many different kinds of anymals for which, through our dealings with them, we have developed many groups with whom we have distinguishable forms of relationships. Consider companion, farm, lab, wild, park, feral, genetically engineered, introduced, endangered, totemic, symbolic, cartoon, and fictionalized anymals. All of these human-anymal relationships, whether real or virtual, historical or contemporary, factual or fictional, beneficial or detrimental, fall within the purview of HAS. Given this broad scope, scholars from many fields within the social sciences (psychology, sociology, political science, anthropology, geography) and the humanities (philosophy, history, religion, literary criticism, feminist studies, postcolonial studies, rhetoric, legal studies) are conducting investigations in HAS.

Here are a few examples: in the social sciences, Gene Myers, a psychologist, through field work in a preschool setting, examined the development of young children's perception, conceptualization, and first relationship with animals; working in an animal shelter, sociologists Janet and Steven Alger described the subculture of a group of cats as they live with humans in this specialized setting; and, on the basis of data from focus groups, geographers showed how a culturally based practice involving anymals (dog-eating) resulted in the mainstream culture viewing a particular ethnic group (Filipinas in Los Angeles) as a subordinate group based on race ("racialization"). In the humanities, philosophers and legal scholars developed arguments that anymals are part of the moral community, that they can be wronged, and that at least some classes of anymals should be given their day in court; Kathy Grier, a historian, showed how companion animals became an important part of the socialization of children in nineteenth-century America; scholars in women's studies show how discrimination against women is built on their prior association through language usage ("nice piece of meat") with a degraded view of anymals; and literary theorists critique authors' treatment of anymals in fiction as property, symbol, or fully developed character.

The breadth of HAS notwithstanding, not all studies involving anymals fall within its scope. A field study of chimpanzees or a veterinary study of feline cancer does not directly investigate human-anymal relationships. Although they can provide limits to the forms of human-anymal relationships possible, they are not part of HAS proper. More subtly, when an investigator exclusively considers anymals as cultural artifacts, symbols, models (stand-ins), or commodities to understand human culture, psychology, literature, or economics, the resulting study is not a direct contribution to HAS. More typically in HAS, the investigator studies an anymal as an individual in a relationship. Of course, sometimes the form of that relationship is one in which the anymal has been commodified or objectified. As feminist studies often expose relationships in which women have been exploited, so do HAS studies reveal disrespectful human-anymal relationships.

In assessing the robustness of an academic field, we can distinguish between scholarship and the institutional infrastructure that supports it. In terms of scholarly output, the growth of HAS in the past two decades is remarkable, as evidenced in the following representative list: devoted peer-reviewed journals (*Anthrozoos*; *Society and Animals*), book series (*Animals, Culture, and Society*, Temple University; *Human-Animal Studies*, Brill Academic), major academic publishers (Harvard University; University of California; University of Chicago), major commercial publishers (Blackwell; Harper-Collins; Wiley), and conferences (Animals and Human Society, held at Princeton; Animals and Religion, held at Harvard).

Another measure of scholarly output is the number of budding scholars choosing HAS as a field for their doctoral dissertation. Kathy Gerbasi has documented that the number of completed doctoral dissertations in HAS in the 1990s was double the number completed in the 1980s, a rate of growth faster than that of dissertations in any other academic discipline.

Gaining a permanent foothold within academic institutions and professional organizations is a more difficult task than sheer scholarly output. The American Philosophical Association and the American Sociological Association both have recognized groups whose focus is HAS. Efforts to install interest groups or sections or divisions within other disciplines, such as psychology, to date have not been successful. About 200 courses in HAS now are being offered in various fields. Since 1999 the Humane Society of the United States has been giving an annual award to the best established, new, and innovative HAS courses. An example of an innovative course is an offering for instructors in various fields. This course has the potential multiplier effect of teaching teachers to develop their own course in their own field. In addition to the fields listed earlier, other disciplines are producing courses, such as a course on animals in the visual arts.

As Kathy Gerbasi found in her study of HAS dissertations, the distribution of university-based scholars mentoring doctoral candidates and, in most instances, developing HAS courses is very scattered. The lack of a critical mass of scholars at most institutions is a barrier to the further development of institutional infrastructure. Although a few fields (e.g., sociology, philosophy, and legal studies) now have a robust core of scholars who publish in HAS and who network via e-mail lists (e.g., the Human-Animal-Studies electronic mailing list), the field is just beginning to establish minors, majors, and programs. Only one major has been established in the United States—"Animals in Human Society" is one of four majors in the sociology department at Notre Dame de Namur. A group of scholars (the Executive Committee for the Development of Human-Animal Studies) has been meeting for over two years to identify and support prospective concentrations, minors, and majors at universities. Two U.S. universities and one Canadian university do offer a master's degree in animal welfare, through their veterinary schools. These programs could potentially develop the broader curriculum of HAS.

The association with veterinary schools (a professional rather than academic school) suggests an advantage of a more robust presence of HAS in universities. The field provides a broad background for animal-related careers. Many careers involve the direct care of animals (e.g., shelters, sanctuaries, refuges, rehabilitation centers, and veterinary facilities). Others involve the use of or scientific study of animals (e.g., research labs, teaching labs, farms, zoos, wildlife management, marine biology, and animal-assisted therapeutics). There also are many careers available in the animal protection movement.

Part of the gap between the growth in scholarship and the limited development of institutional infrastructure is typical of the development of any new field. But some blocks are peculiar to HAS. We already have referred to the linguistic gap. Beneath this is the mainstream culture's view that anymals are categorically distinct from and subordinate to humans.

Within the politics of universities and colleges, any interdisciplinary field is disadvantaged by universities' organization into separate disciplines, each of which vies with the others for a limited budget.

Certain limitations to and challenges of the scholarship of HAS also present barriers to its institutional development. First, the field needs a more solid theoretical base. To date, most of the theories offered are extensions of existing human-based theories rather than theories peculiar to HAS. For examples, both Tom Regan's rights theory and Peter Singer's utilitarian theory are arguments to extend rights and moral claims to anymals; sociologists have argued that symbolic interactionism can extend to anymals, and psychologists offer evidence that attachment theory and self-psychology help us understand human-anymal relationships. These are helpful to the degree that they explain and predict the various forms of human-anymal relationships, but we await the acceptance of a theory especially devised to explain these phenomena.

In the area of method, there are comparable as well as additional problems. Because anymals do not have full-blown language and, in any case, are perhaps not given to introspection in the same way as humans, studies of their subjective world in a particular human-anymal relationship are limited, but not impossible. Anymals do express their experience through various forms of communication, gesture, and behavior. On the other hand, where the independent variable is behavior, studies involving anymals are more reliable than studies of humans because anymals less often hide their intentions for the sake of social desirability or to impress or maintain a particular self-image. What you observe is what you should get. Finally, measurement and understanding of relationships as compared to individual-based phenomena are more difficult.

In summary, HAS is a rapidly growing field that has drawn scholars from many fields. A robust body of scholarship is now available; institutional representation of the field has a ways to go.

See also

Education—*Human-Animal Studies*

Further Resources

Alger, J. M., & Alger, S. F. (2003). *Cat culture: The social world of a cat shelter*. Philadelphia: Temple University Press.

Arluke, A., & Sanders, C. R. (1996). *Regarding animals*. Philadelphia: Temple University Press.

Gerbasi, K. C., Anderson, D. C., Gerbasi, A. M., & Coultis, D. (2002). Doctoral dissertations in human-animal studies. *Society and Animals, 10,* 330–46.

Grier, K. (2006). *Pets in America: A history*. Chapel Hill: University of North Carolina Press.

Griffith, M., Wolch, J., & Lassiter, U. (2002). Animal practices and racialization of Filipinas in Los Angeles. *Society and Animals, 10,* 221–48.

Kemmerer, L. (2006). *Ethics and animals: In search of consistency.* Leiden: Brill Academic.

Myers, O. E. (1998). *Children and animals: Social development and our connections to other species.* Boulder: Westview.

Shapiro, K. (2002). The state of human-animal studies [Special issue]. *Society and Animals, 10*(4).

Waldau, P. (2002). *The specter of speciesism: Buddhist and Christian views of animals.* Oxford: Oxford University Press.

Kenneth Shapiro

■ Education
Human-Animal Studies

Human-animal studies (HAS) is the interdisciplinary study of human-animal relations. At times referred to as animal studies, animal humanities, critical animal studies, or anthrozoology, it examines the complex interactions between the worlds of human and other animals.

First, HAS is a young discipline and one of the fastest-growing areas of research in the academy. Its growth to date is akin to that of environmental studies in the latter decades of the twentieth century. At the beginning of the twenty-first century, HAS is primarily located in established disciplines as either an official or de facto subfield. For example, one can find HAS courses, research, and special interest groups in a variety of the social and interdisciplinary sciences, such as anthropology, geography, political science, psychology, and sociology (to name a few). HAS is equally well represented in the arts and humanities, such as the fine arts, cultural studies, history, literature, philosophy, and religious studies (to name a few more). HAS has also institutionalized itself through journals (e.g., *Anthrozoos; Society and Animals*), book series (e.g., Brill; Temple University Press; Columbia University Press), and international societies and online networks (e.g., International Society of Anthrozoology; Nature in Legend and Story e-network), as well as policy institutes that use the fruits of scholarship (e.g., Institute for Society and Animals). The proliferation of HAS opens up social and conceptual space for the creation and use of academic publications and ensures its long-term viability in higher education. HAS is also emerging as the focus of undergraduate minors and majors in colleges and universities across the world. Currently, the Center for Animals and Public Policy at Tufts University offers the only graduate-level program where students can engage in human-animal studies full-time.

Second, HAS emerged in response to three problematic ways of understanding animals. The first is the failure of the natural and behavioral sciences to adequately address the sentience, sapience, and agency of many animals. The second is the recognition of anthropocentrism and speciesism as prejudicial paradigms that distort our moral relationship with other people, animals, and the rest of nature. The third is a burgeoning interest in the cultural, social and political place of animals in human societies. A clarion response to these problems was the publication of two books by Mary Midgley— *Beast and Man* (1978) and *Animals and Why They Matter* (1984). Both texts were intended in part as responses to the ethical and scientific blinders of behaviorism, genetic determinism, and sociobiology. Midgley is arguably the field's most celebrated scholar, and her incisive critiques of ethical, philosophical, and scientific themes inspired scholars to consider the animal question as a serious subject of study.

Third, the interdisciplinarity of HAS produces a wealth of theories, methods, and topics. Scholars have approached HAS from diverse theoretical positions, ranging from empiricism and positivism to postcolonialism and ecofeminism. They undertake their studies using qualitative, quantitative, mixed, and other methodologies. And their topics touch on wild, companion, farm, and research animals. Although this plurality generates a vibrant dialogue that should be praised, it can also obscure fundamentally different approaches to ethics, science, and society. This is becoming something of an unacknowledged struggle for HAS, as positivists and anti-positivists begin to clash in conferences, faculty meetings, seminars, and publications. This is to be expected because the positivist claim to undertaking value-free and objective science is widely discredited, and the anti-positivist alternatives represent such a diversity of theoretical and methodological points-of-departure that it is both impossible and undesirable to establish a unitary point of view. HAS is laden with incommensurable points of view on naturalistic versus humanistic models of science, quantitative versus qualitative methodologies, and value-free versus value-forming scholarship. These clashes have not become the primary focus of debate as of yet, but they bear watching as sources of rough weather.

Fourth, like any academic field with environmental or social relevance, there is an ongoing tension between scholarship and activism. The perspectives of activists for animal welfare, protection, or rights are a source of inspiration and insight to the academy and society alike. Yet scholarship and activism are neither identical nor inseparable. Some scholars and students have precommitments to animal social movements, and for reasons of academic freedom and social relevance, this is well and good. Even so, the intellectual requirement of social movements frequently engenders dogmatic approaches to moral and political problems. This may serve advocates well as they mobilize support for their positions. It is antithetical, however, to the best norms of scholarship that aspire to theoretical and methodological rigor or the plurality of subject matter that may not be popular in activist circles. The trick to managing this tension is not to privilege one concern and discourse over another, but to allow each to inform and challenge the other.

Fifth, HAS will face crucial challenges in the years ahead. One such challenge has to do with its legitimacy in the academy. HAS (broadly understood) is of obvious interest to a great many people. The popularity of nature videos, animal-focused ecotourism, birdwatching and the like are well established. So too is the emergence over the past decade of "animal art" as a subject for study and creation. Interestingly, I was told several years ago by a curator at the Ashmolean Museum in Oxford, England, that animal art was for children and not a fitting subject for serious study. Even so, HAS is likely to receive a cool reception in many established disciplines. The reasons for this will vary, but will include the following:

- Hostility toward animals as a serious subject of study
- Fears that interdisciplinary fields such as HAS diminish students and resources for established departments
- Theoretical imperialism and the distaste for upstart disciplines that do not toe the theoretical line
- Advocacy concerns that a focus on the well-being of animals will detract from the well-being of humans
- Censorship by university administrators who fear that HAS will jeopardize corporate and government sources of funding

HAS will have to face all these concerns directly if its efforts at institutionalizing itself are to avoid being undermined by ivory tower politics.

Another challenge has to do with creating a global learning community out of interdisciplinary plurality. HAS draws insights from many disciplines, theories, methods, topics, and experiences. These insights are drawn not only from North America or the animal protection movement, but also from places and identity groups across the globe. The globalization of HAS will likely continue in the years ahead. This then raises questions about how academics and others learn to generate a body of knowledge that is open to a wide diversity of perspectives, without lapsing into a lazy relativism about knowledge claims or moral norms. Grappling with the problem of relativism—and its opposite, objectivism—will likely require an ongoing debate over the status of situated knowledge in both science and ethics. It will also require ongoing attention to the forums for dialogue that creates the possibility for such knowledge.

See also

Education—*A History of Human-Animal Studies*
Law—*Public Policy and Animals*

Further Resources

Baker, S. (2001). *Picturing the beast: Animals, identity, and representation.* Urbana: University of Illinois Press.

Balcolmbe, J. P. (1999). Animals and society courses: A growing trend in post-secondary education. *Society and Animals, 7*(3), 229–40.

Bekoff, M. (Ed.). (1998). *Encyclopedia of animal rights and animal welfare.* Westport, CT: Greenwood Press.

———. (Ed.). (2004). *Encyclopedia of animal behavior.* Westport, CT: Greenwood Press.

Franklin, A. (1999). *Animals and modern cultures: A sociology of human-animal relations in modernity.* Thousand Oaks: Sage.

Jamieson, D. (2002). *Morality's progress: Essays on humans, other animals and the rest of nature.* New York: Oxford University Press.

Lavigne, D. (Ed.). (2006). *Gaining ground: In pursuit of ecological sustainability.* Limerick, Ireland: University of Limerick Press.

Lynn, W. S. (2002). *Canis lupus* cosmopolis: Wolves in a cosmopolitan worldview. *Worldviews, 6*(3), 300–27.

———. (2004). Animals. In S. Harrison, S. Pile, & N. Thrift (Eds.), *Patterned ground: Entanglements of nature and culture.* London: Reaktion Press.

Midgley, M. (1995). *Beast and man: The roots of human nature.* London: Routledge.

———. (1998). *Animals and why they matter.* Athens: University of Georgia Press.

———. (2005). *The essential Mary Midgley.* New York: Routledge.

Patton, K., & Waldau, P. (2006). *A communion of subjects: Animals in religion, science and ethics.* New York: Columbia University Press.

Philo, C., & Wilbert, C. (Eds.). (2000). *Animal spaces, beastly places: New geographies of human-animal relations.* London: Routledge.

Rollin, B. E. (2006). *Science and ethics.* Cambridge: Cambridge University Press.

Sax, B. (Ed.). (2001). *The mythical zoo: An A–Z of animals in world myth, legend and literature.* New York: ABC-Clio.

Wolch, J., & Emel, J. (Eds.). (1998). *Animal geographies: Place, politics and identity in the nature-culture borderlands.* London: Verso.

Wolf, C. (Ed.). (2003). *Zootologies: The question of the animal.* Minneapolis: University of Minnesota Press.

William S. Lynn

■ Education
Humane Education

Humane education explores all the challenges facing our planet, from human oppression and animal exploitation to materialism and ecological degradation. It explores how we might live with compassion and respect for everyone—not just our friends and neighbors, but all people; not just our dogs and cats, but all animals; not just our own homes, but the earth itself, our ultimate home. Humane education inspires people to act with kindness and integrity and provides an antidote to the despair that many feel in the face of entrenched and pervasive global problems and persistent cruelty and abuse toward both people and animals. Humane educators cultivate an appreciation for the ways in which even the smallest decisions we make in our daily lives can have far-reaching consequences. By giving students the insight they need to make truly informed, compassionate, and responsible choices, humane education paves the way for them to live according to abiding values that can lend meaning to their own lives while improving the world at the same time.

The term "humane education" originated in the late nineteenth century as founders of societies for the prevention of cruelty to animals (SPCAs) and child-protection organizations (often the same people) realized the importance of teaching children the principles of kindness and respect for others, both human and nonhuman. For many decades in the late twentieth century, humane education became synonymous with elementary-level school programs that primarily taught children about kindness toward and care of companion animals. As the crisis of dog and cat overpopulation grew, humane education began to focus on the importance of spaying and neutering. With the emergence of dog fighting as a popular "sport" among some communities, humane education programs often discussed the cruelty inherent in dog fighting and also offered bite prevention presentations.

In the 1990s, several humane education programs emerged that expanded the then-limited perception of humane education, returning to humane education's roots. These programs focused on the definition of the word "humane" (meaning "having what are considered the best qualities of human beings") and applied this definition to our relationships with everyone: all animals, people, and the earth.

Humane education now encompasses animal-protection education, environmental and sustainability education, media literacy, character education, and social justice education. It is the only educational movement that currently does so. Drawing the connections between all forms of oppression and exploitation, humane education empowers and inspires students to be change-makers who have the skills not only to connect the dots between various problems and forms of abuse, but also to find solutions that work for everyone.

Quality humane education accomplishes its goals through the use of four elements:

1. Providing accurate information—so that students understand the consequences of their decisions as consumers and citizens
2. Fostering the three Cs: curiosity, creativity, and critical thinking—so that students can evaluate information and solve problems on their own
3. Instilling the three Rs: reverence, respect, and responsibility—so that students will act with kindness and integrity
4. Offering positive choices that benefit oneself, other people, the earth, and animals—so that students are able to help bring about a better world

Humane education achieves these goals through interactive and engaging teaching techniques that model compassion, respect, and openness.

Providing Accurate Information

In order to make the kindest and wisest choices, we need knowledge. For example, unless we know about the problem of dog and cat overpopulation, the abuse of farmed animals in factory farms, the plight of women and children working in sweatshops, the dangers of certain products and chemicals to the environment, or the escalating travesty of worldwide slavery (to name a few issues), we cannot make informed, conscious, and humane choices that help solve these growing threats and problems. With knowledge, however, individuals, businesses, and governments are able to make choices that do not cause suffering and destruction. Humane educators empower their students by offering them accurate information so that the students can make wise and compassionate decisions both personally and as emerging members of a democracy.

Fostering the Three Cs: Curiosity, Creativity, and Critical Thinking

Humane educators do more than expose students to hidden truths. They teach the critical thinking skills necessary to evaluate information, as well as foster curiosity and creativity so that students pursue lifelong learning and imaginative, yet practical, solutions to difficult problems. When one visits a school where humane education is in progress, one may find students analyzing popular advertising or reading pamphlets from opposing groups, trying to separate fact from opinion. Students may be working together to develop creative answers to challenges often portrayed in either–or terms; crafting persuasive essays on various issues; tracing the effects on animals, people, and the environment of certain products and behaviors; or coming up with ideas for everything from proposed legislation to meaningful disclosure on product labels. Humane educators inspire their students to think about, consider, and creatively and positively respond to norms and attitudes that are often accepted without question, from what is served in the cafeteria to how and where the school uniforms are produced, to the use and disposal of paper in the school, to dissection in biology classes, and much more.

Instilling the Three Rs: Reverence, Respect, and Responsibility

Without the three Rs of reverence, respect, and responsibility, the acquisition of knowledge and improved critical and creative thinking abilities by themselves will generally fail to inspire a person to take the necessary steps toward solving problems and making kinder and more positive choices in their lives and communities.

Reverence is an emotion akin to awe. What people revere, they tend to honor and protect. If young people have reverence for life, for other humans, for animals, and for the planet Earth, they are more likely to find the will to make choices that diminish harm to others and create more peace. Respect is an attitude people bring to the world; it is reverence manifested in interactions. Responsibility is respect turned into action. When young people are filled with reverence, and when they feel respect for others, taking responsibility for their actions and choices is an inevitable next step.

How do humane educators cultivate the three Rs? Through age-appropriate activities, reflections, field trips, opportunities to meet people and animals who have been exploited or abused, stories, pictures, and films, humane educators awaken the hearts and souls of their students and ignite their love for this earth, its people, and its animals. They spark students' innate empathy, so that respect follows easily and the motivation to take responsibility (again, in age-appropriate ways) is the likely result.

Providing Positive Choices

Humane educators do not tell students what to think or what to do (which would be the opposite of teaching critical and creative thinking), but they do make sure that students know that they have choices that can improve or diminish the world, end suffering or contribute to it, solve problems or perpetuate them.

This fourth element of humane education is the one that makes the rest meaningful. If students are exposed to the problems in the world and the suffering and destruction that abound, but are given no tools or choices to make a difference, they may become cynical and apathetic, exactly the opposite outcome from what humane education tries to achieve. When, instead, humane educators introduce students to innovative ideas and inspiring successes, and they provide examples of ways in which individuals, communities, corporations, and governments can make a lasting positive contribution, they pave the way for young people to become visionary entrepreneurs, leaders, change agents, and engaged citizens in both small and large ways.

When these four elements come into play, young people not only become aware of the challenges facing animals, people, and our planet, but also learn to trust that they can make a difference, and they become more enthusiastic and committed citizens. Their education becomes deeply meaningful, and their lives may take on a purpose greater than simply good grades or a future lucrative career. For those students whose future may appear bleak, humane educators offer hope, meaning, and solidarity, empowering such students to create a better future for themselves as well as for others, drawing links between the oppression of other species and oppressive systems in our society that affect those who are disenfranchised.

Humane education has the capacity to change the world through educating a new generation to be caring, compassionate, and responsible. As humane education is integrated into curricula and as humane educators are hired by schools in the same number as math or language arts teachers, students will gain the knowledge, opportunity, and will to live with more respect for others, be they other humans, other animals, or the ecosystems that support us all.

Further Resources

Bekoff, M. (2000). *Strolling with our kin*. New York: Lantern Books.

Lickona, T. (1991). *Educating for character: How our schools can teach respect and responsibility*. New York: Bantam.

Orr, D. (1994). *Earth in mind*. Washington, DC: Island Press.

Seed, J., Macy, J., Fleming, P., & Naess, A. (1988). *Thinking like a mountain: Toward a council of all beings*. Gabriola Island, British Columbia: New Society Publishers.

Selby, D. (1995). *EarthKind: A teachers' handbook on humane education*. Stoke on Trent, UK: Trentham Books.

Weil, Z. (1990). *Animals in society: Facts and perspectives on our treatment of animals*. Jenkintown, PA: AVS.

———. (1994). *So, you love animals: An action-packed, fun-filled book to help kids help animals*. Animalearn. Jenkintown, PA: AAVS.

———. (2003). *Above all, be kind: Raising a humane child in challenging times.* Gabriola Island, British Columbia: New Society Publishers.

———. (2004). *The power and promise of humane education.* Gabriola Island, British Columbia: New Society Publishers.

Zoe Weil

■ Education
Information Resources on Humans and Animals

Human-animal interactions represent a vast area of study, and any single aspect of this broad subject may warrant further investigation. For those seeking useful information and resources, whether for a class assignment, professional relevance, or general interest, a more detailed search is necessary. Although anecdotal information is readily available via such Internet search engines as Google, relevant research results are not. Identifying and then locating reliable and authoritative information on a specific topic within the human-animal relationship is challenging because the field remains widely dispersed. But a successful search may be accomplished through the use of a variety of databases and Web-based resources.

How does a counselor begin searching for information on how the elderly benefit from a relationship with their pet or on how caring for animals helps rehabilitate prisoners? Where does the potential graduate student begin to search in order to locate information on programs incorporating the human-animal relationship or on what careers exist in animal-related fields? How does the veterinarian learn more about how to help her clients deal with the loss of their pets? In all these cases, the best practice for locating reliable, useful, and quality information requires the searcher to consider both *how* to search, including the approach and terminology, and *where* to search among the many options.

How

It is important to consider what exactly is wanted and to make notes of terms commonly used in the area. By considering all of the synonyms, the search is expanded significantly, improving the likelihood of retrieving more of the relevant information. Search terms are much more than simply the topic: human-animal interactions or human-animal relationships. Other broad search terms might include "human-animal interactions," "animal-human interaction," "human-animal bond," "animal companions," "pet-client relationship," "human-animal relationships," and "pets and companion animals." In order to narrow the search results to those of a particular area or focus, additional specific search terms, or synonyms, are necessary. Using the elderly and their pets as an example, specific search terms might include the following: elder, elderly, aged, geriatric, Alzheimer's, dementia, nursing home, or convalescent; as well as pet, dog, cat, bird, animal companion, or domestic animals. The terms selected will directly affect the quality of the search results.

Where

There are many options for locating relevant high-quality materials, dependent on both interest and availability. Information is published and available in scientific journals and books, as well as on the Internet via Web-based publications and resources. Finding current, relevant journals and identifying which journal article or which book chapter may be of interest remains challenging. In order to learn about the most recent developments in the field, researches need databases. A variety of databases exist to assist in identifying potentially useful books, journal articles, and Web sites.

Books

Large public online catalogs and databases are available to help identify books published in the area, from the human-animal bond to the history of cat domestication. Once specific titles are identified, the next step is to request the material through interlibrary loan from the local library.

Melvyl is the comprehensive catalog of the University of California libraries and the California State Library, containing records for books, journals, movies, dissertations, and government documents. The catalog is available worldwide via the Web and may be searched by title, author, keyword, or subject.

The Library of Congress, as the nation's library, is the largest library in the world. Its online catalog is a database of records representing the vast collection of materials held by the library, containing approximately 14 million records, including books, serials, manuscripts, and visual materials dating from 1898 to the present. It too may be easily searched by title, author, subject, and keyword.

The British Library is the national library of the United Kingdom and is another amazing resource that is available to all. Searching the Integrated Catalogue provides a European perspective on the topic.

Journal Articles

Bibliographic databases allow searching by a topic to identify precise articles that have been published. Rather than browsing through potentially useful journals, searching in particular databases for pet loss or zoonoses will point the way to the exact article title, journal, year, and page number. Although most of these tools are only available with paid subscriptions, a few are available free to the general public. The primary free databases that index in the area of human-animal interactions are AGRICOLA, PubMed, and GoogleScholar.

AGRICOLA is a free database published by the National Agricultural Library, indexing books and journal articles from all areas of agriculture and related disciplines, including animal science and veterinary science. For example, a keyword search of "detection dogs" will retrieve a list of citations to journal articles that may be of interest on that topic.

PubMed, by the U.S. National Library of Medicine, provides citations and abstracts for articles published in journals in the fields of medicine, life sciences, health administration, veterinary medicine, and others. Performing a keyword search such as "pets and mental health" or "pet loss and grief" will identify relevant and authoritative research articles.

GoogleScholar searches for scholarly literature on the Web, including peer-reviewed journal articles, theses, books, and abstracts from a variety of academic publishers, professional societies, and universities. The Advanced Scholar Search function allows searching by phrase, author, or keywords, retrieving resources on such varied topics as animals used in commercials and environmental enrichment for laboratory animals.

Web Sites

Finally, organizations, professional societies, and academic centers often develop resources and provide access to information that might be otherwise unavailable. These may be self-published resources, such as *The Guide to Pet Loss Resources* by David Anderson, or access to materials, such as abstracts from conference proceedings available at the International Society for Anthrozoology (ISAZ) Web site. The UC Center for Animal Alternatives at the University of California–Davis focuses on organizing information related to service animals and animal-assisted therapies, and the Center for the Interaction of Animals and Society (CIAS) at the University of Pennsylvania provides a forum for addressing many practical and moral issues arising from the interactions of animals and society.

Human-animal interactions represent an enormous and diverse area of study. Research and publication in the field is equally widely dispersed, creating a significant challenge to identifying authoritative information and reliable resources, and locating definitive answers. A variety of useful tools and approaches have been outlined, enabling the user to pinpoint particular literature of interest in the field.

Further Resources

Anderson, D. (2006). *The guide to pet loss resources* (3rd ed.). Lincoln, CA: RockyDell Resources. Available at http://rockydellresources.homestead.com/petloss.html

The British Library: The world's knowledge. http://www.bl.uk

Center for the Interaction of Animals and Society, University of Pennsylvania School of Veterinary Medicine. http://www2.vet.upenn.edu/research/centers/cias

GoogleScholar. http://scholar.google.com

International Society for Anthrozoology. http://www.vetmed.ucdavis.edu/CCAB/isaz.htm

The Library of Congress Online Catalog. http://catalog.loc.gov

Melvyl: The Catalog of the University of California Libraries. http://melvyl.cdlib.org

National Agricultural Library. *AGRICOLA*. http://agricola.nal.usda.gov

National Library of Medicine. *PubMed*. http://www.ncbi.nlm.nih.gov/entrez/query.fcgi?db=PubMed

University of California Center for Animal Alternatives. http://www.vetmed.ucdavis.edu/Animal_Alternatives/main.htm

Wood, M. W. (2006). Techniques for searching the animal-assisted therapy literature. In A. Fine (Ed.), *Handbook on animal-assisted therapy: Theoretical foundations and guidelines for practice* (2nd ed., pp. 409–19). New York: Elsevier.

Mary W. Wood

■ |Education
|Nurturing Empathy in Children

Fostering the development of a caring person is a major goal of many child-education programs, including humane education, which helps people treat animals humanely. There are more unknowns than knowns in the development of empathy. In addition, very little research has been conducted into the development of empathy toward other animals per se. Nonetheless, enough is know to furnish some insight into how empathy develops and how to help nurture its growth into prosocial behavior. It is also reasonable to expect that the development of empathy toward other animals follows a similar track to development of empathy toward other people, although research into this is sparse.

Humans have a natural tendency to be empathic (Zahn-Waxler et al., 1992). Even young children appear to react with concern toward others' distress, largely independent of caregiver influences. As children develop, however, their inborn empathy can flourish or atrophy, depending on what they do and how others react. Early in life, the most important relationship for developing empathy is that with one's primary caregiver. When this relationship is healthy—that is, when a child gains comfort from the caregiver and has learned enough trust to withstand the caregiver leaving—then empathy, cooperation, social competence, sociability, and personal resilience develop easily (Ainsworth et al., 1978; Bowlby, 1988; Bryant & Crockenberg, 1980; Yarrow et al., 1973).

Another important component of the caregiver–child relationship is the way morals are enforced and the child is disciplined. The optimal strategy appears to be setting and compassionately enforcing age-appropriate standards of behavior (Baumrind, 1971). Neither overly permissive nor overly punitive caregiving instills social responsibility. When one child hurts another, psychologist Martin Hoffman and others contend that the most effective way of disciplining a child (to help them internalize morals) is to bring attention both to the victim's distress and to the child's role in causing it. Done correctly, this helps children develop a sense of guilt for hurting others.

Indeed, asking children to develop empathy means exposing them to more suffering. To lead children to empathize and then leave them there helpless to it, so to speak, may be counterproductive and even inhumane. The children may simply learn to turn away from other's sorrow to protect the tender feelings they have developed. Therefore, children should also be taught effective ways of helping others out of distress. Learning that they can effectively help others does more than let children endure empathy and guilt; it shows children through their own actions that they are caring, competent individuals. Indeed, children from cultures where they are routinely required to help others are more prosocial than children from cultures in which children are not expected to do this (Whiting & Whiting, 1973, 1975).

As children grow into a larger world, more factors influence their empathy and prosocial behaviors. Role models become important (e.g., Elliot & Vasta, 1970; Rushton, 1975), especially successful ones who are close to the child (Eisenberg & Geisheker, 1979). The overall community and its norms for the way people treat each other is another powerful influence on young people's prosocial behavior (Ianni, 1992).

A mother teaches her young son the importance (and joy) of caring for a small rabbit. Courtesy of Shutterstock.

A veterinarian teaches children how to care for a hamster. School performance among these children increased significantly with animals present. Courtesy of Ceres Faraco, PAN's South American Advisor and President of the Brazilian Veterinarian Welfare and Ethics Committee, Porto Allegro, Brazil.

People who are empathic in one situation do tend to be empathic in other situations (Elliot & Vasta, 1970). In addition, nurturing empathy in one situation can encourage it in others. For example, professor Frank Ascione et al. (1992, 1996) found that teaching children empathy toward animals can generalize to humans. However, empathy does not always generalize. Hitler, after all, cared lovingly for his dogs. It seems prudent, then, not to rely on empathy generalizing from where it is found to where one wants it to go: to create kindness toward animals, it is surely best to cultivate it there.

Developing this sort of relationship with animals should help develop empathy toward them as it does toward people. Nurturing empathy toward animals is not exactly the same as nurturing it toward people, however. One barrier that is probably higher in developing empathy toward animals is that people must learn how animals express their emotions—or even that they have emotions. In addition, cultural differences in the treatment of animals must be respectfully considered and realistic objectives set.

Humane educators have successfully nurtured empathy toward animals by employing the strategies previously described in brief. Especially with younger children, humane educators help students understand that animals can feel pleasure, pain, and other emotions and sensations; somewhat older children (e.g., fourth graders) can begin to identify some of the ways animals evince their emotions. Beginning in late elementary school, students not only can benefit from frequent exposure to role models, but also can start to be models themselves. Being a role model may foster prosocial behavior even more powerfully than seeing models.

It is not reliable to expect empathy to generalize. In any case, children (and adults) learn best when they learn what they will actually be expected to do. Therefore, empathy and prosocial behavior can flourish during hands-on experiences in which children witness their actions directly benefiting animals in contexts very similar to those in which the children will interact with animals in their daily lives. A model that incorporates several of these components was proffered by scientists Barbara Rogoff, Eugene Matusov, and Cynthia White (1996) for creating a community of learners (see diagram below). In this model, teachers begin by modeling the target behavior. The students then take on progressively more of the task until the teachers are simply overseeing what students do. Note that "teachers" can be knowledgeable students who lead other students under an actual teacher's supervision.

Model for Creating Community of Learners

I DO	→	I DO	→	YOU DO	→	YOU DO
YOU WATCH		YOU HELP		I HELP		I WATCH

For example, students can research and plan a service-learning project that directly benefits animals in their immediate area. During the entire project, the teacher sets and fairly, compassionately enforces rules of conduct. This example is one type of ideal that incorporates much of what is known to nurture empathy; of course, certain situations dictate different strategies. In all cases, however, integrating a variety of strategies should work better than any one tactic alone.

See also

> Children—*The Appeal of Animals to Children*
> Children—*Children and Animals*

Further Resources

Ainsworth, M. D. S., Blehar, M. C., Waters, E., & Wahl, S. (1978). *Patterns of attachment*. Hillsdale, NJ: Erlbaum.

Ascione, F. R. (1992). Enhancing children's attitudes about the humane treatment of animals: Generalization to human-directed empathy. *Anthrozoös, 5*(3), 176–81.

Ascione, F. R., Claudia, V., & Weber, M. S. (1996). Children's attitudes about the humane treatment of animals and empathy: One-year follow up of a school-based intervention. *Anthrozoös, 9*(4), 188–95.

Bar-Tal, D., Nadler, A., & Blechman, N. (1980). The relationship between Israeli children's helping behavior and their perception on parents' socialization practices. *Journal of Social Psychology, 111*, 159–67.

Baumrind, D. (1971). Current patterns of parental authority. *Developmental Psychology Monographs, 4*(1).

Bowlby, J. (1988). *A secure base: Parent-child attachment and healthy human development*. New York: Basic Books.

Bryant, B. K., & Crockenberg, S. B. (1980). Correlates and dimensions of prosocial behavior: A study of female siblings with their mothers. *Child Development, 51*, 529–44.

Eisenberg, N. (1993). *Special report: The socialization and development of empathy and prosocial behavior*. East Haddam, CT: The National Association for Humane and Environmental Education.

Eisenberg, N., & Geisheker, E. (1979). Content of preachings and power of the model/preacher: The effect on children's generosity. *Developmental Psychology, 15*, 168–75.

Elliot, R., & Vasta, R. (1970). The modeling of sharing: Effects associated with vicarious reinforcement, symbolization, age, and generalization. *Journal of Experimental Child Psychology, 10*, 8–15.

Hoffman, M. L. (1998). Varieties of empathy-based guilt. In J. Bybee (Ed.), *Guilt and children*. San Diego: Academic Press.

Ianni, F. A. J. (1992). Meeting youth needs with community programs. ERIC Digest, Number 86.

Rogoff, B., Matusov, B., & White, S. (1996). Models of teaching and learning: Participation in a community of learners. In D. Olson & N. Torrance (Eds.), *The handbook of cognition and human development* (pp. 388–414). Oxford: Blackwell.

Rushton, J. P. (1975). Generosity in children: Immediate and long-term effects of modeling, preaching, and moral judgment. *Journal of Personality and Social Psychology, 31*, 459–466.

Whiting, B. B., & Whiting, J. W. M. (1975). Children of six cultures: A psychocultural analysis. Cambridge, MA: Harvard University Press.

Whiting, J. W. M., & Whiting, B. B. (1973). Altruistic and egotistic behavior in six cultures. In L. Nader & T. Maretzki (Eds.), *Cultural illness and health*. Washington, DC: American Anthropological Association.

Yarrow, M. R., Scott, P. M., & Waxler, C. Z. (1973). Learning concern for others. *Developmental Psychology, 8,* 240–60.

Zahn-Waxler, C., Robinson, J. L., & Emde, N. E. (1992). The development of empathy in twins. *Developmental Psychology, 28,* 1038–47.

William Ellery Samuels

■ Education
Teaching Human-Animal Interactions

Why should human-animal interactions be taught as a university course? Mostly because we share the planet with so many nonhuman animal species and we live intimately in so many different relationships with them. Yet we don't realize the extent of these relationships until we look more closely, and that is what such a course could do.

Who should take a course in human-animal relationships? It could be useful to students in many different areas. Of course, biology students should see the interrelationships between the animals they study and us humans. Similarly, psychology students would enjoy looking at our complex and diverse set of interactions with the species they study and other animals. Students with applied interests can particularly benefit. Agricultural students with an animal husbandry concentration have a strong need to understand what they are involved in, as do environmental studies students. Pre-veterinary students will gain a lot by this overview, and given the ubiquity of animal research, it would be useful for premed students as well. Philosophy students might enjoy looking more deeply at the morals and ethics of our relationships with animals too.

Who should teach a course on human-animal relationships? Here there is a good argument for a comparative psychologist like me. Someone with training in this area has a good view of animal behavior but also an extensive background in understanding how humans think, believe, and reason. Of course people who teach this course should also be involved with animals themselves. It is hard to imagine that anyone on this planet is not involved with animals, but surely someone with a broad enough overview to teach the course should have pets (for example, we have kept cats, dogs, mice, parakeets, turtles, and guinea pigs) and should have experience with animals in research (for example, I have studied six cephalopod species, reptiles, and humans). Of course, such a person could have all these qualifications without being in psychology, but he or she should have a more broad background than, say, someone who specializes in agricultural animal care. There are theoretical as well as practical issues here.

What areas should such a course cover? The first basic area is information about what the interrelationships between animals and humans are. The range of these interrelationships is very wide, stretching all the way from hunting to keeping animals for food, from the intimate relationships of pet owners to the less intimate and perhaps exploitative ones of pet breeders and handlers. Study in this area must cover our cooperative relationships with hunting and herding dogs and ranch horses as well as guide and companion dogs. And it must touch on animals for display in aquariums and zoos and the assets and liabilities of bringing wild animals to where the general public can watch them in safety.

A second area of study must address the type of relationships involved, which of the partners is in power, and what each gains from the interaction. Some relationships, such as those of the agriculture industry, are clearly exploitation of the animals for the bene-fit of the humans. Some bring myth and custom to the interactions, such as big game hunting. Nonhuman animals and humans are on a much more equal footing in some other relationships; it is easy to argue that a border collie or a sled dog is doing what it loves and that the humans involved are fostering its success. Other relationships are harder to see in terms of power and gain. The zoo animal is kept, often cosseted, but it has lost its freedom. The animals in national parks are assumed to be "wild and free," but ecologists argue that there is no "true wilderness" any more, that every landscape has been altered by humans. So how does this more subtle interaction work? Think of "problem bears" that come to rely on our garbage and end up moved or shot.

A third and most important area to cover in a class on human-animal interactions is the morality of the interaction and the values we humans bring to it. No decision about animals is value-free, and we must ask who gains and loses and on what basis we decide to exploit or foster animals. A Cartesian approach that values human good and sees animals as tools to be exploited is one end of a continuum of value. The other is surely that of the vegetarian who refuses to eat meat because it would involve the slaughter of animals, or the individual who argues against the spaying of female cats because it would deprive them of the experience of motherhood. In between we have the moral issue of deciding which animals we should value, so that a treatment that would be seen as abhorrent for a kitten is judged reasonable for a cockroach. Another point of view is to decide which animal might have consciousness and undergo suffering.

How should a course in human-animal interactions be taught? First, as a specialty course, it should probably be taught at the third-year level, with students having several biology or psychology courses behind them so that they can understand the basic material. Then to some extent, the approach depends on the number of students. What can be done with a small class of twenty is impossible, or at least difficult, for a large one of 180. But let us imagine a class of under fifty.

Lecturing is probably an important component of such a course, particularly in the first part of a semester or trimester, because students need to get an idea of the three basic areas mentioned earlier. But then they need to interact with the material, and this can be done in several ways. Students can explore and think about these issues by contacting the people involved. This could be as formal as an interview with a veterinarian, an ani-mal shelter worker, a rancher, or a zookeeper. Asking questions and getting the person's point of view can bring the depth of the relationships home to the class.

Second, students should be asked to look at the ethical and moral issues involved. Each can do an investigation of the issues involved in raising veal calves, sport fishing, breeding an endangered species in a zoo or aquarium, or altering the appearance of dogs to keep their confirmation to a standard for its breed (such as tail docking). Actually looking at the issues involved helps the students to see that there are no clear and easy answers, yet also that we hold the upper hand and make the decisions in most interac-tions. We handle a large animal like a horse with some care, but we ultimately decide its feeding, housing, and even opportunity to mate.

Third, students can observe these interactions themselves. In any city, large or small, there are lots of opportunities to do this. Being from a small city, I worried about limited opportunities for watching, but students found them. Students can follow a veterinarian or an animal rescue specialist at work. They can go to the local park and watch the inter-actions between dogs and their owners. They can go to aquariums and zoos, which welcome volunteers and often give them opportunities for research. They can watch animals in training, see the interactions between farm animals and their handlers—think of lambing or calving time on a farm.

Of course, if the class is small enough, it is a great benefit to give the students a chance to show the results of their investigations to each other. PowerPoint is almost ubiquitous in upper-level classes, and presentations to the class give class members a chance to practice their oral skills. Who knows—they might use those skills in the future to talk about their research or advocate for new laws or good treatment of the animals they are talking about. Poster presentations are a particularly interesting way to show and persuade. I remember one student, after looking at the poster of another on the treatment of veal calves, saying, "That's it! I'm not eating veal again." Another thoughtfully scanned a poster in which a classmate had pointed out that big game hunting was a way for small isolated towns to keep a viable economy going. Posters and oral presentations can easily be shared between several students, which makes them more possible for larger classes. Evaluating them is a bit tougher. What is the relative importance of the student's knowledge, the student's ability to answer questions, and the visual appearance of the poster? That takes time to work out and to judge. If it can be done, having a research component to the class is beneficial; for mine it was observational research. We humans are curious creatures, and having students try to look and see and then report what they saw is a useful component of their education. It may help the future vet to look carefully at his or her patients, the horse handler to know what the animal feels, or even the city dweller to know when that dog at the front of its yard is dangerous. Again, there are lots of opportunities. One of my students watched to see how veterinary assistants alleviated pain, another how Canada geese so common in parks today knew when someone wanted to feed them. Again, getting involved, looking, and thinking about what you see is such a valuable learning experience.

What should students take away from a class on human-animal relationships? Of course, the first thing they should take away is factual knowledge about the animals and how we interact with them in all different ways. A second thing they should take away is the extent to which we humans control and dictate the lives of animals—we are in charge. And hopefully, they will exit the class having thought about the morality of what we do every day, the extent to which, like it or not, we are making complex decisions of costs and benefits that need a morality behind them, even if we do not realize it. Hopefully, these will all sum to a simple thing: respect for all the animals that we share the planet with. That would be enough.

Jennifer A. Mather

■ Enrichment for Animals
Animal Boredom

The suffering of animals from chronic boredom is a widespread, but largely unrecognized, problem in our relationship with animals. Animals in many ways are our friends and companions—they enrich our lives with their presence, and we hope we do the same for them. However, there is a deep ambiguity to that relationship, in that we also use and manipulate animals as an industrial resource. We keep enormous numbers of animals confined in small, enclosed areas for the sake of food production, scientific research, and leisure entertainment, and we control almost every aspect of these animals' lives. As a result, they are safe from predators, food scarcity, extreme temperatures, and other natural threats—but their lives have also become highly predictable and uneventful, leaving them with very little of interest to do. Even though an animal's cage environment may be full of noise, smells, and things to see, it mostly lacks the space and variety of provisions that would allow the animal to lead an active, self-motivated life. In human beings, such conditions are known to create chronic passivity, boredom, and depression, and we should ask whether the same may be true for the animals in our care.

For decades it was assumed that animals in captivity are basically all right as long as they stay physically healthy and readily reproduce. However, in recent times, we have discovered so much about the intelligence and awareness of animals that we have begun to realize they can suffer in ways that go well beyond physical health. It is now acceptable to speak of "psychological" or "mental" well-being in animals and to ask what is required to give them a psychologically wholesome life. To address this question, it is of primary importance to recognize that animals value interaction and communication with companions and physical surroundings not only to secure survival, but also for interaction and communication's own sake. Together or alone, animals explore, play, and experiment with their environment to create opportunities for trying out and experiencing new things. Play occurs in a wide variety of species, but particularly birds and mammals are known to develop inventive rituals and games. Common ravens, for example, will fly upside down, slide down snowy slopes on their backs, play tug of war, or play "pass–the stick" in midair. Such behavior typically appears spontaneous, "in-the-moment," and relaxing, and it is hard to avoid the impression that playing animals, like children, are having fun. When interpreting the behavior of animals, it is important to realize that different species have different habits and expressions and that to understand these correctly, we must know the animals well. Scientists are often concerned that we attribute emotions to animals that are essentially human and that in reality they do not have. However, the opposite problem is perhaps more likely to occur, namely that we fail to recognize an animal's emotions because we do not know its expressions well enough. Only if we study animals and communicate with them close-up and for long periods of time can we begin to understand what they feel.

How then can we recognize boredom in captive animals when it occurs? How do we know that an animal sitting or lying down in a corner for hours on end is not simply resting or quietly content? One important answer to this question lies in the close observation of the overall style, or manner, in which the animal behaves throughout the day. When housed in small, barren cages, animals are much less active than animals in larger, more enriched cages, and they spend much of their time lying down, sleeping, or dozing. They can also sit or stand quite still for long periods of time, with a hunched back, drooping heads and ears, half-closed eyes, and abnormally bent limbs. Such prolonged drowsy, half-awake behavior indicates that the animals are not relaxing and taking a break, but have become more chronically lethargic and listless. This also comes to expression in the unease with which they respond to events around them. They grow wary of disturbance, irritable, and even aggressive, and they rarely play or positively express themselves. They wander around, sniffing or nibbling different objects in their cages, but never stay with anything for long; unsure of what they are doing, they are easily provoked or spooked. In all this, the animal appears rather forlorn, never fully absorbed in a task or a game. Animals can be physically active at times during the day, but that activity is not necessarily focused or creative; it is listless, tense, restless, despondent, anxious, hostile, all at the same time—all signs of a chronic absence of meaningful things to do, which in their totality suggest that the animal is bored or, in severe cases, depressed.

To be able to recognize a transition from lively to listless styles of behaving requires patience, good knowledge of individual animals, and above all, well-honed observational skills. Exactly how an animal expresses symptoms of boredom is likely to differ between species. Some animals swim and float rather than walk or sit, and others may rely on sensors other than their eyes, so the signs of lethargy and irritability in such species will assume different forms. We may, for example, quite easily recognize "listless sitting" in a pig, but the detection of "listless swimming" in fish requires thorough familiarity with fish behavior. When looking for such signs, it is important to know an animal's biological background and to understand how it prefers to spend its time under more natural conditions. This gives an idea of the animal's priorities and needs and suggests where we might look for signs that those needs are not met. Pigs in the wild, for example, spend long hours rooting for, and chewing, food; in small industrial pens, however, where they are given all their food for the day at once, all they can do is rub their noses against the concrete floor and chew the bars of their pen and parts of each other's bodies. Such substitute behaviors can gradually take on a fixed, compulsive, "stereotyped" form, accompanied by drowsy postures and tense, agitated, anxious expressions. These behaviors indicate that for pigs it is important to spend much of the day actively searching for food; other species, however, according to their ecological niche, may have different needs.

Apart from suffering the monotony of confined cages, animals can also be subjected by trainers and caretakers to monotonous tasks. Animals in the entertainment industry (circuses, dolphinaria, film studios), laboratory animals, rescue or police dogs, or dogs that guide the blind may all be asked to perform the same task again and again. Little study has yet been made of how such routines affect the animals' welfare, and there may certainly also be positive effects. In circuses and dolphinaria, for example, learning to perform routines may bring animals temporary relief from the boredom of their living quarters, especially when they are positively rewarded through communication with their caretakers. However, when animals are not sufficiently motivated to perform a task, they have to be forced to pay attention and stay vigilant against their will and are likely to become tense, anxious, and bored. The provision of spacious and enriched housing conditions would help to compensate for the negative impact of such tasks.

As these descriptions indicate, boredom is a complex notion that encompasses fluctuating levels of other emotions and states. Acute emotions such as fear and anger mostly have clear expressions and are not easily misunderstood, but the expression of boredom is multifaceted and spread out over time and so potentially is easily missed. To observe and describe shifts in behavioral style, scientists increasingly adopt qualitative methods of assessment, which focus on the animal as a whole rather than on separate elements of its behavior. Such assessments already play a prominent role in studies of animal temperament and personality, describing animals, for example, as friendly, hostile, anxious, or relaxed. However, to make such assessments in the context of animal welfare, and use them to describe an animal's emotional state, is more controversial and still relatively rare. As indicated previously, scientists often fear that qualitative assessments are subjective judgments that fall outside the scientific domain. However, we should ask ourselves why we should not put our sophisticated human observational skills to good scientific and practical use. Those who work with animals on a daily basis (e.g., farmers, veterinarians, trainers, laboratory technicians) very much depend on this skill; they need to evaluate their animals' "body language" to check whether they are ill, languishing, or doing well. Animals may not give us verbal reports, but that does not mean they do not express themselves in an intelligible way that, if we pay attention, we can learn to understand.

Thus the notion of boredom has inspired many keepers and caretakers to provide their animals with more complex, interesting, and challenging environments. When animals are given the chance to lead an active life, by searching for food, building nests, finding shelter, and communicating with other animals and humans, their energy and liveliness return. They appear relaxed, inquisitive, and alert and are less likely to develop stereotyped behaviors. It is not enough to provide them with a few objects that stimulate their senses or physical movements (e.g., television screens, brightly colored objects, exercise wheels, or arenas). Although these are better than nothing and may capture the animal's attention for a while, they will not sustain the sort of self-generated activities and interests that keep the animal busy and prevent boredom. This seems true for all forms of enrichment that are too far removed from the animal's natural environment—they are unlikely to appeal to the animal as organic materials do, and they will not keep the animal occupied in the long run.

To allow animals to create a meaningful life, we must, first of all, give them space to explore, move around, run, and play when they wish. In addition, we must provide companions and materials that enable them to fulfill their social and physical needs in inventive and varied ways. It is inevitable that in more complex environments, animals will endure a certain amount of stress, through aggressive conflicts, competition for food, or greater vulnerability to physical illness and harm. However, a great deal can be done to alleviate such stress through good environmental design—for example, by creating places where the animal can hide or withdraw from others, rest, or sleep. The solution is to give animals more choice in how to deal with the challenges of daily life, not to take those challenges away. If animals are given the means to cope, they may, because of a complexity of factors, be temporarily stressed; however, they will never be chronically bored.

See also

Sentience and Cognition—*Animal Pain*
Zoos and Aquariums—*Enrichment and Well-Being for Zoo Animals*

Further Resources

Bekoff, M., & Byers, J. A. (Eds). (1998). *Animal play: Evolutionary, comparative, and ecological perspectives.* Cambridge: Cambridge University Press.

Fagen, R. (1992). Play, fun, and communication of well-being. *Play & Culture, 5*(1), 40–58.

Newberry, R. C. (1995). Environmental enrichment—Increasing the biological relevance of captive environments. *Applied Animal Behaviour Science, 44*(2/4), 229–43.

Wemelsfelder, F. (1990). Boredom and laboratory animal welfare. In B. E. Rollin (Ed.), *The experimental animal in biomedical research* (pp. 243–72). Boca Raton, FL: CRC-Press.

———. (1993). The concept of animal boredom and its relationship to stereotyped behaviour. In A. B. Lawrence & J. Rushen (Eds.), *Stereotypic animal behaviour: Fundamentals and applications to animal welfare* (pp. 65–95). Wallingford: CAB-International.

———. (2005). Animal boredom: Understanding the tedium of confined lives. In F. McMillan (Ed.), *Mental health and well-being in animals* (pp. 79–93). Oxford: Blackwell.

Young, R. J. (2003). *Environmental enrichment for captive animals.* Oxford: Blackwell.

Françoise Wemelsfelder

■ Enrichment for Animals
Emotional Enrichment of Captive Big Cats

A large number of exotic felines live in captivity. In the wild most species of big cats are designated as either threatened or endangered, mainly because of loss of habitat from human incursion. It is therefore likely that big cats will for the foreseeable future face existence in captivity of some form. Their prospect for continued survival is dependent on interaction with humans both in the wild and in captive settings. Various enrichment techniques are suitable for captive big cats. Of particular interest is the enrichment that enhances emotional well-being through human interactions. Sanctuaries such as the International Exotic Feline Sanctuary (IEFS) in Boyd, Texas, that specialize in the long-term care of a limited number of species share a common goal with other zoological institutions of providing state-of-the-art care and scientific innovation in the field of zoo biology.

Big Cats in Captivity

In ancient Rome, when captive lions and other big cats were kept for gladiatorial games in the Circus Maximus, presumably there was little concern for the well-being of animals intended for public amusement. In contrast to ancient practice, big cats today are mainly kept for scientific, conservation, and educational purposes. The health and well-being of felines are carefully monitored in modern zoos and wild animal parks. Although these felines are well cared for, other felines are kept in private menageries where the owners' practices are often less humane. The exact population of captive big cats worldwide is unknown, but estimates range as high as ten thousand in the United States (Nyhus, Tilson, & Tomlinson, 2003), where regulations controlling the possession of exotic felines were limited or nonexistent until recently. Therefore, a large number of big cats are being kept by private individuals with insufficient knowledge, skill, and ability, as well as insufficient financial means to properly care for a dangerous predatory animal. Private owners of big cats may consider them as pets, and consequently, rearing practices often resemble the hand-raising of a domesticated cat or dog.

The emergence of captive animal sanctuaries is in large part a response to the failings of private ownership in states such as Texas, which has one of the highest populations of privately kept big cats (Siderius, 2002). The International Exotic Feline Sanctuary is a fully accredited American Zoological Association facility that provides for the long-term care of approximately seventy big cats. The big cats housed at IEFS typically have had many interactions with humans prior to their arrival at the sanctuary. Unfortunately, some were inhumanely treated or poorly hand-reared by unknowledgeable owners. Such poor treatment prior to sanctuary living has often resulted in the development of maladaptive behavioral problems in captivity. The IEFS has developed many practices for big cat enrichment involving human and feline interactions.

Captive Feline Enrichment

Enrichment is a general term that refers to the modification of the environment or behavior of an animal typically in held in a human-made environment. Approximately four decades ago, scientists discovered that simply enriching the sensory environment of the normal impoverished lab-animal enclosure could have profound behavioral and physiological effects on laboratory animals. Since the early scientific studies, the beneficial effect of enrichment has been investigated in a wide variety of academic disciplines and species. Because of this solid basis in science, enrichment has emerged as a common practice in applied animal behavior settings such as zoos.

Environmental enrichment and behavioral enrichment are terms used interchangeably in most scientific literature. However, we wish to distinguish between the two terms, based on whether the enclosure (i.e., environment) is enhanced or whether a behavior is altered through the introduction of a device or operant conditioning. Thus, in environmental enrichment, the context affords additional behavioral opportunities for the captive feline, whereas in behavioral enrichment, the additional behavior opportunities are afforded through interaction with a toy or a device such as an operant manipulandum (e.g., a lever that can be pressed to deliver a food reward). A third form of enrichment can also occur when the purpose is to enhance the animal's emotional state through social interaction with humans. We refer to this as emotional enrichment because of its heavy reliance on human interaction to enhance the animals' well-being. All three types of enrichment are employed at the International Exotic Feline Sanctuary.

Environmental Enrichment

The American Zoological Association defines environmental enrichment as "a process for improving or enhancing zoo animal environments and care within the context of their inhabitant's behavioral biology and natural history." It is a dynamic process in which changes to structures and husbandry practices are made with the goal of increasing the behavioral choices available to animals and drawing out their species-appropriate behaviors and abilities, thus enhancing their welfare. As the term implies, enrichment typically involves the identification and subsequent addition to the zoo environment of a specific stimulus or characteristic that the occupant(s) needs but that was not previously present. In other words, enrichment is primarily concerned with physical objects and structures that seem to make life a bit more interesting for the species.

Sanctuaries such as IEFS practice environmental enrichment in many ways. Very large naturalistic habitats are built with ample shade, terrain changes, and vegetation (see example of IEFS enclosure in the figure below). Such elements provide for the cats' needs for prospect and refuge (i.e., having characteristics of a visual vantage point and a sense of security). Additionally, ramps and high perches are provided for the climbing cats such as cougars, leopards, bobcats, lynx, and jaguars. There are perches for the lions, whereas large dirt mounds serve the same purpose in cheetah habitats. There are also perches and pools with running water for the tigers and jaguars, as well as other cats. The complexity of the habitat can reduce stereotypy (mechanical repetition of the same movement, such as pacing, for example), decrease aggression, and increase species-typical behaviors.

Behavioral Enrichment

Some argue that environmental enrichment by itself does not provide an incentive for increasing species-typical behaviors of the animals within a so-called enriched environment. Advocates of behavioral enrichment, such as Dr. Hal Markowitz, have pointed to the lack of behavioral contingencies inherent in captive settings. They argue that creating enrichment activities that afford opportunities to self-control schedules of feeding, for example, is far more mentally stimulating than a quasi-naturalistic enclosure. An operant manipulandum that

Environmental enrichment of a typical big cat habitat at IEFS includes pools, variable terrain, and climbing structures. Courtesy of Scott Coleman.

activates artificial prey or that delivers food will encourage hunting-like behavior, and these have been successfully used for zoo felines. Providing the opportunity to capture

Behavioral enrichment devices that encourage play are used at IEFS. For example, pumpkins are a seasonal treat with a strong appeal to tigers and other big cats. Courtesy of Scott Coleman.

live fish has also been shown to be effective feline enrichment. The current enrichment practice at IEFS minimizes associations between humans and food and thus does not employ food reward for training. However, there are enrichment devices that elicit species-typical prey behaviors. For instance, we at IEFS give large and small balls to the cats to paw, chase, and chew. We give boat buoys and large plastic "pickles" to the larger cats to carry and bite. In the fall, we give the cats pumpkins to play with (see figure above), and in the summer, cats receive ice blocks embedded with food objects to play with.

One limitation of an enrichment approach based on operant learning theory may be its dependence on motivation factors such as hunger in order to be most effective. The cat's interest in toys may also be limited by short-term motivational factors, and the approach therefore becomes less effective when the novelty wears off. A system of periodically exchanging toys may be required to minimize the cat's boredom. Considering the apparent inadequacy of environmental and behavioral enrichment to provide sufficient mental stimulation, another enrichment technique based on social contact with humans may serve to enhance the captive feline's total well-being.

Emotional Enrichment

Emotional enrichment is defined as the enhancement of an animal's psychological well-being through social interaction with humans. The premise for emotional enrichment stems largely from scientific research involving animal personality theory, together with the acceptance of animal emotion currently flourishing in modern animal cognition theory. Furthermore, recent research on the underlying neurological mechanism and behavioral benefits of early handling suggests that emotional enrichment exists separately from environmental enrichment. Early handling is simply the human contact and taming of domestic and laboratory animals during early phases of development. The effects of social contact with people for caged domestic house cats has been shown to have an ameliorating effect on stress, which implies the existence of an identical response in captive big cats. Contrary to a prevailing attitude that denies the need for human social interaction, a human-feline relationship with captive exotic felines has been shown to increase reproductive success in some small exotic felines in a zoo setting. Thus, taken together, there is emerging evidence to suggest that enhancing the emotional state of a captive big cat is a distinct form of enrichment.

The purpose of enrichment in sanctuaries is often different than that set forth in zoos because it is employed as a therapeutic technique for felines. Thus, emotional enrichment is based on individual differences in big cat personalities rather than prompting species-typical behavior. Developing a theory that explains environmental and behavioral enrichment has been difficult. One problematic reason is that the behavioral variation within a species is usually unknown. Thus, we cannot fully know what is or is not species-typical behavior. In addition, common behavioral measures of enrichment often lack relevancy to overarching biological theory. The ability to link enrichment with evolutionary biology theory is important because it could provide a valid assessment of the outcome in terms of Darwinian fitness. On the other hand, the emotional enrichment program for captive big cats at IEFS, where there is awareness of the need to create behavioral variation as well, could be described as less aligned with an evolutionary biology model and more likened to a biomedical model of animal health using a protocol to establish emotional contentment.

Emotional enrichment is practiced in a variety of ways at IEFS. One frequent mistake is categorizing a species in general terms vis-à-vis its personality and emotional characteristics. At IEFS, we work with each individual according to its own personality and emotional needs. In general, we have our staff and volunteers treat each cat with the respect and dignity that would be accorded another human. Care is taken not to agitate, irritate, or unduly excite any cat. No demands are placed on it, other than the necessary movement into and out of its separated area to be locked down for feeding and cleaning of the habitat. Volunteers and staff sit outside the habitats of various cats and give them companionship and company. When the keepers are working in a cat's area, they take the time to softly talk to the cat and reassure the cat before moving on, thereby having protected interaction that is beneficial to the cat without risk to the keeper. Keepers also take the time to relax the animal and make it comfortable with the keeper's presence before initiating maintenance or other activities (e.g., operating gates and feeding and moving the animal).

The reverse might be more appropriate: to consider behavioral enrichment programs based on the animal's instinctual and emotional desires rather than what a human would consider enriching, and to recognize the possibility that the animal's emotional ranges are more far more complex than we choose to recognize. For instance, facilities frequently give exotic cats items filled with food that the cats have to dig out or hang something just out of reach for the cat to jump at. In truth, exotic cats, when not needing to obtain food for existence, want nothing as much as to be relaxed and free of stress. The enjoyment of seeing a cat working at a frustrating item is perhaps pleasurable to the human, but not to the cat. They enjoy relaxing or sleeping without any stress. A relaxed cat is considered to be a happy cat, whereas an excited cat is quite often an agitated cat. Altering their behavior to increase activity may be to their detriment.

The point is that we find not only that emotional enrichment heightens the behavioral enrichment and acts as an important adjunct to it, but also that the two activities in concert substantially improve the emotional and physical lives of the individuals involved. Enrichment items that would otherwise not be given much attention are used much more when they are a source of play and interaction with a human, and the cats seem to derive much more pleasure from the activity. Also, our experience is that the whole attitude of our cats is much more positive and peaceful as a result of our emotional enrichment program.

Most of the cats in our facility were either intentionally or unintentionally mistreated or abused. They come to the sanctuary with an antagonistic attitude toward humans. We have found that the great majority of them change that attitude as they come to appreciate our feelings and conduct toward them. Obvious behavioral changes indicate that their lives have improved. Their stress level is minimized. They almost never act defensively or aggressively toward humans. Cats at our facility are rarely seen pacing, baring their fangs, or charging at humans. Pacing is very minimal and is generally associated with specific causal stimuli, such as a new piece of machinery or an unusual group of people.

In certain cases, unprotected physical interaction is initiated with some of our large cats that have had direct contact prior to arriving at the sanctuary. In most cases, the prior interaction was adversarial because training methods of domination and control were apparently utilized. Most of the cats respond surprisingly well to using only affection, trust, and respect in direct interaction. In emotional enrichment, sitting near the cats and becoming a companion through softly talking to these big cats gives them reassurance and a sense of security. Many of the larger cats become so relaxed that they fall asleep as a consequence of human comforting. They obviously enjoy and want positive emotional interaction with humans. With emotional enrichment, even the most solitary of cats in the wild, such as leopards, cougars, and tigers, can be among the most affectionate of the sanctuary's cats. This again varies from individual to individual. Some cats respond simply to standing near people who are talking. Other cats want proximity and will sit next to a person during the interaction. In some cases, an individual cat would rather play interactively with humans by using some of its enrichment items. For example, interacting with a cat by rolling a ball back and forth and holding one of its enrichment items or tossing the item for the cat will direct its attention to the item. Often, simply holding a dry twig for the animal to take in its mouth and break pieces off bit by bit (this is a favorite game for most of the cats, actually) can be stimulating interaction. Some also enjoy direct physical contact, such as being rubbed or scratched. Some cats become contented enough that they will fall asleep in the presence of humans and enjoy placing a paw over a person's hand or arm. The purpose of human presence is to be a reassuring influence on the cats' state of mind and their environment.

This program does not necessitate actual direct physical contact with a potentially dangerous big cat to be successful. Many of the elements of our program could be used to improve an animal's life without unprotected human contact. Indeed, we have several cats that derive a great deal of benefit from our program without having any direct interaction. In fact, unless unprotected contact can be safely employed without any adversarial consequences, it is better to confine the technique to protected contact. The requirements for interaction in an unprotected environment with an individual animal are that it must (1) substantially benefit that individual animal's quality of life, and (2) be accomplished safely without danger to either the animal or the human keeper. Emotional enrichment is a vital and helpful addition to any behavioral enrichment program and would substantially benefit the lives of the affected animals.

Do Big Cats Purr?

Scott L. Coleman

Cougars purr, frequently for the same reason domestic cats do, because they are apparently contented. Bobcats and most cats smaller than cougars also purr. Cougars are the largest cats that purr in the same manner as domestic cats, but the larger cats have their version of a purr too. The "chuff" of a tiger means somewhat the same as a purr, and leopards make a sound somewhat like a growl that is their version of a purr. Lions vocalize a sort of moan that has the same meaning. In short, all cats have a vocalization that has the same meaning as a domestic cat's purr, and all big cats have affectionate emotions for selected individuals— human and nonhuman—that are very similar to those of domestic cats

See also

Zoos and Aquariums—*Enrichment and Well-Being for Zoo Animals*

Further Resources

Bekoff, M. (2001). Human-carnivore interactions: Adopting proactive strategies for complex problems. In J. L. Gittleman, S. M. Funk, D. W. Macdonald, & R. K. Wayne (Eds.), *Carnivore conservation*. New York: Cambridge University Press.

Hauser, M. (2000). *Wild minds: What animals really think*. New York: Holt.

Markowitz, M. (1981). *Behavioral enrichment in the zoo*. New York: Reinhold.

Newberry, R. (1995). Environmental enrichment: Increasing the biological relevance of captive environments. *Applied Animal Behaviour Science, 44*, 229—43.

Nyhus, P. J., Tilson, R. L., & Tomlinson, J. L. (2003). Dangerous animals in captivity: Ex situ tiger conflict and implications for private ownership of exotic animals. *Zoo Biology, 22*, 573–86.

Shepard, D., Mellen, J., & Hutchins, M. (1998). Second nature: Environmental enrichment for captive animals. Washington, DC: Smithsonian Books.

Siderius, C. (2002, February 28). Catch those tigers: Years of little or no regulation have made Texas a place where big cats prowl—and sometimes kill. *Dallas Observer*.

Young R. J. (2003). *Environmental enrichment for captive animals*. Oxford: Blackwell.

Scott L. Coleman and Louis Dorfman

■ Enrichment for Animals
Pets and Environmental Enrichment

Environmental enrichment involves changing a captive animal's environment and changing husbandry practices to benefit the animal. Physically or psychologically stimulating an animal, allowing behavioral choices, and increasing species-specific behaviors enables an animal to cope better in its environment and thus ultimately improve its welfare.

In pet species, improving welfare is likely to have a positive impact on the human-animal relationship. Animals with improved welfare are likely to be easier to train and interact with. Distress and anxiety can impact negatively on learning, making training and interaction difficult. Arnold Chamove suggests that environmental enrichment should increase desirable and reduce undesirable behavior. Implementing appropriate environmental enrichment can reduce negative states such as distress, anxiety, frustration, and boredom. These states are often the underlying motivation for undesirable or problematic behavior that owners or handlers may find upsetting or distressing (e.g., aggression, eating of feces) or for those behaviors that impact the welfare of the animal (e.g., self-injury, over-grooming). Environmental enrichment can additionally increase behaviors that the owner is likely to find desirable and behaviors that are likely to increase the welfare of the animals (e.g., play, exploration, social behaviors). Environmental enrichment therefore facilitates a better relationship for both human and animal.

Many pets have behavioral problems, and therefore their welfare may be compromised in some way, suggesting a need for environmental enrichment. Behavioral problems include separation-related problems, aggression, and inappropriate toileting. In addition, the home environment pets live in is greatly varied, from the urban to the rural to the feral environment. For example, a dog's environment can vary from a flat to a country estate to the local dump. Many pets are kept in captive situations—for example, horses are often stabled for long periods of the day; cats are often restricted or prevented from outdoor access; dogs are often kept in social isolation for many hours a day in runs, kennels, or the home; and smaller pet species such as guinea pigs, birds, and rabbits are often kept in small cages and runs, often with little social interaction. Ultimately, pets are kept as captive animals, and the home environment can greatly affect the animal's well-being.

The domestic setting of the home is not an environment every pet species encounters. Many dogs, cats, rabbits, rats, and guinea pigs live in laboratories where they are often restricted in space, preventing a lot of natural behaviors such as foraging, hunting, and social behaviors. Other common captive environments include rescue shelters, quarantine kennels, veterinary hospitals, boarding catteries or kennels, police and army kennels or stables, aviaries, and pet shops. These varied and often restrictive environments can be manipulated in many ways to improve the lives of their inhabitants.

To understand how environmental enrichment works, it is often better to consider its main goals. The first goal of environmental enrichment is to increase behavioral diversity. For a rabbit that spends most of its time inactive, one goal of environmental enrichment would be to increase behaviors such as exploring the environment, foraging for food, and hopping and jumping, all natural species behaviors.

The second goal is to reduce the frequencies of abnormal behavior. Some problem behaviors in pets are abnormal for the animal (e.g., self-injury—parrots who pluck out their feathers; route-pacing—dogs in captive environments such as kennels will often trace the same route continuously; over-grooming—cats who are distressed will often over-groom, causing baldness). The reduction of abnormal behavior will ultimately improve the animal's welfare.

The third goal is to increase the range or number of normal (species-specific) behavior patterns. For example, hunting is a species-specific behavior for a cat, but not for a rabbit. Environmental enrichment aims to increase these species-specific behaviors in a manner that is acceptable for each animal's situation. For example, it may not be possible or acceptable for an indoor cat to be able to hunt live prey. Instead, by using certain types of environmental enrichment, the behavior of hunting can be carried out on an inanimate object. Hence, the actual behavior the animal performs is natural, but the substrate the cat performs it on is not.

The fourth goal is to increase positive utilization of the environment, thus encouraging the animal to use its entire environment in a species-specific manner. It would not be that beneficial to increase the size of a dog's run if it only ever remained in one small section of it. Environmental enrichment can be used to encourage the dog to use all of the space available to it in a positive way (e.g., exploring and playing in its entire run rather than pacing or jumping repeatedly in only one area).

The final goal is to increase the ability to cope with challenges in a more normal way. Donald Broom describes the welfare of an animal as its ability to cope with its environment. Providing enrichment that allows an animal to cope better in its environment can therefore directly improve its welfare and well-being.

The environment of an animal consists of a wide range of stimuli, including social, sensory, and physical elements. The social environment includes the company of the animal's species and other species, including humans. Many of our pet species such as rabbits and dogs would live in naturally occurring social groups, and providing this in the captive environment can be enriching. Karen Overall and Donna Dyer suggest that cats (who are now known to be not as solitary as once thought) can benefit from being housed in maternal groups or by affiliation, with these groups being most likely to produce the most normal feline social behaviors. For pets that have been socialized to people, human company can be extremely enriching. Frequent handling and playing with socialized pets such as dogs and cats can provide the animal with both mental and physical stimulation while increasing their confidence and enabling them to cope better in a variety of circumstances. Training programs can give pets opportunities to encounter the levels of mental stimulation and problem solving seen in and used by their wild relatives.

Pet species very often have better senses (vision, audition, and olfaction) than humans. Well-positioned outdoor enclosures and runs can visually stimulate pets—for example, allowing dogs and cats to observe other species, such as prey items, in their natural habitat; and for horses and rabbits, providing them with as vast a visual view as they would encounter in the wild. Even for indoor pets such as cats, a window with a view can be enriching.

Captive and domestic settings can prevent a lot of auditory stimulation encountered in the natural environment, and efforts have been made to rectify this. Recordings are now available of species-specific sounds, such as rainforest sounds for parrots and sounds of mice and birds for cats. Recently, there has been a lot of interest in the potential use of human music as enrichment for pets. Dr. Deborah Wells and her colleagues found that barking decreased and resting increased in shelter dogs exposed to classical music; both behaviors are potentially indicative of improved welfare.

Hidden food that can be found by smell is likely to be enriching for many pets. The introduction of both novel scents and species-specific scents may also be enriching, although scientific evidence should be obtained before considering different scents.

Finally, much enrichment can be carried out in the physical environment (the animal's living area and its contents). For example, toys can provide a great degree of physical and psychological stimulation. Toys that need to be manipulated to gain a food reward have been seen to improve welfare in a number of pet species. Carder and Berkowitz even found that rats preferred earned food even when free food was available, suggesting that animals may find working for their food enriching. For cats, providing food in devices that have to be manipulated to gain the food helps to mimic the cats' natural feeding schedule of small and frequent meals. In addition, toys that are utilized with other animals or a human can have the added benefit of being socially enriching.

Providing physical space is important. Areas of different heights can be enriching for cats, dogs, birds, and rabbits, providing opportunity for exercise, visual stimulation, and social interaction while minimizing the effects of distress. For prey animals such as rabbits and guinea pigs, providing spaces to hide in, such as tubes and boxes, can greatly reduce distress and improve welfare. Enclosure complexity (whether the enclosure is a house, cage, or other) is likely to encourage more natural behaviors and therefore increase welfare.

The options for enrichment are vast, and only the basics have been covered in the scope of this essay. The main points to remember when considering environmental enrichment are that there are differences between species, breeds, and even individuals and that, therefore, one type of enrichment is not necessarily going to suit a whole species or breed, but possibly only a few individuals. The greatest benefits from environmental enrichment will be seen when enrichment is designed on a tailor-made basis for each individual, taking into consideration both the animal's natural environment and behavior and what that individual animal can physically and psychologically cope with.

See also

Bonding—*Companion Animals*

Further Resources

Broom, D. M. (1986). Indicators of poor welfare. *British Veterinary Journal, 142,* 524–26.

Chamove, A. S. (1989). Environmental enrichment: A review. *Animal Technology, 40*(3), 155–78.

Gunn, D., & Morton, D. B. (1995). Rabbits. In C. P. Smith & V. Taylor (Eds.), *Environmental enrichment information resources for laboratory animals: 1965–1995: Birds, cats, dogs, farm animals, ferrets, rabbits and rodents* (pp. 127–143). Hertfordshire: UFAW.

Hubrecht, R. (1995). Dog housing. In C. P. Smith & V. Taylor (Eds.), *Environmental enrichment information resources for laboratory animals: 1965–1995: Birds, cats, dogs, farm animals, ferrets, rabbits and rodents* (pp. 49–62). Hertfordshire: UFAW.

McCune, S. (1995). Enriching the environment of the laboratory cat. In C. P. Smith & V. Taylor (Eds.), *Environmental enrichment information resources for laboratory animals: 1965–1995: Birds, cats, dogs, farm animals, ferrets, rabbits and rodents* (pp. 27–42). Hertfordshire: UFAW.

Overall, K. L., & Dyer, D. (2002). Enrichment strategies for laboratory animals from the viewpoint of clinical veterinary behaviour medicine: Emphasis on cats and dogs. *Institute for Laboratory Animal Research, 46*(2), 202–16.

Van de Weerd, H. A., & Baumans V. (1995). Environmental enrichment in rodents. In C. P. Smith & V. Taylor (Eds.), *Environmental enrichment information resources for laboratory animals: 1965–1995: Birds, cats, dogs, farm animals, ferrets, rabbits and rodents* (pp. 145–212). Hertfordshire: UFAW.

Wells, D., Graham, L., & Hepper, P. (2002). The influence of auditory stimulation on the behaviour of dogs housed in a rescue shelter. *Animal Welfare, 11*(4), 385–93.

Young, R. J. (2003). *Environmental enrichment for captive animals.* Oxford: Blackwell.

Sarah Ellis

■ Enrichment for Animals
Sanctuary Construction

As the world fills up with people and their crops, there is less room for many animal species, especially those that are large and those that are tasty to humans. As zoos breed more babies to attract visitors, places need to be found for those babies when they become adults and no longer draw visitors. In addition, as pet animals grow from cuddly infants to not-so-cuddly adults or into old age, some people lose interest in their pets and stop caring for them. Fortunately, many people have a desire to help those unwanted animals. The answer is often to develop sanctuaries—places where the animals can live in approximately natural conditions and where people can see and study them. This entry provides information to start your quest if you wish to provide or develop an animal sanctuary.

Sanctuaries for large animals are normally quite expensive to build. The enclosures are expensive because they use expensive material for few animals, and they are commonly sited on expensive real estate. And because of cost, the usual size of any enclosure is kept small, putting pressure on the animals, which in turn puts pressure on the fencing, which necessitates even more expensive fencing. Give small enclosure size, only small groups can be accommodated because of fighting when closely confined. And because there are a number of small groups that require individual maintenance, feeding and cleaning animals is time-consuming and consequently costly.

For most sanctuaries, costs to humans are high; benefits to animals are low. To reduce costs, one needs to minimize the following:

- Expensive materials
- Expensive real estate
- Expensive fencing
- Small groups
- Individual feeding and cleaning
- Small impoverished areas for animals

Small-Group Housing

Often the animals taken to sanctuaries have been reared in ways that make group housing difficult, such as being reared singly by humans in their homes. Many have never seen another animal before in their lives. Because they cannot be grouped, they are put into individual or pair housing, in small cages. Small cages are the most expensive enclosures, using the highest proportion of the most expensive materials, and on costly premises. Costs for the enclosure alone can exceed US$65,000 to house one or two chimpanzees. More economic than individual or pair housing is housing in larger groups.

Zoo-Style Enclosures

Larger enclosures with more animals, such as those seen in zoos, are less expensive overall. But they still use expensive materials for the floor, walls, and housing; and they keep a relatively small number of animals on valuable land. Although maintenance of group-housed animals in small enclosures makes feeding and cleaning more economic, the animals live for long periods on the same land, with conditions that can lead to the problem of parasite burdens, a problem effectively addressed by farmers. Farmers deal with such parasite problems by rotating species, but this is uncommon in zoos, laboratories, or sanctuaries. With groups, animals defecate in localized places, reducing cleaning time. The use of self-cleaning substrates such as coarse gravel or wood-chips can also reduce cleaning times.

As an example of costs of a zoo-style enclosure, Carole Noon's Save the Chimps program is constructing twelve 1.2 hectares (three-acre) enclosures on an 80-hectare (200-acre) site in Alamogordo, New Mexico. The twenty-one existing chimpanzees live in one enclosure. The project costs about US$1 million per enclosure, or US$70 per yard of fencing. Also, the annual operating budget is estimated at US$2 million.

Island Sanctuaries

The least expensive enclosure is one that is naturally formed, normally an island. These natural enclosures are uncommon and still expensive. Uninhabited islands are usually small in size. Animals are difficult to feed and maintain by boat. Examples include the Tiwai Island Wildlife Sanctuary in Sierra Leone; the Ngamba Island Chimp Sanctuary, a 100-acre island twenty-five kilometers offshore in Uganda's Lake Victoria; and Dennis Rasmussen's Primate Refuge and Sanctuary of Panama on the Tiger and Bruja Islands in Gatun Lake.

Exclusion Sanctuaries

An enclosure fence keeps some animals in and others out. When the target animals are susceptible to predation, keeping animals out assumes more importance. New Zealand has many such animals, some of which exist only in single-digit numbers. Because of the historical absence of mammals and predators, a number of flightless or ground-nesting birds and slow-moving reptiles live there. Predators introduced by humans, such as mice, rats, cats, dogs, and pigs, eat either these indigenous animals or their food—German wasps, possums, deer, tahr. In New Zealand, sanctuaries exclude predators from threatened populations.

Such exclusion sanctuaries need to keep out a variety of animals, ranging in size from rats to deer. Mice are too difficult! The New Zealand fence is a substantial and expensive construction, even extending below ground. Custom-made components add to the cost. The approximate cost of the Karori fence was US$95–$145 (NZ$160–$240) per square meter, depending on the terrain. The fence is about three meters high, so the cost per running meter is three times the previously noted value, or US$300 per yard. The perimeter-to-area ratio or "bang for the buck" is another important consideration. The Karori 8.5 kilometer fence, costing NZ$240/m, protects 230 hectares at a relative cost of about US$6,000 per hectare (NZ$8,850/ha). This cost includes engineering design, permits, track excavation, retaining, drainage, gates, stream crossings, and so on. The cost of the fence materials may be only 50 percent of the total cost.

Optimal Sanctuaries

For an inexpensive sanctuary, it is desirable to keep the cost–benefit ratio as low as possible, to reduce the costs of building material, land, and animal maintenance while increasing the benefits to the animals of space and complexity.

The optimal sanctuary, therefore, has inexpensive fencing and a large but inexpensive area. For restricting most animals, the electric fence is the most cost-effective. Such a fence contains or excludes animals *not* on the basis of the fence's strength, but instead by its aversive nature. It is the cheapest fence that can be built per animal or per hectare. It is inexpensive because it utilizes a relatively small amount of expensive material per animal or per hectare.

The land used can be inexpensive, needing only the presence of a small amount of water. The area of the enclosure can be large and still be enclosed economically. A large area of land has several benefits. Animals can avoid their excrement, it provides a more complex environment, and it puts less pressure on the perimeter fence.

Animals need to be able to escape from other animals, of both the same and different species. To escape, they need space, enough space so that the pursuer stops its pursuit. The lack of sufficient space is the main reason animals escape from cages and enclosure. They are driven out during fights and will usually return if allowed. As animal numbers increase, subgroups form, each attempting to obtain and retain their territory. So the increasing territorial demands of subgroups put pressure on a fence, almost exclusively during fights.

The cheapest land is that which cannot be traditionally farmed because it is too dry or too steep and land that is sufficiently far from a city so as to be perceived as beyond daily commuting distance. Such a distance may be well under one hour in countries with low population densities where commuting is not usually necessary.

A 400-hectare (1,000 acre) site is the size of Golden Gate Park in San Francisco, California, and covers an area of 4 square kilometers (1.4 square miles or 1.3 miles on a side). As an example, such land in New Zealand would sell for US$460 per hectare (US$200 per acre). In Australia, similar land would sell for half that price. A 400-hectare site would cost US$200,000 in New Zealand. Such land would graze 285 cattle, 200 horses, or 2,000 sheep year-round. If the land was too dry to be farmed, then the stock numbers would be reduced in the absence of irrigation, and the land cost would be reduced.

Electric Fencing

The least expensive option to enclose a large area is an electric fence. The cost of a five-wire electric fence, including posts, wire, and insulators, is about US$4 per yard (NZ$6/m). Electric fences must be regularly checked. Their integrity is dependent on electricity being transmitted along the whole length of a wire, so interruption by a break in the wire or by an earth conductor being applied to a wire renders that particular wire ineffective. Checking is not a problem because equipment is readily available that tests out the fence automatically and at a distance and warns of problems.

In climates where the grass is green year-around, the simple all-live-wire system is adequate; where the ground dries or becomes frozen, the earth-wire system is needed. In the former, all of the wires are live, and the moist earth provides the "ground" to complete the electrical circuit; in the latter, alternate wires in the fence are connected to the earth to provide the "ground" to complete the circuit. Where the animals can jump onto the fence and avoid contacting both the ground and the fence simultaneously, the earth-wire system is necessary because the animals are insulated from a shock if all wires are live, similar to what is seen when birds perch on an electric power wire.

There is considerable expertise available in the use of electric wire enclosures. In countries where electric fencing was developed and is extensively used, such as New Zealand, it is commonly used for enclosing not only cattle and sheep, but also pigs, deer, horses, and raites and used for excluding wild deer and possums. More recently, it has been used to enclose primates (e.g., Jim Cronan's Monkey World Ape Rescue Centre in the United Kingdom and Japanese macaques in the 26-hectare (65-acre) enclosure at Arashiyama West Primate Center near Dilley, Texas). It is being used more frequently for excluding even the largest mammals in Africa. An electric fence charger of 5,000 V, 32 mA is commonly used worldwide.

To see what equipment is commercially available for electric fencing, the best Web site is the Gallagher Group Limited's Power Fencing site: http://www.gallagher.co.nz. It includes equipment for fencing over streams that flood and that require automatic shutting off the power for that section; a handheld device that shows the direction of any electrical short in the fence; lights that flash at a distance for reassurance that the fence is working; and a gizmo to turn off the power unit to allow repairs when the person repairing the fence is far from the energizer.

An electric fence is constructed using heavy posts that carry the weight of the wire and that are located at the corners of the enclosure or where any straight line ends. The posts are commonly braced at corners given that they are under considerable pressure when the wires are strained to 90 kilograms (200 pounds) under tension. Consequently, the cheapest electric fences enclose a rectangle and the most costly a polygon. Where terrain is irregular, a level track on which to build the fence is commonly bulldozed along the fence line prior to construction of the fence.

The bottom wire is parallel with the ground and at a height of fifteen centimeters (six inches) for sheep, goats, pigs, and dogs. If the wires are not parallel with the ground, young animals can wander under the fence where there is a depression in the ground and are often unable to find their way back. This is the primary cause of death with electric fencing. The only other cause of death is when an animal becomes entangled in the fence because the fence is sited at the bottom of a steep slope where the animal can slip into the fence.

For a five-wire fence for goats, one of the more difficult animals to keep fenced, the spacing from the ground is 16, 16, 16, 20, and 25 centimeters (6, 6, 6, 8, and 10 inches). For macaque-sized primates who put more pressure on the upper levels of the fence, spacing should not exceed 20 centimeters (8 inches). Arthur Hunt of the Vervet Monkey Sanctuary in Tzaneen, South Africa, recommends electrified woven mesh. For animals that burrow, such as rabbits, pigs, and wombats, a ground wire 5 centimeters (2 inches) from the ground is added to the previously described cocktail.

We also recommend that an additional wire, or preferably a warning tape, be placed at a point before the animal will come into contact with the electric fence. For most animals, that is at nose height. Electric tape with an aposematic warning pattern would alert an animal to the proximity of the fence.

Gallagher Power Fencing Company sells a tape that warns animals of the presence of electric shock. According to research by Arnold S. Chamove, this tape uses the natural warning patterns of animals such as honeybees, coral snakes, and poison-arrow frogs to increase the aversive nature of the tape, so that even when the electricity is off, animals avoid the warning tape. Animals will avoid the tape even if they have never seen it before or before the animals have had any contact with electric tape, just like some animals will instinctively avoid animals that have warning patterns.

Few animals will challenge an electrified fence unless under immediate and considerable pressure, such as the pressure of being chased with nowhere to escape. About 95 percent of monkeys in a group will contact an electric fence about twice throughout the first week of exposure when no warning of electricity is present. Thereafter, the fence is never touched. Using a warning tape reduces all contacts by 75 percent.

The height of an electric fence is determined by the height the animals can normally jump and still contact the fence or by face height for non-jumpers. For sheep, goats, cattle, and horses, that height is 0.9 meters (3 feet); for deer 1.6 meters (5 feet); and for elephants 2 meters (6.5 feet). For leaping animals such as primates, the fence should be at such a height so that the animal, when jumping onto the fence, would grab the next-to-top wire. Fail-safe chimpanzee fencing at Monkey World Ape Rescue Centre uses 15-centimeter (6-inch) spacing to a height of 2.4 meters (8 feet), a local government requirement, but neither chimps nor baboons have ever been observed contacting the fence above 1.2 meters (4 feet) high. Primates do not jump *over* things but rather jump *onto* things. It is likely that 1.2 meters (4 feet) is the minimal but adequate height for an electric fence for primates in large enclosures, but some recommend 2 meters.

To further keep climbing animals from scaling fences, offset brackets, standoff insulators, or "outriggers" are often used. Each outrigger supports an additional live wire for the fence, a wire that juts out from the fence on the inside of the fence. Now the animals cannot simply dash quickly up the fence but must climb part of the fence, climb over the overhanging wire(s), and continue up the electrified fence, slowing the animal's rate of climb. Between one and four outriggers are placed near the top of the fence so that the fence forms an inverted "L" shape.

The power for fences can be supplied by commercial energizer units powered by a battery, by the sun, or by a conventional electric supply. The conventional electric supply is the most common source of power for the commercial energizer. Because the power burst is so brief that it does not move an electricity meter, the power it uses is free. There are several sizes of power energizer units, and the size to choose is determined by the length of the fence (actually the length of the fence times the number of live wires) and not by the jolt you wish to deliver to the animal—the jolt is always the same.

The quality of an electric fence relies on its earth or ground. Surveys show that 80 percent of fences have inadequate ground systems. A good ground system uses at least three interconnected galvanized rods driven 2 meters (6.5 feet) into the earth and at least 4.5 meters (15 feet) apart in soil that is moist all year around. Check the ground every year during the driest season by touching the ground wire while having your other hand on the ground. If you feel a shock, the ground is inadequate.

A sanctuary need not be expensive if the land used is inexpensive, the enclosures are large, and the fence is electric. There is much land that is not suitable for farming but is suitable for sanctuaries. For those who wish to help unwanted animals in our crowded world, sanctuaries enclosed by an electric fence provide an opportunity to do so.

Further Resources

Chamove, A. S., & Rohrhuber, B. (1989). Moving callitrichid monkeys from cages to outside areas. *Zoo Biology, 8*(1), 151–63.

Gallagher, W. M. *Gallagher power fence manual* (10th ed.). Hamilton, New Zealand: Gallagher. http://www.gallagherusa.com/pf.manual.aspx

Williams, J. (1986). How do warning colours work? *Animal Behaviour, 34*(1), 286–88.

Arnold S. Chamove and Carol J. Chamove

■ Environment
See Conservation and Environment

■ Ethics and Animal Protection
Animals in Disasters

Heart-wrenching photos of oil-soaked seabirds after the *Exxon Valdez* accident and footage of stranded pets after Hurricane Katrina illustrate that any incident that affects people is likely to affect animals as well. Pets, wildlife, livestock, and captive animals face risks from such natural disasters as floods, tornados, hurricanes, and earthquakes. They are also at risk in such technological disasters as nuclear accidents, oil spills, terrorist attacks, and chemical leaks. Wildlife and their habitat can be affected by disasters, including fire, drought, and disease threats. In addition, large-scale disease outbreaks, including avian flu, SARS, and foot and mouth disease, can devastate livestock populations and local economies. Moreover, many diseases are *zoonotic,* meaning they can spread between humans and animals. The intensive agriculture practices widely used today present ideal environments for the rapid spread of livestock diseases. The close confinement and transportation of birds and animals destined for slaughter means that a disease outbreak in one facility can quickly escalate into a regional or national disaster. Animal stakeholders of all kinds, including pet owners, breeders, zoo keepers, farmers, veterinarians, and others, face unique challenges in planning and response.

The difference between an emergency and a disaster is a matter of scale. In an emergency, the existing local response structures (police and fire departments) can take action and meet the immediate needs created by the event. In contrast, a disaster overwhelms local resources and often makes it difficult for outside help to arrive. The response to a disaster that affects animals involves numerous agencies and stakeholders. Some of these agencies may also be involved in the response to human needs, whereas others have responsibility solely for animals. The response activates a complex network of government and nonprofit agencies at the federal, state, and local levels. In the case of a naturally occurring incident, such as a flood or tornado, local agencies have primary control of the emergency response. Animal issues will usually be addressed within the local framework by animal control, animal shelters, veterinary associations, and livestock organizations. Depending on the type of incident and the numbers and species of animals affected, various agencies will assist with the response. At the request of local government, the Federal Emergency Management Agency (FEMA) might activate veterinary medical assistance teams (VMATs) in a response involving animals. Veterinarians who volunteer with VMAT assist when a disaster compromises the area's veterinary infrastructure. The Department of Agriculture and the Fish and Wildlife Service each have many branches that may contribute resources when animals are involved. The Department of Health and Human Services, which oversees the Centers for Disease Control and Urban Search and Rescue, could also play a large role. At the state level, offices of emergency management and departments of agriculture and wildlife could enter the picture. However, state and federal agencies would get involved only after requests from the local level. In most events that affect animals, local animal control and law enforcement agencies would seek the help of national nonprofit animal welfare groups that have disaster response programs, such as the American Humane Association and the Humane Society of the United States. Local animal shelters, rescue groups, and livestock organizations, as well as local veterinary associations, would also assist in the response. In most events, large numbers of volunteers donate time and money.

In disasters, animal issues are associated with public safety, the human-animal bond, public health, the economy, and ethical and moral issues.

Public Safety

People will risk their lives to protect their animals, and will consequently risk the lives of others when they fail to evacuate or they reenter evacuated areas. A common reason for evacuation failure (along with fear of looting) is the inability or unwillingness to evacuate animals. Animal concerns accounted for several cases of evacuation failure following the post-Katrina flooding in New Orleans. When people remain in unsafe buildings or reenter them to rescue pets, emergency responders often have to rescue them, using time and resources that are always in short supply during a disaster. This public safety risk is not limited to pet owners but occurs with those who own and work with livestock, as well.

Numerous issues surround the evacuation of animals, including property rights, contamination, evidence preservation, and infrastructural hazards. Following Hurricane Katrina, rescuers entered many properties without permission to rescue stranded pets. Some homeowners objected to what they saw as breaking and entering. Moreover, rescuers encountered sewage, oil, gas, and other chemicals because of their efforts to save stranded pets, which were also contaminated. After a disaster, the scene must be maintained for insurance documentation. When people enter damaged areas, they can compromise the integrity of the evidence needed for insurance claims through their movement and by moving debris.

A dramatic example of the public safety risk when people reenter evacuated areas comes from a 1996 chemical spill in Weyauwega, Wisconsin. At 5:30 AM on March 4, 1996, thirty-five cars of a train derailed while passing through the town. Fifteen of the cars carried propane, and five of these caught fire. At 7:30 AM, residents of 1,022 households were ordered to evacuate because of the risk of explosion. Emergency managers anticipated that the response would take several hours. The effort instead took over two weeks, reflecting the unpredictability of disaster response. Half of the 241 pet-owning households left their pets behind. Others who were not at home at the time had little choice. Shortly after the evacuation, pet owners began to reenter the evacuation zone illegally to rescue their pets, at considerable risk to their own safety. Following protocol, emergency managers prevented residents from entering their own homes. In response, a group of citizens made a bomb threat "on behalf" of the animals, which directed considerable negative media attention at the response. Four days after the evacuation, the emergency operations center organized an official pet rescue, supervised by the National Guard and using armored vehicles.

The Human-Animal Bond

Sixty percent of American households now include pets, which exceeds the numbers that include children. The majority of pet owners consider their pets as members of the family. The human-animal bond is a powerful presence in our society. Interaction with animals has positive effects on people's mental health and physical well-being. During disasters, the human-animal bond can be either a source of support for victims of disaster or a source of significant stress, anxiety, and even depression. People who have lost everything find solace in an animal companion. Failure to consider this bond in disaster response creates substantial concerns among the public. Consequently, disaster planning at all levels must take animals into account. In 2005 Hurricane Katrina brought this need to public attention.

During Hurricane Katrina, many people were forced to leave their animals behind, and adequate resources were not readily available for those who did evacuate with their animals. Even some who left their homes with their animals had to abandon them when evacuating for a second time, as was the case with residents who went to the Louisiana Superdome. Most animals left behind were rescued and brought to temporary shelters, such as the one at the Lamar-Dixon Expo Center in Gonzalez, Louisiana. The animal shelter in New Orleans, where stray animals would normally have gone, was destroyed in the flood. Evacuated residents had no way of knowing where their animals had been taken or even whether they had been rescued or remained in the city. Moreover, after Katrina, many people had to relocate to other cities and states before locating lost pets. The emotional impact of losing and, in some cases, finding pets adds to the trauma of the disaster's physical losses. Through the innovative efforts of such Internet resources as *petfinder.com,* many families were reunited with their dogs and cats. However, many could not reclaim their pets because they were simply unable to house and care for them. Shelters all over the country took in homeless animals from Louisiana, Mississippi, and Texas. Overall, the 2005 hurricane season called attention to the importance of including pets in evacuation plans.

Public Health

Animal and human health issues are closely interwoven. The complex roles that animals play in public health seldom come to mind when people think of disasters. However, people and animals can often contract the same diseases, and they can transmit diseases from one to the other as well. Diseases that can be transmitted between humans and animals are known as *zoonoses.* They can become epidemics or pandemics. Examples include rabies, severe acute respiratory syndrome (SARS), avian flu, Lyme disease, West Nile virus, monkeypox, and various types of influenza. In addition, some animal diseases, including anthrax (*bacillus anthracis*) and plague (*yersinia pestis*), could serve as weapons of mass destruction. Zoonoses spread when people or animals travel from endemic areas to other parts of the world.

There are over 200 zoonoses, and many have been known for centuries. Zoonoses can involve bacteria, parasites, viruses, and other agents. Millions of people can become ill through such foodborne bacterial zoonoses as *Salmonellosis* and *E. coli,* which cause fever, diarrhea, abdominal pain, and nausea. *Influenza* is a viral zoonosis that causes respiratory disease in humans, and some influenza viruses also infect a wide variety of other animals, including poultry, horses, swine, and marine mammals. Avian influenza viruses, resorted and transmitted to humans, were responsible for the pandemics of 1918, 1957, and 1968. In poultry and ducks, the infection affects the gastrointestinal tract (rather than the respiratory tract), and is therefore subclinical, or without symptoms. The subclinical nature of these infections, combined with the birds' migratory behavior and the virus's ability to persist in cold water, makes birds an enormous natural reservoir for influenza viruses.

Zoonoses can also be transmitted through unconventional means. A type of protein known as a *prion* causes bovine spongiform encephalopathy (BSE), or mad cow disease. In infected cattle, abnormal prions appear in the small intestines, the tonsils, and the tissues of the central nervous system. Eating BSE-contaminated beef products has been associated with a variant form of Creutzfeldt-Jakob disease (vCJD).

The Economy

Animal issues in disasters have economic impact not only because of the costs of the recovery efforts themselves, but also because of the role that animals play in the economy. For example, when wildfires and drought affect wildlife, local and state economies feel the impact on their tourism, hunting, and fishing industries. The cost of rescuing and sheltering pets abandoned during the 2005 hurricane season was tremendous. The economic impact is particularly notable with livestock disasters, such as widespread disease outbreaks. In the United States, livestock production directly contributes over $100 billion to the economy annually, and many times that value indirectly. Disease threats to livestock, either accidental or intentional (as when a disease agent is used as a weapon), could have devastating impacts on the economy and the nation's food supply.

Few countries understand the economic impact of livestock disease to the extent that Great Britain does. The first cases of foot and mouth disease (FMD) appeared only five years after the outbreak of mad cow disease. The 2001 outbreak of FMD paralyzed the agricultural infrastructure and cost the equivalent of $12 billion. The outbreak resulted in the depopulation of over 4 million cows, pigs, and sheep, the majority of which lived in the affected areas but did not have the disease. The economic impact included such direct costs as lost animals, carcass disposal, and response and eradication efforts. Slaughterhouse workers lost jobs as the production of beef products was suspended. The outbreak affected peripheral industries as well: hauling companies reported a large downturn, and the rendering industry, which had previously produced economically valuable raw materials, essentially became a waste disposal industry in response to the massive slaughter. The outbreak caused significant indirect costs to tourism and trade in Britain and Western Europe. The UK experienced significant economic impact because of the decline of international tourism. Travel was significantly restricted in an effort to control the spread of the disease. Many small businesses, such as pubs and inns, closed down in the affected areas. In addition, the outbreak brought significant nonmonetary and moral consequences. Some herds in Great Britain were "legacy" herds, raised by particular families for generations. The outbreak meant the loss of lifestyle. Farm families were ostracized, and over eighty suicides were reported among farmers and other animal stakeholders affected by the outbreak of FMD.

Ethical and Moral Issues

Ethical and moral issues enter into the disaster response through at least two related factors. First, humans are responsible for animals in many ways; we bring them into our homes and include them in our families. Therefore, they depend on us when they are in danger. We house food animals in extremely crowded conditions with little chance of escape when barns flood, catch fire, or collapse. On a more basic level, human beings are responsible for bringing many species of animals into existence in the first place. When we domesticated animals, we took on the responsibility for their care.

We bear responsibility for wild animals, as well. Wildlife species are a critical component of the environment. Birds and wild animals are at risk in human-caused fires, oil spills, and other technological disasters. For example, oil is carcinogenic to fish, birds, and mammals, and many animals die from ingesting it. Seals and sea lions often drown because of the weight of oil on their coats, and oil exposure also decreases seals' reproductive rate.

Second, disaster responders have always put human health and safety first. Evacuation plans have long involved getting humans out of danger and going back for animals later. This tenet of emergency management failed during the evacuation following Hurricane Katrina, when people would not leave without their pets, and when the animals left behind required massive rescue and sheltering efforts. The lesson of Katrina was that one of the best ways to save people in a disaster is to save their animals at the same time. When Rita, the subsequent hurricane, forced evacuations in Galveston, Texas, emergency managers allowed people to take their animals with them.

Planning for Disasters

Disaster planning on a large scale takes place at the governmental level, but individual households must also prepare. Moreover, such animal stakeholders as veterinary clinics, breeding facilities, boarding kennels, shelters, and farms must prepare. Whereas some disasters require evacuation, others call for what is known as "sheltering in place," or staying put until the risk has passed. Depending on the disaster, animal stakeholders might have to evacuate or take in evacuated animals. Consequently, the preparations must consider various scenarios.

Disaster planning begins with assessing the risks in a given area. A region that is vulnerable to hurricanes and flooding probably faces little risk of blizzards and ice storms. Wildfires obviously do not threaten highly urban areas. In short, the type of response necessary will depend on the potential threat, and planning must take the likely threats into consideration. However, there are many "equal opportunity" risks. For example, railroad tracks intersect most regions, and there are numerous homes within a mile of tracks. Trains regularly carry flammable or hazardous chemicals, putting wide areas at risk in the event of derailment, such as the incident in Weyauwega.

Planning at the household level usually means anticipating the needs of pets. Designate a cupboard, shelf, or a container for emergency supplies for pets. At minimum, households should have sufficient food, water, litter, bedding, and other necessities to last at least seventy-two hours. Pets should have up-to-date identification and vaccinations. A waterproof plastic bag can hold copies of vaccination records and any licenses. It can be helpful to include one or two photos of the pets, ideally with family members, in the emergency supply kit. If an animal is lost, the photo can supplement a description and also verify that a found animal belongs with a particular family. If the incident requires evacuation rather than sheltering in place, dogs' leashes must be easily located. Cats and smaller dogs must have travel carriers. An adequate supply of any medications must accompany the animals. Because most emergency shelters do not allow pets to be housed with people, animals and their guardians will most likely be separated during the evacuation period. This highlights the importance of current identification.

Horse and livestock bring additional issues to consider in planning. Owners must have sufficient operable trailers and transporters. Moreover, horse owners should practice loading their horses into trailers so that they can do it quickly and safely when necessary. Horses and livestock are often evacuated to local farms and ranches, but they are also housed at fairgrounds and similar facilities that have barns. As with pets, the need for identification also arises with livestock. Brands on livestock and tattoos on horses are common means of linking owners and animals, and owners should ensure that their animals have current identification.

In addition to preparing to shelter in place, individuals and families should locate animal-friendly accommodations outside the immediate area in case emergency managers call for evacuation. Knowing where to find pet-friendly motels before the incident occurs, or having friends and family who can house animals, can save lives and prevent separation from pets.

Further Resources

Crisp, T. (1997). *Out of harm's way: The extraordinary true story of one woman's lifelong devotion to animal rescue.* New York: Pocket Books.

Heath, S. (1999). *Animal management in disasters.* West Lafayette, IN: Purdue University Press.

Heath, S., & O'Shea, A. (1999). *Rescuing rover: A first aid and disaster guide for dog owners.* West Lafayette, IN: Purdue University Press.

Leslie Irvine